STUDENT'S SOLUTIONS MANUAL

BRENT GRIFFIN
Georgia Highlands College

COLLEGE ALGEBRA IN CONTEXT WITH APPLICATIONS FOR THE MANAGERIAL, LIFE, AND SOCIAL SCIENCES

SECOND EDITION

Ronald J. Harshbarger
University of South Carolina–Beaufort

Lisa S. Yocco
Georgia Southern University

PEARSON

Addison
Wesley

Boston San Francisco New York
London Toronto Sydney Tokyo Singapore Madrid
Mexico City Munich Paris Cape Town Hong Kong Montreal

Reproduced by Pearson Addison-Wesley from electronic files supplied by the author.

Copyright © 2007 Pearson Education, Inc.
Publishing as Pearson Addison-Wesley, 75 Arlington Street, Boston, MA 02116.

ISBN   0-321-36977-7

2 3 4 5 6 BB 09 08 07 06

TABLE OF CONTENTS

Chapter 1   Functions, Graphs, and Models; Linear Functions

Algebra Toolbox — 1

1.1 Functions and Models — 4

1.2 Graphs of Functions — 9

1.3 Linear Functions — 15

1.4 Equations of Lines — 20

1.5 Algebraic and Graphical Solutions of Linear Equations — 27

1.6 Fitting Lines to Data Points; Modeling Linear Functions — 34

1.7 Systems of Linear Equations in Two Variables — 41

1.8 Solutions of Linear Inequalities — 51

Chapter 1 Skills Check — 60

Chapter 1 Review — 66

Chapter 1 Group Activities/Extended Applications — 76

Chapter 2   Quadratic and Other Non-Linear Functions

Algebra Toolbox — 77

2.1 Quadratic Functions; Parabolas — 83

2.2 Solving Quadratic Equations — 90

2.3 A Library of Functions — 101

2.4 Transformations of Graphs and Symmetry — 107

2.5 Quadratic and Power Models — 113

2.6 Combining Functions; Composite Functions — 121

2.7 Inverse Functions — 128

2.8 Additional Equations and Inequalities — 134

Chapter 2 Skills Check — 145

Chapter 2 Review — 151

Chapter 2 Group Activities/Extended Applications — 164

Chapter 3   Exponential and Logarithmic Functions

   Algebra Toolbox   **167**

   3.1 Exponential Functions   **170**

   3.2 Logarithmic Functions   **175**

   3.3 Solving Exponential Equations; Properties of Logarithms   **180**

   3.4 Exponential and Logarithmic Models   **186**

   3.5 Exponential Functions and Investing   **190**

   3.6 Annuities; Loan Repayment   **194**

   3.7 Logistic and Gompertz Functions   **198**

   Chapter 3 Skills Check   **202**

   Chapter 3 Review   **206**

   Chapter 3 Group Activities/Extended Applications   **215**

Chapter 4   Higher Degree Polynomials and Rational Functions

   Algebra Toolbox   **217**

   4.1 Higher Degree Polynomial Functions   **222**

   4.2 Modeling Cubic and Quartic Functions   **230**

   4.3 Solution of Polynomial Equations   **236**

   4.4 Polynomial Equations Continued; Fundamental Theorem of Algebra   **241**

   4.5 Rational Functions and Rational Equations   **248**

   4.6 Polynomial and Rational Inequalities   **255**

   Chapter 4 Skills Check   **262**

   Chapter 4 Review   **271**

   Chapter 4 Group Activities/Extended Applications   **278**

Chapter 5 Systems of Equations and Matrices

    Algebra Toolbox   **281**

    5.1 Systems of Linear Equations in Three Variables   **284**

    5.2 Matrix Solution of Systems of Linear Equations   **294**

    5.3 Matrix Operations   **305**

    5.4 Inverse Matrices; Matrix Equations   **312**

    5.5 Systems of Nonlinear Equations   **322**

    Chapter 5 Skills Check   **329**

    Chapter 5 Review   **340**

    Chapter 5 Group Activities/Extended Applications   **352**

Chapter 6 Special Topics: Systems of Linear Inequalities and Linear Programming; Sequences and Series; Preparing for Calculus

    6.1 Systems of Linear Inequalities   **357**

    6.2 Linear Programming: Graphical Methods   **375**

    6.3 Sequences and Discrete Functions   **388**

    6.4 Series   **392**

    6.5 Preparing for Calculus   **396**

    Chapter 6 Skills Check   **399**

    Chapter 6 Review   **412**

    Chapter 6 Group Activities/Extended Applications   **420**

CHAPTER 1
Functions Graphs, and Models; Linear Functions

Algebra Toolbox Exercises

1. $\{1,2,3,4,5,6,7,8\}$ and
   $\{x \mid x < 9, x \in \mathbb{N}\}$
   Remember that $x \in \mathbb{N}$ means that x is a natural number.

2. Yes.

3. Yes. Every element of B is also in A.

4. No. $\mathbb{N} = \{1,2,3,4,...\}$. Therefore, $\frac{1}{2} \notin \mathbb{N}$.

5. Yes. Every integer can be written as a fraction with the denominator equal to 1.

6. Yes. Irrational numbers are by definition numbers that are not rational.

7. Integers. Note this set of integers could also be considered a set of rational numbers. See question 5.

8. Rational numbers

9. Irrational numbers

10. $x > -3$

11. $-3 \le x \le 3$

12. $x \le 3$

13. $(-\infty, 7]$

14. $(3, 7]$

15. $(-\infty, 4)$

16.
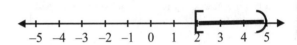

17.

Note that $5 > x \ge 2$ implies $2 \le x < 5$, therefore
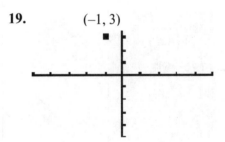

18.
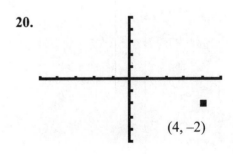

19.

(−1, 3)

20.

(4, −2)

21.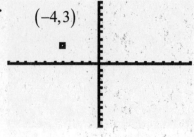

$(-4, 3)$

22. $|-6| = 6$

23. $|7 - 11| = |-4| = 4$

24. The x^2 term has a coefficient of -3. The x term has a coefficient of -4. The constant term is 8.

25. The x^4 term has a coefficient of 5. The x^3 term has a coefficient of 7. The constant term is -3.

26. $\left(z^4 - 15z^2 + 20z - 6\right) + \left(2z^4 + 4z^3 - 12z^2 - 5\right)$

$= \left(z^4 + 2z^4\right) + \left(4z^3\right) + \left(-15z^2 - 12z^2\right) +$

$\quad \left(20z\right) + \left(-6 - 5\right)$

$= 3z^4 + 4z^3 - 27z^2 + 20z - 11$

27. $-2x^3y^4 + \left(2y^4 + 5y^4\right) - 3y^2 +$

$\quad \left(3x - 5x\right) + \left(-119 + 110\right)$

$= -2x^3y^4 + 7y^4 - 3y^2 - 2x - 9$

28. $4\left(p + d\right)$

$= 4p + 4d$

29. $-2\left(3x - 7y\right)$

$= \left(-2 \cdot 3x\right) + \left(-2 \cdot -7y\right)$

$= -6x + 14y$

30. $-a\left(b + 8c\right)$

$= -ab - 8ac$

31. $4\left(x - y\right) - \left(3x + 2y\right)$

$= 4x - 4y - 3x - 2y$

$= 1x - 6y$

$= x - 6y$

32. $4\left(2x - y\right) + 4xy - 5\left(y - xy\right) - \left(2x - 4y\right)$

$= 8x - 4y + 4xy - 5y + 5xy - 2x + 4y$

$= \left(8x - 2x\right) + \left(-4y - 5y + 4y\right) + \left(4xy + 5xy\right)$

$= 6x - 5y + 9xy$

33. $2x\left(4yz - 4\right) - \left(5xyz - 3x\right)$

$= 8xyz - 8x - 5xyz + 3x$

$= \left(8xyz - 5xyz\right) + \left(-8x + 3x\right)$

$= 3xyz - 5x$

34. $\dfrac{3x}{3} = \dfrac{6}{3}$

$x = 2$

35. $\dfrac{x}{3} = 6$

$\left(\dfrac{3}{1}\right)\left(\dfrac{x}{3}\right) = \left(\dfrac{3}{1}\right)\left(\dfrac{6}{1}\right)$

$x = \dfrac{18}{1}$

$x = 18$

36. $x + 3 = 6$

$x + 3 - 3 = 6 - 3$

$x = 3$

37. $4x - 3 = 6 + x$

$4x - x - 3 = 6 + x - x$

$3x - 3 = 6$

$3x - 3 + 3 = 6 + 3$

$3x = 9$

$\dfrac{3x}{3} = \dfrac{9}{3}$

$x = 3$

38. $3x - 2 = 4 - 7x$

$3x + 7x - 2 = 4 - 7x + 7x$

$10x - 2 = 4$

$10x - 2 + 2 = 4 + 2$

$10x = 6$

$\dfrac{10x}{10} = \dfrac{6}{10}$

$x = \dfrac{6}{10}$

$x = \dfrac{3}{5}$

39. $\dfrac{3x}{4} = 12$

$4\left(\dfrac{3x}{4}\right) = 4(12)$

$3x = 48$

$x = 16$

40. $2x - 8 = 12 + 4x$

$2x - 4x - 8 = 12 + 4x - 4x$

$-2x - 8 + 8 = 12 + 8$

$-2x = 20$

$\dfrac{-2x}{-2} = \dfrac{20}{-2}$

$x = -10$

Section 1.1 Skills Check

1. Using Table A

   a. -5 is an x-value and therefore is an input into the function $f(x)$.

   b. $f(-5)$ represents an output from the function.

   c. The domain is the set of all inputs. D:$\{-9,-7,-5,6,12,17,20\}$. The range is the set of all outputs. R:$\{4,5,6,7,9,10\}$

   d. Every input x into the function f yields exactly one output $y = f(x)$.

3. $f(-9) = 5$
   $f(17) = 9$

5. No. In the given table, x is not a function of y. If y is considered the input variable, one input will correspond with more than one output. Specifically, if $y = 9$, then $x = 12$ or $x = 17$.

7. a. $f(2) = -1$

   b. $f(2) = 10 - 3(2)^2$
      $= 10 - 3(4)$
      $= 10 - 12$
      $= -2$

   c. $f(2) = -3$

9. Recall that $R(x) = 5x + 8$.

   a. $R(-3) = 5(-3) + 8 = -15 + 8 = -7$

   b. $R(-1) = 5(-1) + 8 = -5 + 8 = 3$

   c. $R(2) = 5(2) + 8 = 10 + 8 = 18$

11. Yes. Every input corresponds with exactly one output. The domain is $\{-1,0,1,2,3\}$. The range is $\{-8,-1,2,5,7\}$.

13. No. The graph fails the vertical line test. Every input does not match with exactly one output.

15. No. If $x = 3$, then $y = 5$ or $y = 7$. One input yields two outputs. The relation is not a function.

17. a. Not a function. If $x = 4$, then $y = 12$ or $y = 8$.

    b. Yes. Every input yields exactly one output.

19. a. Not a function. If $x = 2$, then $y = 3$ or $y = 4$.

    b. Function. Every input yields exactly one output.

21. No. If $x = 0$, then
    $(0)^2 + y^2 = 4 \Rightarrow y^2 = 4 \Rightarrow y = \pm 2$. So, one input of 0 corresponds with 2 outputs of -2 and 2. Therefore the equation is not a function.

23. $C = 2\pi r$, where C is the circumference and r is the radius.

25. A function is a correspondence that assigns to each element of the domain exactly one element of the range.

27. The range of a function is the set of all possible outputs from the function.

29. a. No. Every input (x, given day) would correspond with multiple outputs (p, stock prices). Stock prices fluctuate throughout the trading day.

b. Yes. Every input (x, given day) would correspond with exactly one output (p, the stock price at the end of the trading day).

31. Yes. Every input (month) corresponds with exactly one output (cents per pound).

33. Yes. Every input (education level) corresponds with exactly one output (average income).

35. Yes. Every input (depth) corresponds with exactly one output (pressure). The graph of the equation passes the vertical line test.

37. a. Yes. Every input (day of the month) corresponds with exactly one output (weight).

b. The domain is $\{1,2,3,4,\ldots,13,14\}$.

c. The range is $\{171,172,173,174,175,176,177,178\}$.

d. The highest weights were on May 1 and May 3.

e. The lowest weight was May 14.

f. Three days from May 12 until May 14.

39. a. $B(3) = 16,115$

b. $B(2) = 23,047$. $B(2)$ represents the balance owed by the couple at the end of two years.

c. Year 2.

d. $t = 4$

41. a. When $t = 2005$, the ratio is approximately 4.

b. $f(2005) = 4$. For year 2005 the projected ratio of working-age population to the elderly is 4.

c. The domain is the set of all possible inputs. In this example, the domain consists of all the years, t, represented in the figure. Specifically, the domain is $\{1995, 2000, 2005, 2010, 2015, 2020, 2025, 2030\}$.

d. As the years, t, increase, the projected ratio of the working-age population to the elderly decreases. Notice that the bars in the figure grow smaller as the time increases.

43. a. $f(1990) = 492,671$

b. The domain is the set of all possible inputs. In this example, the domain is all the years, t, represented in the table. Specifically, the domain is $\{1985, 1986, 1987, \ldots, 1997, 1998\}$.

c. The maximum number of firearms is 581,697, occurring in year 1993. Note that $f(1993) = 581,687$.

45. a. Yes. Every year, t, corresponds with exactly one percentage, p.

b. $f(1840) = 68.6$. $f(1840)$ represents the percentage of U.S. workers in a farm occupation in the year 1840.

c. If $f(t) = 27$, then $t = 1920$.

d. $f(1960) = 6.1$ implies that in 1960, 6.1% of U.S. workers were employed in a farm occupation.

e. As the time, t, increases, the percentage, p, of U.S. workers in farm occupations decreases. Note that the graph slopes down if it is read from left to right.

47. a. $f(1990) = 3.4$. In 1990 there are 3.4 workers for each retiree.

b. 2030

c. As the years increase, the number of workers available to support retirees decreases. Therefore, funding for social security into the future is problematic. Workers will need to pay larger portion of their salaries to fund payments to retirees.

49. a. $R(200) = 32(200) = 6400$. The revenue generated from selling 200 golf hats is $6400.

b. $R(2500) = 32(2500) = \$80,000$

51. a. $\begin{aligned} P(500) &= 450(500) - 0.1(500)^2 - 2000 \\ &= 225,000 - 25,000 - 2000 \\ &= 198,000 \end{aligned}$
The profit generated from selling 500 ipod players is $198,000.

b. $\begin{aligned} P(4000) \\ &= 450(4000) - 0.1(4000)^2 - 2000 \\ &= 1,800,000 - 1,600,000 - 2000 \\ &= \$198,000 \end{aligned}$

53. a. $\begin{aligned} f(1000) &= 0.105(1000) + 5.80 \\ &= 105 + 5.80 \\ &= 110.80 \end{aligned}$

The monthly charge for using 1000 kilowatt hours is $110.80.

b. $f(1500) = 0.105(1500) + 5.80$

$\qquad = 157.5 + 5.80$

$\qquad = \$163.30$

55. a. $h(1) = 6 + 96(1) - 16(1)^2$

$\qquad = 6 + 96 - 16$

$\qquad = 86$

The height of the ball after one second is 86 feet.

b. $h(3) = 6 + 96(3) - 16(3)^2$

$\qquad = 6 + 288 - 144$

$\qquad = 150$

After three seconds the ball is 150 feet high.

c. Test $t = 5$.

$h(5) = 6 + 96(5) - 16(5)^2$

$\qquad = 6 + 480 - 400$

$\qquad = 86$

After five seconds the ball is 86 feet high. The ball does eventually fall, since the height at $t = 5$ is lower than the height at $t = 3$. Considering the following table of values for the function, it seems reasonable to estimate that the ball stops climbing at $t = 3$.

X	Y1
0	6
1	86
2	134
3	150
4	134
5	86
6	6

X=0

57. Y1=4√(4X+1)

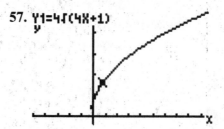

a. Yes. The graph seems to pass the vertical line test.

b. Any input into the function must not create a negative number under the radical. Therefore, the radicand, $4s + 1$, must be greater than or equal to zero.

$\qquad 4s + 1 \geq 0$

$4s + 1 - 1 \geq 0 - 1$

$\qquad 4s \geq -1$

$\qquad s \geq -\dfrac{1}{4}$

Therefore, based on the equation, the domain is $\left[-\dfrac{1}{4}, \infty\right)$.

c. Since s represents wind speed in the given function and wind speed cannot be less than zero, the domain of the function is restricted based on the physical context of the problem. Even though the domain implied by the function is $\left[-\dfrac{1}{4}, \infty\right)$, the actual domain in the given physical context is $[0, \infty)$.

59. a. The domain is $[0, 100)$.

b. $C(60) = \dfrac{237,000(60)}{100 - 60} = 355,500$

$\quad\ C(90) = \dfrac{237,000(90)}{100 - 90} = 2,133,000$

61. a.

$$V(12) = (12)^2(108 - 4(12))$$
$$= 144(108 - 48)$$
$$= 144(60)$$
$$= 8640$$

$$V(18) = (18)^2(108 - 4(18))$$
$$= 324(108 - 72)$$
$$= 324(36)$$
$$= 11,664$$

b. First, since x represents a side length in the diagram, then x must be greater than zero. Second, to satisfy postal restrictions, the length plus the girth must be less than or equal to 108 inches. Therefore,

$$\text{Length} + \text{Girth} \leq 108$$
$$\text{Length} + 4x \leq 108$$
$$4x \leq 108 - \text{Length}$$
$$x \leq \frac{108 - \text{Length}}{4}$$
$$x \leq 27 - \frac{\text{Length}}{4}$$

Since x is greatest if the length is smallest, let the length equal zero to find the largest value for x.

$$x \leq 27 - \frac{0}{4}$$
$$x \leq 27$$

Therefore the conditions on x and the corresponding domain for the function $V(x)$ are $0 \leq x \leq 27$ or $x \in [0, 27]$.

c.

X	Y1
10	6800
12	8640
14	10192
16	11264
18	11664
20	11200
22	9680

X=18

The maximum volume occurs when $x = 18$. Therefore the dimensions that maximize the volume of the box are 18 inches by 18 inches by 36 inches.

Section 1.2 Skills Check

1. a.

x	$y = x^3$	(x, y)
−3	−27	(−3, 27)
−2	−8	(−2, −8)
−1	−1	(−1, −1)
0	0	(0, 0)
1	1	(1, 1)
2	8	(2, 8)
3	27	(3, 27)

b. Y1=X^3

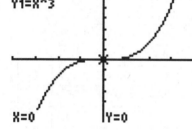

[−4,4] by [−30,30]

c. The graphs are the same.

3. Y1=X2−5

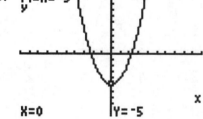

5. Y1=X^3−3X^2

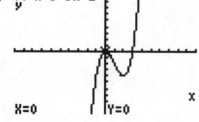

7. Y1=9/(X2+1)

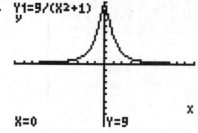

9. a. Y1=X+20

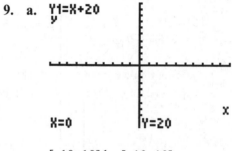

[−10, 10] by [−10, 10]

b. Y1=X+20

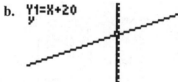

[10, 10] by [−10, 30]

View b) is better.

11. a. Y1=(.4*(X−.1))/(X^2+300)

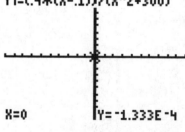

[−10, 10] by [−10, 10]

b. Y1=(.4*(X-.1))/(X^2+300)

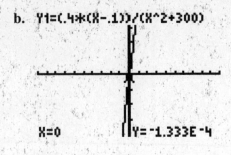

[−20, 20] by [−0.02, 0.02]

View b) is better.

13. When $x = -3$ or $x = 3$, $y = 59$. When $x = 0$, $y = 50$. Therefore, $[-3,3]$ by $[0,70]$ is an appropriate viewing window. {Note that answers may vary.}

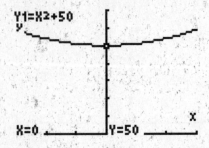

15. When $x = -10, y = 250$. When $x = 10, y = 850$. When $x = 0, y = 0$. Therefore, $[-10,10]$ by $[-250,1000]$ is an appropriate viewing window. {Note that answers may vary.}

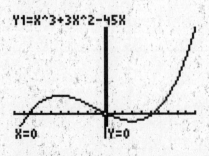

[−10,10] by [−250,1000]

17. Y1=10X²−90X+300

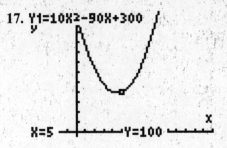

[−5, 15] by [−10, 300]

{Note that answers may vary.}

19.

t	$S(t) = 5.2t - 10.5$
12	51.9
16	72.7
28	135.1
43	213.1

21.

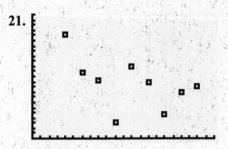

[0, 110] by [0, 600]

23. a.

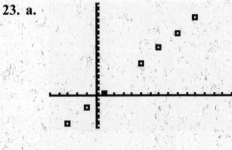

[−5, 15] by [−50, 150]

b. Y1=12X-6

X=5   Y=54

[–5, 15] by [–50, 150]

c. Yes. Yes. Compare the following table of points generated by $f(x) = 12x - 6$ to the given table of points:

X	Y1
-3	-42
-1	-18
1	6
3	30
5	54
7	78
9	102

X= -3

25. a. $f(20) = (20)^2 - 5(20)$
$= 400 - 100$
$= 300$

b. $x = 20$ implies 20 years after 2000. Therefore the answer to a) yields the millions of dollars earned in 2020.

Section 1.2 Exercises

27. a. $x = \text{Year} - 1990$
For 1994, $x = 1994 - 1990 = 4$
For 1998, $x = 1998 - 1990 = 8$

b. $y = -112(8)^2 - 107(8) + 15056 = 7032$
7032 represents the number of welfare cases in Niagara, Canada in 1998.

c. For 1995, $x = 5$. Therefore,
$y = -112(5)^2 - 107(5) + 15,056$
$= 11,721$

There were 11,721 welfare cases in Niagara, Canada in 1995.

29. a. $t = \text{Year} - 1995$
For 1996, $t = 1996 - 1995 = 1$
For 2004, $t = 2004 - 1995 = 9$

b. $P = f(8) = 6.02(8) + 3.53 = 51.69$.
51.96 represents the percentage of households with Internet access in 2003.

c. $x_{\min} = 1995 - 1995 = 0$
$x_{\max} = 2005 - 1995 = 10$

31. $S = 100 + 64t - 16t^2$

a. Y1=100+64X-16X²

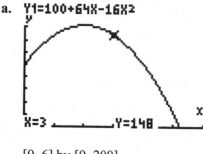

X=3   Y=148

[0, 6] by [0, 200]

b.

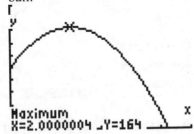

Considering the table, $S = 148$ feet when x is 1 or when x is 3. The height is the same for two different times because the height of the ball increases, reaches a maximum height, and then decreases.

c. The maximum height is 164 feet, occurring 2 seconds into the flight of the ball.

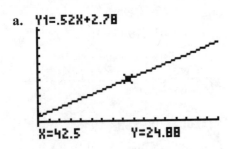

33. $F = 0.52M + 2.78$

a.

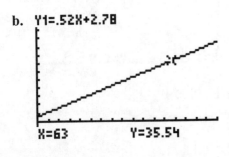

[0, 85] by [0, 65]

b. Y1=.52X+2.78

[0, 85] by [0, 65]

When $x = 63$, $y = 35.54$. Therefore, when the median male salary is \$63,000, the median female salary is \$35,540.

35. $S = 0.027t^2 - 4.85t + 218.93$

a.

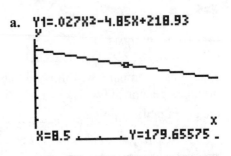

[0, 17] by [0, 300]

b.

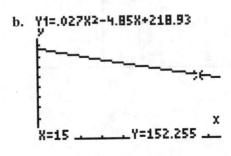

[0, 17] by [0, 300]

When $t = 15$, $S = 152.255$

c. 1995 corresponds to $t = 1995 - 1980 = 15$. When $t = 15$,
$$S = 0.027(15)^2 - 4.85(15) + 218.93$$
$$= 152.255$$
See the graph in part *b* above. The estimated number of osteopathic students in 1995 is 152,255.

37. $B(t) = 20.37 + 1.83t$

a.

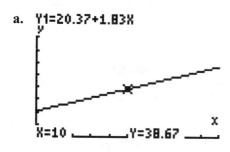

[0, 20] by [0, 100]

b. The tax burden increased. Reading the graph from left to right, as t increases so does $B(t)$.

39. $C(x) = 15,000 + 100x + 0.1x^2$

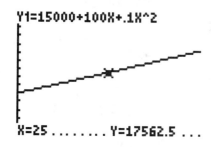

[0, 50] by [10,000, 25,000]

41. $P(x) = 200x - 0.01x^2 - 5000$

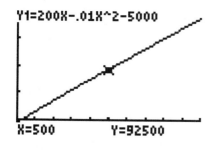

[0, 1000] by [−25,000, 200,000]

43. $f(t) = 982.06t + 32,903.77$

a. Since the base year is 1990, 1990-2005 correspond to values of t between 0 and 15 inclusive.

b. For 1990:
$$f(0) = 982.06(0) + 32,903.77$$
$$= 32,903.77$$
For 2005:
$$f(15) = 982.06(15) + 32,903.77$$
$$= 47,634.67$$

c.

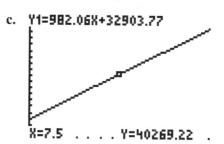

[0, 15] by [30,000, 50,000]

{Note that answers may vary.}

45.

x (years since 1990)	y (number of near-hits)
0	281
1	242
2	219
3	186
4	200
5	240
6	275
7	292
8	325
9	321
10	421

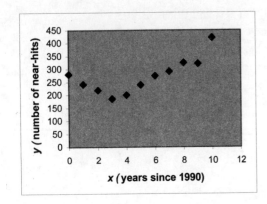

b.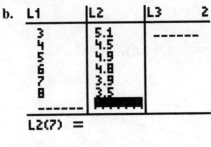

L2(7) =

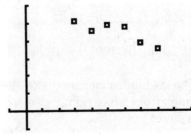

47. a.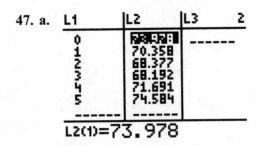

L2(1)=73.978

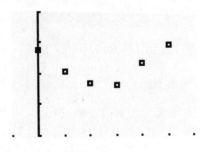

[–1, 6] by [60, 80]

[–1, 10] by [–1, 6]

c. Y1=.0583X^4–1.315X^3+10...

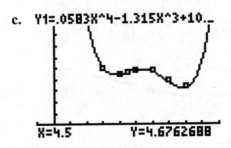

X=4.5   Y=4.6762688

[–1, 10] by [–1, 10]

Yes. The fit is reasonable but not perfect.

b. Y1=.973X^2–4.667X+73.950

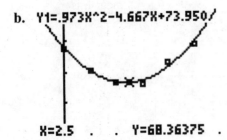

X=2.5   Y=68.36375

[–1, 6] by [60, 80]

Yes. The fit is reasonable but not perfect.

49. a. In 2003 the unemployment rate was 3.5%.

Section 1.3 Skills Check

1. Recall that linear functions must be in the
   form $f(x) = ax + b$.

   a. Not linear. The equation has a 2nd
      degree (squared) term.

   b. Linear.

   c. Not linear. The x-term is in the
      denominator of a fraction.

3. $m = \dfrac{y_2 - y_1}{x_2 - x_1}$

   $= \dfrac{4 - (-10)}{8 - 8}$

   $= \dfrac{14}{0}$

   $=$ undefined

   Zero in the denominator creates an
   undefined expression.

5. a. x-intercept: Let $y = 0$ and solve for x.

   $5x - 3(0) = 15$

   $5x = 15$

   $x = 3$

   y-intercept: Let $x = 0$ and solve for y.

   $5(0) - 3y = 15$

   $-3y = 15$

   $y = -5$

   x-intercept: (3, 0), y-intercept: (0, –5)

b.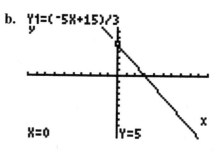

$Y1=(-5X+15)/3$

$X=0$ $Y=5$

[–10, 10] by [–10, 10]

7. a. x-intercept: Let $y = 0$ and solve for x.

   $3(0) = 9 - 6x$

   $0 = 9 - 6x$

   $0 - 9 = 9 - 9 - 6x$

   $-6x = -9$

   $x = \dfrac{-9}{-6}$

   $x = \dfrac{3}{2} = 1.5$

   y-intercept: Let $x = 0$ and solve for y.

   $3y = 9 - 6(0)$

   $3y = 9$

   $y = 3$

   x-intercept: (1.5, 0), y-intercept: (0, 3)

b.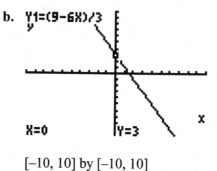

$Y1=(9-6X)/3$

$X=0$ $Y=3$

[–10, 10] by [–10, 10]

9. Horizontal lines have a slope of **zero**.
   Vertical lines have an **undefined** slope.

11. $m = 4, b = 8$

13. $5y = 2$

$y = \dfrac{2}{5}$, horizontal line

$m = 0, \; b = \dfrac{2}{5}$

15. a. $m = 4, \; b = 5$

   b.   Rising. The slope is positive

   c.   Y1=4X+5

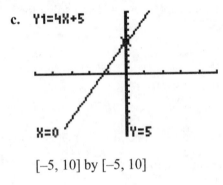

X=0   Y=5

[–5, 10] by [–5, 10]

17. a. $m = -100, \; b = 50{,}000$

   b.   Falling. Slope is negative.

   c.

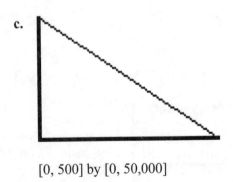

[0, 500] by [0, 50,000]

19. For a linear function, the rate of change is equal to the slope. $m = 4$.

21. For a linear function, the rate of change is equal to the slope. $m = -15$.

23. For a linear function, the rate of change is equal to the slope.

$$m = \frac{y_2 - y_1}{x_2 - x_1} = \frac{-7 - 3}{4 - (-1)} = \frac{-10}{5} = -2.$$

25. The lines are perpendicular. The slopes are negative reciprocals of one another.

27. a.   The identity function is $y = x$. Graph *ii* represents the identity function.

   b.   The constant function is $y = k$, where k is a real number. In this case, $k = 3$. Graph *i* represents a constant function.

29. a.   The slope of a constant function is zero ($m = 0$).

   b.   The rate of change of a constant function equals the slope, which is zero.

Section 1.3 Exercises

31. Linear. Rising—the slope is positive.
$m = 0.155$.

33. Linear. Falling—the slope is negative.
$m = -0.762$.

35. a. x-intercept: Let $p = 0$ and solve for x.

$$30p - 19x = 30$$
$$30(0) - 19x = 30$$
$$-19x = 30$$
$$x = -\frac{30}{19}$$

The x-intercept is $\left(-\dfrac{30}{19}, 0\right)$.

b. p-intercept: Let $x = 0$ and solve for p.

$$30p - 19x = 30$$
$$30p - 19(0) = 30$$
$$30p = 30$$
$$p = 1$$

The y-intercept is $(0,1)$.
In 1990, the percentage of high school students using marijuana daily is 1%.

c. $x = 0$ corresponds to 1990, $x = 1$ corresponds 1992, etc.

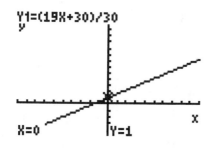

Y1=(19X+30)/30

X=0   Y=1

$[-10, 10]$ by $[-5, 15]$

37. a. The data can be modeled by a constant function. Every input x yields the same output y.

b. $y = 11.81$

c. A constant function has a slope equal to zero.

d. For a linear function the rate of change is equal to the slope. $m = 0$.

39. a. $m = 26.5$

b. Each year, the percent of Fortune Global 500 firms recruiting via the Internet increased by 26.5%.

41. a. For a linear function, the rate of change is equal to the slope. $m = \dfrac{12}{7}$. The slope is positive.

b. For each one degree increase in temperature, there is a $\dfrac{12}{7}$ increase in the number of cricket chirps.

43. a. Yes, it is linear.

b. $m = 0.959$

c. For each one dollar increase in white median annual salaries, there is a 0.959 dollar increase in minority median annual salaries.

45. a. To determine the slope, rewrite the equation in the form $f(x) = ax + b$ or $y = mx + b$.

$$30p - 19x = 30$$

$$30p = 19x + 30$$

$$\frac{30p}{30} = \frac{19x + 30}{30}$$

$$p = \frac{19}{30}x + 1$$

$$m = \frac{19}{30} \approx .633$$

b. Each year, the percentage of high school seniors using marijuana daily increases by approximately 0.63%.

47. x-intercept: Let $R = 0$ and solve for x.

$$R = 3500 - 70x$$

$$0 = 3500 - 70x$$

$$70x = 3500$$

$$x = \frac{3500}{70} = 50$$

The x-intercept is $(50, 0)$.

y-intercept: Let $x = 0$ and solve for R.

$$R = 3500 - 70x$$

$$R = 3500 - 70(0)$$

$$R = 3500$$

The y-intercept is $(0, 3500)$.

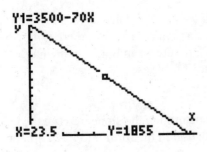

Y1=3500-70X

X=23.5   Y=1855

[0, 52] by [0, 3500]

49. a. $m = 5.74$

$y - \text{intercept} = b = 14.61$

b. The y-intercept represents the percentage of the population with Internet access in 1995. Therefore in 1995, 14.61% of the U.S. population had Internet access.

c. The slope represents the annual change in the percentage of the population with Internet access. Therefore, the percentage of the population with Internet access increased by 5.74% each year.

51. a. $m = \dfrac{y_2 - y_1}{x_2 - x_1}$

$$= \frac{700,000 - 1,310,000}{20 - 10}$$

$$= \frac{-610,000}{10}$$

$$= -61,000$$

b. Based on the calculation in part a), the property value decreases by $61,000 each year. The annual rate of change is –61,000.

53. Marginal profit is the rate of change of the profit function.

$$m = \frac{y_2 - y_1}{x_2 - x_1}$$

$$= \frac{9000 - 4650}{375 - 300}$$

$$= \frac{4350}{75}$$

$$= 58$$

The marginal profit is $58 per unit.

55. a. $m = 0.56$

b. The marginal cost is $0.56 per unit.

    c. Manufacturing one additional golf ball each month increases the cost by $0.56 or 56 cents.

57. a. $m = 1.60$

    b. The marginal revenue is $1.60 per unit

    c. Selling one additional golf ball each month increases revenue by $1.60.

59. The marginal profit is $19 per unit. Note that $m = 19$.

Section 1.4 Skills Check

1. $m = 4$, $b = \dfrac{1}{2}$. The equation is $y = 4x + \dfrac{1}{2}$.

3. $m = \dfrac{1}{3}$, $b = 3$. The equation is $y = \dfrac{1}{3}x + 3$.

5. $m = -\dfrac{3}{4}$, $b = 2$. The equation is

$y = -\dfrac{3}{4}x + 2$.

7. $y - y_1 = m(x - x_1)$

$\quad y - 4 = 5(x - (-1))$

$\quad y - 4 = 5(x + 1)$

$\quad y - 4 = 5x + 5$

$\quad\quad y = 5x + 9$

9. $y - y_1 = m(x - x_1)$

$\quad y - (-6) = -\dfrac{3}{4}(x - 4)$

$\quad y + 6 = -\dfrac{3}{4}x + \left(\dfrac{3}{4} \cdot \dfrac{4}{1}\right)$

$\quad y + 6 = -\dfrac{3}{4}x + 3$

$\quad\quad y = -\dfrac{3}{4}x - 3$

11. $y - y_1 = m(x - x_1)$

$\quad y - 4 = 0(x - (-1))$

$\quad y - 4 = 0$

$\quad\quad y = 4$

13. $x = 9$

15. Slope: $m = \dfrac{y_2 - y_1}{x_2 - x_1} = \dfrac{1 - 7}{-2 - (4)} = \dfrac{-6}{-6} = 1$

Equation: $y - y_1 = m(x - x_1)$

$\quad\quad\quad\quad y - 7 = 1(x - 4)$

$\quad\quad\quad\quad y - 7 = x - 4$

$\quad\quad\quad\quad\quad y = x + 3$

17. Slope: $m = \dfrac{y_2 - y_1}{x_2 - x_1} = \dfrac{6 - 3}{2 - (-1)} = \dfrac{3}{3} = 1$

Equation: $y - y_1 = m(x - x_1)$

$\quad\quad\quad\quad y - 6 = 1(x - 2)$

$\quad\quad\quad\quad y - 6 = x - 2$

$\quad\quad\quad\quad\quad y = x + 4$

19. Slope: $m = \dfrac{y_2 - y_1}{x_2 - x_1} = \dfrac{5 - 2}{-3 - (-4)} = \dfrac{3}{1} = 3$

Equation: $y - y_1 = m(x - x_1)$

$\quad\quad\quad\quad y - 5 = 3(x - (-3))$

$\quad\quad\quad\quad y - 5 = 3x + 9$

$\quad\quad\quad\quad\quad y = 3x + 14$

21. Slope: $m = \dfrac{y_2 - y_1}{x_2 - x_1} = \dfrac{5 - 2}{9 - 9} = \dfrac{3}{0} = $ undefined

The line is vertical. The equation of the line is $x = 9$.

23. With the given intercepts, the line passes through the points $(-5, 0)$ and $(0, 4)$. The slope of the line is

$m = \dfrac{y_2 - y_1}{x_2 - x_1} = \dfrac{4 - 0}{0 - (-5)} = \dfrac{4}{5}$.

Equation: $y - y_1 = m(x - x_1)$

$$y - 0 = \frac{4}{5}\left(x - (-5)\right)$$

$$y = \frac{4}{5}(x + 5)$$

$$y = \frac{4}{5}x + 4$$

b. $f(x + h) - f(x)$

$$= 45 - 15x - 15h - \left[45 - 15x\right]$$

$$= 45 - 15x - 15h - 45 + 15x$$

$$= -15h$$

c. $\dfrac{f(x + h) - f(x)}{h} = \dfrac{-15h}{h} = -15$

25. Slope: $m = \dfrac{y_2 - y_1}{x_2 - x_1} = \dfrac{13 - (-5)}{4 - (-2)} = \dfrac{18}{6} = 3$

Equation: $y - y_1 = m(x - x_1)$

$$y - 13 = 3(x - 4)$$

$$y - 13 = 3x - 12$$

$$y = 3x + 1$$

27. For a linear function, the rate of change is equal to the slope. Therefore, $m = -15$. The equation is

$$y - y_1 = m(x - x_1)$$

$$y - 12 = -15(x - 0)$$

$$y - 12 = -15x$$

$$y = -15x + 12.$$

29. $\dfrac{f(b) - f(a)}{b - a}$

$$= \dfrac{f(2) - f(-1)}{2 - (-1)}$$

$$= \dfrac{(2)^2 - (-1)^2}{3} = \dfrac{4 - 1}{3} = \dfrac{3}{3} = 1$$

The average rate of change between the two points is 1.

31. a. $f(x + h) = 45 - 15(x + h)$

$$= 45 - 15x - 15h$$

33. a. $f(x + h) = 2(x + h)^2 + 4$

$$= 2\left(x^2 + 2xh + h^2\right) + 4$$

$$= 2x^2 + 4xh + 2h^2 + 4$$

b. $f(x + h) - f(x)$

$$= 2x^2 + 4xh + 2h^2 + 4 - \left[2x^2 + 4\right]$$

$$= 2x^2 + 4xh + 2h^2 + 4 - 2x^2 - 4$$

$$= 4xh + 2h^2$$

c. $\dfrac{f(x + h) - f(x)}{h}$

$$= \dfrac{4xh + 2h^2}{h}$$

$$= \dfrac{h\left(4x + 2h\right)}{h}$$

$$= 4x + 2h$$

35. a.   The difference in the *y*-coordinates is consistently 30, while the difference in the *x*-coordinates is consistently 10. Note that 615–585 = 30, 645 – 630 = 30, etc. Considering the scatter plot below, a line fits the data exactly.

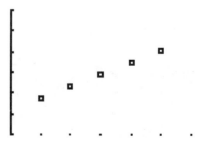

[0, 60] by [500, 800]

b. Slope:

$$m = \frac{y_2 - y_1}{x_2 - x_1}$$

$$= \frac{615 - 585}{20 - 10}$$

$$= \frac{30}{10}$$

$$= 3$$

Equation:

$$y - y_1 = m(x - x_1)$$

$$y - 585 = 3(x - 10)$$

$$y - 585 = 3x - 30$$

$$y = 3x + 555$$

Section 1.4 Exercises

37. Let x = KWh hours used and let y = monthly charge in dollars. Then the equation is $y = 0.0935x + 8.95$.

39. Let t = number of years, and let s = value of the machinery after t years. Then the equation is $s = 36,000 - 3,600t$.

41. a. Let x = the number of years since 1996, and let P = the population of Del Webb's Sun City Hilton Head community. The linear equation modeling the population growth is $P = 705x + 198$.

  b. To predict the population in 2002, let $x = 2002 - 1996 = 6$. The predicted population is $P = 705(6) + 198 = 4428$.

43. a. From year 0 to year 5, the automobile depreciates from a value of $26,000 to a value of $1,000. Therefore, the total depreciation is 26,000–1000 or $25,000.

  b. Since the automobile depreciates for 5 years in a straight-line (linear) fashion, each year the value declines by $\frac{25,000}{5} = \$5,000$.

  c. Let t = the number of years, and let s = the value of the automobile at the end of t years. Then, based on parts a) and b) the linear equation modeling the depreciation is $s = -5000t + 26,000$.

45. Notice that the x and y values are always match. That is the number of deputies always equals the number of patrol cars. Therefore the equation is $y = x$, where x

represents the number of deputies, and y represents the number of patrol cars.

47. $m = \dfrac{y_2 - y_1}{x_2 - x_1}$

$ = \dfrac{9000 - 4650}{375 - 300}$

$ = \dfrac{4350}{75} = 58$

Equation:

$y - y_1 = m(x - x_1)$

$y - 4650 = 58(x - 300)$

$y - 4650 = 58x - 17,400$

$ y = 58x - 12,750$

49. $m = \dfrac{y_2 - y_1}{x_2 - x_1}$

$ = \dfrac{700,000 - 1,310,000}{20 - 10}$

$ = \dfrac{-610,000}{10}$

$ = -61,000$

Equation:

$y - y_1 = m(x - x_1)$

$y - 1,920,000 = -61,000(x - 0)$

$y - 1,920,000 = -61,000x$

$y = -61,000x + 1,920,000$

$v = -61,000x + 1,920,000$

51. $m = \dfrac{y_2 - y_1}{x_2 - x_1}$

$ = \dfrac{32.1 - 32.7}{26 - 6}$

$ = \dfrac{-0.6}{20}$

$ = -0.03$

Equation:

$y - y_1 = m(x - x_1)$

$y - 32.7 = -0.03(x - 6)$

$y - 32.7 = -0.03x + 0.18$

$y = -0.03x + 32.88$

$p = 32.88 - 0.03t$

where t = number of years beyond 1975, y = percentage of cigarette use

53. a. Notice that the change in the x-values is consistently 1 while the change in the y-values is consistently 0.05. Therefore the table represents a linear function. The rate of change is the slope of the linear function.

$m = \dfrac{\text{vertical change}}{\text{horizontal change}} = \dfrac{0.05}{1} = 0.05$

b. Let x = the number of drinks, and let y = the blood alcohol content. Using points $(0, 0)$ and $(1, 0.05)$, the slope is

$m = \dfrac{y_2 - y_1}{x_2 - x_1}$

$ = \dfrac{0.05 - 0}{1 - 0}$

$ = \dfrac{0.05}{1} = 0.05.$

Equation:

$y - y_1 = m(x - x_1)$

$y - 0 = 0.05(x - 0)$

$y = 0.05x$

55. a. Let x = the year at the beginning of the decade, and let y = average number of men in the workforce during the decade. Using points $(1890, 18.1)$ and $(1990, 68.5)$ to calculate the slope yields:

$m = \dfrac{y_2 - y_1}{x_2 - x_1}$

$= \dfrac{68.5 - 18.1}{1990 - 1890}$

$= \dfrac{50.4}{100} = 0.504$

Equation:

$y - y_1 = m(x - x_1)$

$y - 18.1 = 0.504(x - 1890)$

$y - 18.1 = 0.504x - 952.56$

$y - 18.1 + 18.1 = 0.504x - 952.56 + 18.1$

$y = 0.504x - 934.46$

b. Yes. Consider the following table of values based on the equation in comparison to the actual data points.

x	y (Equation Values)	Actual Values
1890	18.1	18.1
1900	23.14	22.6
1910	28.18	27
1920	33.22	32
1930	38.26	37
1940	43.3	40
1950	48.34	42.8
1960	53.38	47
1970	58.42	51.6
1980	63.46	61.4
1990	68.5	68.5

c. It is the same since the points $(1890, 18.1)$ and $(1990, 68.5)$ were used to calculate the slope of the linear model.

57. a. $\dfrac{f(b) - f(a)}{b - a}$

$= \dfrac{f(2001) - f(1996)}{2001 - 1996}$

$= \dfrac{40.1 - 23}{5}$

$= \dfrac{17.1}{5}$

$= 3.42$

The average rate of change is $3.42 billion dollars per year.

b. $m = \dfrac{y_2 - y_1}{x_2 - x_1}$

$= \dfrac{40.1 - 23}{2001 - 1996}$

$= \dfrac{17.1}{5}$

$= 3.42$

c. No. Note that change in education spending from one year to the next is not constant. It varies.

d. No. Since the change in the y-values is not constant for constant changes in the x-values, the data can not be modeled exactly by a linear function.

59. a. $m = \dfrac{y_2 - y_1}{x_2 - x_1}$

$= \dfrac{76 - 15}{46 - 10}$

$= \dfrac{61}{36} = 1.69\overline{4} \approx 1.69$

b. It is the same as part a).

c. Each year between 1960 and 1996, the percentage of out-of-wedlock teenage births increased by approximately 1.69%.

d. $y - y_1 = m(x - x_1)$

$$y - 15 = \frac{61}{36}(x - 10)$$

$$y - 15 = \frac{61}{36}x - \frac{610}{36}$$

$$y = \frac{61}{36}x - \frac{610}{36} + 15$$

$$y = \frac{61}{36}x - \frac{610}{36} + \frac{540}{36}$$

$$y = \frac{61}{36}x - \frac{70}{36}$$

$$y \approx 1.69x - 1.94$$

61. a. $\dfrac{f(b) - f(a)}{b - a}$

$$= \frac{f(1997) - f(1960)}{1997 - 1960}$$

$$= \frac{1,197,590 - 212,953}{1997 - 1960}$$

$$= \frac{984,637}{37}$$

$$\approx 26,611.8$$

b. The slope of the line connecting the two points is the same as the average rate of change between the two points. Based on part a), $m \approx 26,611.8$.

c. The equation of the secant line is given by:

$$y - y_1 = m(x - x_1)$$

$$y - 212,953 = \frac{984,637}{37}(x - 1960)$$

$$y - 212,953 \approx 26,611.8x - 52,159,149$$

$$y = 26,611.8x - 51,946,196$$

d. No. The points on the scatter plot do not approximate a linear pattern.

e. Points corresponding to 1990 and 1997. The points between those two years do approximate a linear pattern.

63. a. No.

b. Yes. The points seem to follow a straight line pattern for years between 2010 and 2030.

c. $\dfrac{f(b) - f(a)}{b - a}$

$$= \frac{f(2030) - f(2010)}{2030 - 2010}$$

$$= \frac{2.2 - 3.9}{2030 - 2010}$$

$$= \frac{-1.7}{20}$$

$$= -0.085$$

d. $y - y_1 = m(x - x_1)$

$$y - 3.9 = -0.085(x - 2010)$$

$$y - 3.9 = -0.085x + 170.85$$

$$y = -0.085x + 174.75$$

65. a. Let x = the number of years since 1950, and let y = the U.S. population in thousands. Then, the average rate of change in U.S. population, in thousands, between 1950 and 1995 is given by:

$$\frac{f(b) - f(a)}{b - a}$$

$$= \frac{f(45) - f(0)}{45 - 0}$$

$$= \frac{263,044 - 152,271}{45 - 0}$$

$$= \frac{110,773}{45}$$

$$\approx 2461.6$$

Changing the units from thousands to millions yields $\dfrac{2,461.8}{1,000} = 2.4616$ million per year.

b. Remember to change the units into millions:

$$y - y_1 = m(x - x_1)$$
$$y - 152.271 = 2.4616(x - 0)$$
$$y - 152.271 = 2.4616x$$
$$y = 2.4616x - 152.271$$

c.   1975 corresponds to x = 25.
$$y = 2.4616(25) + 152.271$$
$$y = 213.811 \text{ or } 213,811,000$$

No. The values are different.

d.   The table can not be represented exactly by a linear function.

67. a.   Yes. The *x*-values have a constant change of $50, while the *y*-values have a constant change of $14.

b.   Since the table represents a linear function, the rate of change is the slope.

$$m = \frac{y_2 - y_1}{x_2 - x_1}$$
$$= \frac{5217 - 5203}{30,050 - 30,000}$$
$$= \frac{14}{50} = 0.28$$

For every $1.00 in income, taxes increase by $0.28.

c.   Equation:
$$y - y_1 = m(x - x_1)$$
$$y - 5,203 = 0.28(x - 30,000)$$
$$y - 5,203 = 0.28x - 8400$$
$$y = 0.28x - 3,197$$

d.   When $x = 30,100$,
$$y = 0.28(30,100) - 3197 = 5231.$$
When $x = 30,300$,
$$y = 0.28(30,300) - 3197 = 5287.$$

Yes. The results from the equation match with the table.

Section 1.5 Skills Check

1.
$$5x - 14 = 23 + 7x$$
$$5x - 7x - 14 = 23 + 7x - 7x$$
$$-2x - 14 = 23$$
$$-2x - 14 + 14 = 23 + 14$$
$$-2x = 37$$
$$\frac{-2x}{-2} = \frac{37}{-2}$$
$$x = -\frac{37}{2}$$
$$x = -18.5$$

Applying the intersections of graphs method, graph $y = 5x - 14$ and $y = 23 + 7x$. Determine the intersection point from the graph:

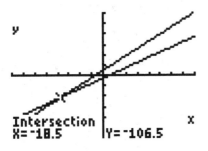

[−35, 35] by [−200, 200]

3.
$$3(x - 7) = 19 - x$$
$$3x - 21 = 19 - x$$
$$3x + x - 21 = 19 - x + x$$
$$4x - 21 = 19$$
$$4x - 21 + 21 = 19 + 21$$
$$4x = 40$$
$$\frac{4x}{4} = \frac{40}{4}$$
$$x = 10$$

Applying the intersections of graphs method yields:

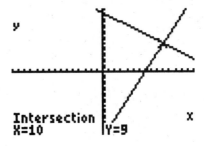

[−15, 15] by [−20, 20]

5. $\quad x - \dfrac{5}{6} = 3x + \dfrac{1}{4}$

$LCM : 12$

$$12\left(x - \frac{5}{6}\right) = 12\left(3x + \frac{1}{4}\right)$$
$$12x - 10 = 36x + 3$$
$$12x - 36x - 10 = 36x - 36x + 3$$
$$-24x - 10 = 3$$
$$-24x - 10 + 10 = 3 + 10$$
$$-24x = 13$$
$$\frac{-24x}{-24} = \frac{13}{-24}$$
$$x = -\frac{13}{24}$$

Applying the intersections of graphs method yields:

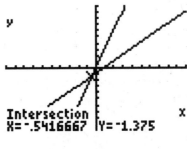

[−10, 10] by [−10, 10]

7. $\dfrac{5(x-3)}{6} - x = 1 - \dfrac{x}{9}$

$LCM : 18$

$18\left(\dfrac{5(x-3)}{6} - x\right) = 18\left(1 - \dfrac{x}{9}\right)$

$15(x-3) - 18x = 18 - 2x$

$15x - 45 - 18x = 18 - 2x$

$-3x - 45 = 18 - 2x$

$-1x - 45 = 18$

$-1x = 63$

$x = -63$

Applying the intersections of graphs method yields:

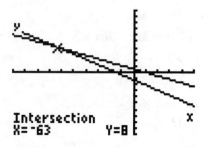

Intersection
X=-63   Y=8

$[-100, 50]$ by $[-20, 20]$

9. $5.92t = 1.78t - 4.14$

$5.92t - 1.78t = -4.14$

$4.14t = -4.14$

$\dfrac{4.14t}{4.14} = \dfrac{-4.14}{4.14}$

$t = -1$

Applying the intersections of graphs method yields:

Intersection
X=-1   Y=-5.92

$[-10, 10]$ by $[-20, 10]$

11. $\dfrac{3}{4} + \dfrac{1}{5}x - \dfrac{1}{3} = \dfrac{4}{5}x$

$LCM = 60$

$60\left(\dfrac{3}{4} + \dfrac{1}{5}x - \dfrac{1}{3}\right) = 60\left(\dfrac{4}{5}x\right)$

$45 + 12x - 20 = 48x$

$-36x = -25$

$x = \dfrac{-25}{-36} = \dfrac{25}{36}$

13. Answers a), b), and c) are the same. Let $f(x) = 0$ and solve for x.

$32 + 1.6x = 0$

$1.6x = -32$

$x = -\dfrac{32}{1.6}$

$x = -20$

The solution to $f(x) = 0$, the x-intercept of the function, and the zero of the function are all -20.

15. Answers a), b), and c) are the same. Let $f(x) = 0$ and solve for x.

$\dfrac{3}{2}x - 6 = 0$

$LCM : 2$

$2\left(\dfrac{3}{2}x - 6\right) = 2(0)$

$3x - 12 = 0$

$3x = 12$

$x = 4$

The solution to $f(x) = 0$, the x-intercept of the function, and the zero of the function are all 4.

17. a. The x-intercept is 2, since an input of 2 creates an output of 0 in the function.

b. The y-intercept is -34, since the output of -34 corresponds with an input of 0.

c. The solution to $f(x) = 0$ is equal to the x-intercept position for the function. Therefore, the solution to $f(x) = 0$ is 2.

19. The answers to a) and b) are the same. The graph crosses the x-axis at $x = 40$.

21. Applying the intersections of graphs method yields:

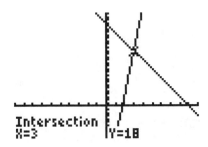

$[-10, 10]$ by $[-10. 30]$
The solution is the x-coordinate of the intersection point or $x = 3$.

23. Applying the intersections of graphs method yields:

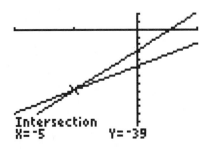

$[-10, 5]$ by $[-70, 10]$

The solution is the x-coordinate of the intersection point or $s = -5$.

25. Applying the intersections of graphs method yields:

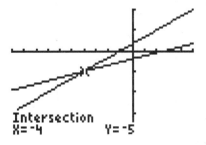

$[-10, 5]$ by $[-20, 10]$

The solution is the x-coordinate of the intersection point. $t = -4$.

27. Applying the intersections of graphs method yields:

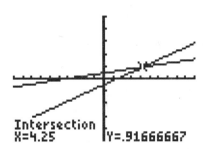

$[-10, 10]$ by $[-5, 5]$

The solution is the x-coordinate of the intersection point, which is $x = 4.25 = \dfrac{17}{4}$.

29. a. $A = P(1 + rt)$
$A = P + Prt$
$A - P = P - P + Prt$
$A - P = Prt$
$\dfrac{A - P}{Pr} = \dfrac{Prt}{Pr}$
$\dfrac{A - P}{Pr} = t$ or $t = \dfrac{A - P}{Pr}$

b. $A = P(1 + rt)$

$$\frac{A}{1 + rt} = \frac{P(1 + rt)}{1 + rt}$$

$$\frac{A}{1 + rt} = P \quad \text{or} \quad P = \frac{A}{1 + rt}$$

31. $5F - 9C = 160$

$5F - 9C + 9C = 160 + 9C$

$5F = 160 + 9C$

$$\frac{5F}{5} = \frac{160 + 9C}{5}$$

$$F = \frac{9}{5}C + \frac{160}{5}$$

$$F = \frac{9}{5}C + 32$$

33. $\dfrac{P}{2} + A = 5m - 2n$

$LCM : 2$

$$2\left(\frac{P}{2} + A\right) = 2(5m - 2n)$$

$P + 2A = 10m - 4n$

$P + 2A - 10m = 10m - 4n - 10m$

$$\frac{P + 2A - 10m}{-4} = \frac{-4n}{-4}$$

$$\frac{P + 2A - 10m}{-4} = n$$

$$n = \frac{P}{-4} + \frac{2A}{-4} - \frac{10m}{-4}$$

$$n = \frac{5m}{2} - \frac{P}{4} - \frac{A}{2}$$

35. $5x - 3y = 5$

$-3y = -5x + 5$

$$y = \frac{-5x + 5}{-3}$$

$$y = \frac{5}{3}x - \frac{5}{3}$$

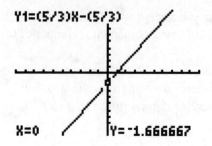

$[-10, 10]$ by $[-10, 10]$

37. $x^2 + 2y = 6$

$2y = 6 - x^2$

$$y = \frac{6 - x^2}{2}$$

$$y = 3 - \frac{1}{2}x^2 \quad \text{or}$$

$$y = -\frac{1}{2}x^2 + 3$$

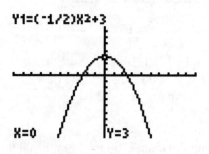

$[-10, 10]$ by $[-10, 10]$

Section 1.5 Exercises

39. Let $y = 690,000$ and solve for x.

$$690,000 = 828,000 - 2300x$$
$$-138,000 = -2300x$$
$$x = \frac{-138,000}{-2300}$$
$$x = 60$$

After 60 months or 5 years the value of the building will be $690,000.

41. $S = P(1 + rt)$
$$9000 = P(1 + (0.10)(5))$$
$$9000 = P(1 + 0.50)$$
$$9000 = 1.5P$$
$$P = \frac{9000}{1.5} = 6000$$

$6000 must be invested as the principal.

43. $M = 0.959W - 1.226$
$$50,560 = 0.959W - 1.226$$
$$0.959W = 50,561.226$$
$$W = \frac{50,561.226}{0.959} \approx 52,723$$

The median annual salary for whites is approximately $57,723.

45. Recall that $5F - 9C = 160$. Let $F = C$, and solve for C.

$$5C - 9C = 160$$
$$-4C = 160$$
$$C = \frac{160}{-4}$$
$$C = -40$$

Therefore, $F = C$ when the temperature is $-40°$.

47. Let $y = 259.4$, and solve for x.

$$259.4 = 0.155x + 255.37$$
$$259.4 - 255.37 = 0.155x$$
$$4.03 = 0.155x$$
$$x = \frac{4.03}{0.155}$$
$$x = 26$$

An x-value of 26 corresponds to the year 1996. The average reading score is 259.4 in 1996.

49. Let $B(x) = 35.32$, and calculate x.

$$35.32 = -3.963x + 51.172$$
$$-3.963x = -15.852$$
$$x = \frac{-15.852}{-3.963} = 4$$

Based on the model, the average monthly mobile phone bill is $35.32 in 1999.

51. Note that p is in thousands. A population of 258,241,000 corresponds to a p-value of 258,241. Let $p = 258,241$ and solve for x.

$$258,241 = 2351x + 201,817$$
$$258,241 - 201,817 = 2351x$$
$$83,424 = 2351x$$
$$x = \frac{56,424}{2351}$$
$$x = 24$$

An x-value of 24 corresponds to the year 1994. Based on the model, in 1994 the population is estimated to be 258,241,000.

53. When the number of prisoners is 797,130, then $y = 797.130$. Let $y = 797.130$, and calculate x.

$$797.130 = 68.476x + 728.654$$

$$797.130 - 728.654 = 68.476x$$

$$68.476 = 68.476x$$

$$x = \frac{68.476}{68.476} = 1$$

An x-value of one corresponds to the year 1991. The number of inmates was 797,130 in 1991.

55. a. Let $p = 49\%$, and solve for x.

$$49 = 65.4042 - 0.3552x$$

$$49 - 65.4042 = -0.3552x$$

$$-61.4042 = -0.3552x$$

$$x = \frac{-16.4042}{-0.3552}$$

$$x \approx 46.2$$

An x-value of 46 corresponds to the year 1996. The model predicts that in 1996 the percent voting in a presidential election is 49%

b. Year 2000 corresponds with an x-value of 50. Let $x = 50$, and solve for p.

$$p = 65.4042 - 0.3552(50)$$

$$p = 65.4042 - 17.76$$

$$p = 47.6442$$

Based on the model the percentage of people voting in the 2000 election was approximately 47.6%. The prediction is different from reality. Models do not always yield accurate predictions.

57. Let $y = 6000$, and solve for x.

$$6000 = 277.318x - 1424.766$$

$$277.318x = 7424.766$$

$$x = \frac{7424.766}{277.318} \approx 26.77$$

An x-value of 27 corresponds to the year 1997. Cigarette advertising exceeds $6 billion in 1997.

59. Let x represent the score on the fifth exam.

$$90 = \frac{92 + 86 + 79 + 96 + x}{5}$$

LCM: 5

$$5(90) = 5\left(\frac{92 + 86 + 79 + 96 + x}{5}\right)$$

$$450 = 353 + x$$

$$x = 97$$

The student must score 97 on the fifth exam to earn a 90 in the course.

61. Let x = the company's 1999 revenue in billions of dollars.

$$0.94x = 74$$

$$x = \frac{74}{0.94}$$

$$x \approx 78.723$$

The company's 1999 revenue was approximately $78.723 billion.

63. Commission Reduction

$$= (20\%)(50,000)$$

$$= 10,000$$

New Commission

$$= 50,000 - 10,000$$

$$= 40,000$$

To return to a $50,000 commission, the commission must be increased $10,000. The percentage increase is now based on the $40,000 commission. L

Let x represent the percent increase from the second year.

$$40,000x = 10,000$$
$$x = 0.25 = 25\%$$

65. Total cost = Original price + Sales tax

Let x = original price.

$$29,998 = x + 6\%x$$
$$29,998 = x + 0.06x$$
$$29,998 = 1.06x$$
$$x = \frac{29,998}{1.06} = 28,300$$

Sales tax = $29,998 - 28,300 = \$1698$

Section 1.6 Skills Check

1. No. The data points do not lie approximately in a straight line.

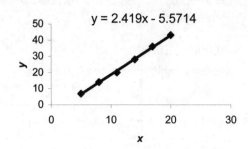

3.
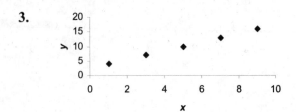

$$y = 2.419x - 5.571$$

5. Exactly. The first differences are constantly three.

7. Using a spreadsheet program yields

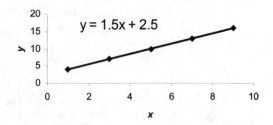

13.
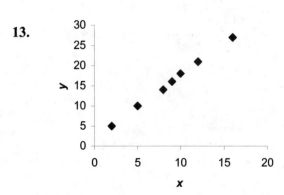

15. Using a spreadsheet program yields

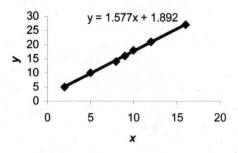

9.
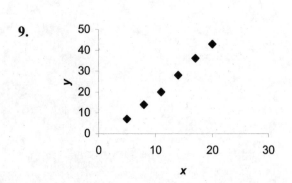

11. The points appear to lie approximately along a line. Using a spreadsheet program yields

17.
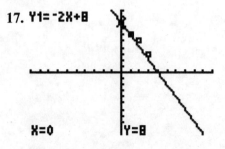

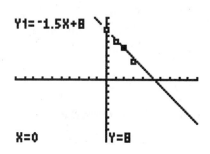

The second equation, $y = -1.5x + 8$, is a better fit to the data points.

19. a. Exactly linear. The first differences are constant
  b. Nonlinear. The first differences increase continuously.
  c. Approximately linear. The first differences vary, but don't grow continuously.

Section 1.6 Exercises

21. a. Discrete. The ratio is calculated each year, and the years are one unit apart.

  b. No. A line would not fit the points on the scatter plot.

  c. Yes. Beginning in 2010, a line would fit the points on the scatter plot well.

23. a. Yes. There is a one unit gap between the years and a constant 60 unit gap in future values.

  b. Yes. Since the first differences are constant, the future value can be modeled by a linear function

  c. Using the graphing calculator yields

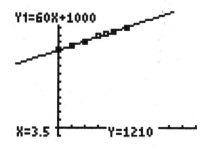

[−3, 10] by [−100, 1500]

25. a. Let $x = 23$. Then,
$$y = 15.910x + 242.758$$
$$= 15.910(23) + 242.758$$
$$= 365.93 + 242.758$$
$$= 608.688$$

Based on the model, 608,688 people were employed in dentist's offices in 1993. Since 1993 is within the range of the data used to generate the model (1970-1998), this calculation is an interpolation.

b. Let $x = 30$. Then,

$$y = 15.910x + 242.758$$
$$= 15.910(30) + 242.758$$
$$= 477.3 + 242.758$$
$$= 720.058$$

Based on the model, 720,058 people were employed in dentist's offices in 2000. Since 2000 is not within the range of the data used to generate the model (1970-1998), this calculation is an extrapolation.

27. a. Using a spreadsheet program yields

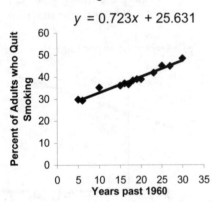

Smoking Cessation

$y = 0.723x + 25.631$

b. Solve $0.723x + 25.631 = 39$. Using the intersections of graphs method to determine x when $y = 39$ yields

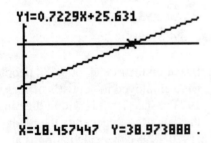

Y1=0.7229X+25.631

X=18.457447 Y=38.973888

Based on the model, 39% of adults had quit smoking 18.46 years past 1960. Therefore, the year was approximately 1978.

c. Since the data is not exactly linear (see the scatter plot in part a) above), the model will yield only approximate solutions.

29. a. Using a spreadsheet program yields

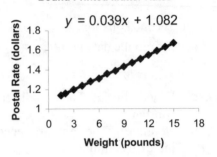

Bound Printed Matter Rates

$y = 0.039x + 1.082$

b. The model fits the data very well. Notice the small residuals in the following table.

Weight (pounds)	Actual postal rate (dollars)	Predicted postal rate based on the regression equation (dollars)	Residual (difference between the actual and predicted values) (dollars)
2	1.16	1.159675	0.000325
3	1.20	1.198734	0.001266
4	1.24	1.237794	0.002206
5	1.28	1.276853	0.003147
6	1.31	1.315913	−0.005913
7	1.36	1.354972	0.005028
8	1.39	1.394031	−0.004031
9	1.43	1.433091	−0.003091
10	1.47	1.472150	−0.002150
11	1.51	1.511210	−0.001210
12	1.55	1.550269	−0.000269
13	1.59	1.589328	0.000672
14	1.63	1.628388	0.001612
15	1.67	1.667447	0.002553

31. a. Let $x = 16$, and solve for y.

$$y = 0.039(16) + 1.082$$
$$y = 0.624 + 1.082$$
$$y = 1.706 \approx 1.71$$

Using the unrounded model yields

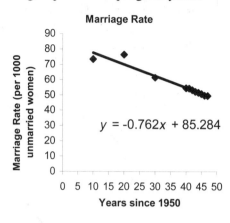

[–3, 20] by [0, 3]

b. Let $y = 1.55$, and solve for x.

$$1.55 = 0.039x + 1.082$$
$$1.55 - 1.082 = 0.039x$$
$$0.039x = 0.468$$
$$x = \frac{0.468}{0.039} = 12 \text{ pounds}$$

c. The slope of the linear model is 0.039. Therefore, for when the weight changes by one pound, the postal rate changes by approximately 3.9 cents.

d. Considering the data, the change in postal rate between 9 pounds and 14 pounds is 4 cents per pound.

33. a. Using a spreadsheet program yields

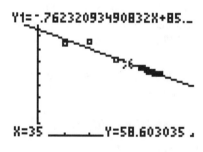

$y = -0.762x + 85.284$

b. Let $x = 1985 - 1950 = 35$, and solve for y.

$$y = -0.7622(35) + 85.284$$
$$y = 58.607 \approx 58.6$$

The model predicts 58.6 marriages per 1000 unmarried women.

Using the unrounded model yields

[–10, 60] by [0, 100]

c. Let $y = 50$, and solve for x.

$$50 = -0.762x + 85.284$$
$$50 - 85.284 = -0.762x$$
$$-0.762x = -35.284$$
$$x = \frac{-35.284}{-0.762} = 46.304 \approx 46.3$$

50 marriages per 1000 unmarried women occurs in approximately 1996.

d. The answer in part c) is an approximation based on the model. Considering the data, the marriage rate is 50 between 1995 and 1996.

35. a. Using a spreadsheet program yields

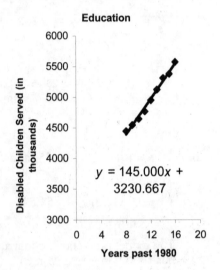

Education

$y = 145.000x + 3230.667$

b. Based on the model, when the time increases by one year, the number of disabled children served increases by 145,000.

c. In 2005, $x = 25$. Using the unrounded model yields

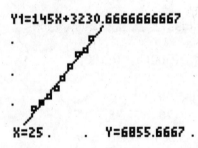

Y1=145X+3230.6666666667

X=25 .   . Y=6855.6667 .

[5, 30] by [4000, 6000]

Approximately 6,856,000 children will be served in 2005 based on the model. This is an extrapolation, since 2005 is beyond the scope of the available data.

37. a. Using a spreadsheet program yields

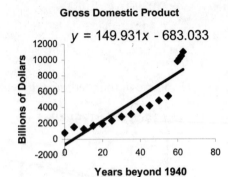

Gross Domestic Product

$y = 149.931x - 683.033$

b. See 37 a) above. The equation is $y = 149.931x - 683.003$.

c. Considering the scatter plot, the model does not fit the data very well.

39. a. Using a spreadsheet program yields

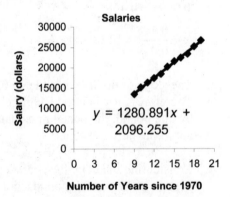

Salaries

$y = 1280.891x + 2096.255$

b. See 39 a) above. The equation is $y = 1280.891x + 2096.255$.

c. See 39 a) above. The linear function fits the data points very well. The line is very close to the data points on the scatter plot.

d. i.   Discrete
   ii. Discrete
   iii. Continuous. The model is not limited to data from the table.

41. a.   Using a spreadsheet program yields

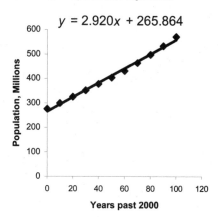

U.S. Population Projections

$y = 2.920x + 265.864$

b.   Using the unrounded model,
$f(65) \approx 455.64$.

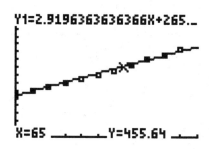

$Y1 = 2.9196363636366X + 265.\_$

$X = 65$ _____ $Y = 455.64$ ____

[0, 100] by [0, 800]

In 2065, the projected U.S. population is 455.64 million or 455,640,000.

c.   In 2080, $x = 80$.

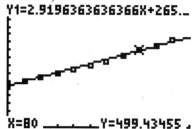

$Y1 = 2.9196363636366X + 265.\_$

$X = 80$ _____ $Y = 499.43455$.

[0, 100] by [0, 800]

Based on the model, the U.S. population in 2080 will be 499.4 million. The prediction in the table is slightly smaller, 497.8 million.

43. a.   Using a spreadsheet program yields

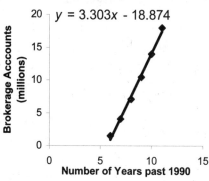

Internet Brokerage Accounts

$y = 3.303x - 18.874$

b.   Considering the slope of the linear model, a one year increase in time corresponds to a 3.3 million unit increase in Internet brokerage accounts.

c.   Let $y = 20$, and solve for x.

$$20 = 3.303x - 18.874$$

$$20 + 18.874 = 3.303x$$

$$38.874 = 3.303x$$

$$x = \frac{38.874}{3.303} = 11.7693 \approx 11.8$$

1.

Based on the model, the number of Internet brokerage accounts will be 20 million near the end of 2001.

d.   The model may no longer be valid due to economic and social instability resulting from the terrorist attacks on September 11, 2001.

45. a. Using a spreadsheet program yields

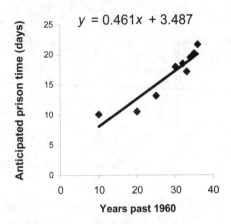

b. Let $x = 15$, and solve for y.

$$y = 0.461(15) + 3.487$$

$$y = 10.402$$

Based on the model the anticipated prison time in 1975 is approximately 10.4 days.

Section 1.7 Skills Check

1. Applying the substitution method
   $y = 3x - 2$ and $y = 3 - 2x$

   $3 - 2x = 3x - 2$

   $3 - 2x - 3x = 3x - 3x - 2$

   $3 - 5x = -2$

   $-5x = -5$

   $x = 1$

   Substituting to find y

   $y = 3(1) - 2 = 1$

   $(1,1)$ is the intersection point.

3. Applying the intersection of graphs method

   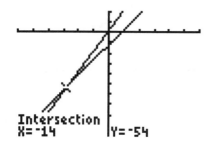

   [–30, 30] by [–100, 20]

   The solution is $(-14, -54)$.

5. Solving the equations for y
   $4x - 3y = -4$

   $-3y = -4x - 4$

   $y = \dfrac{-4x - 4}{-3}$

   $y = \dfrac{4}{3}x + \dfrac{4}{3}$

   and

   $2x - 5y = -4$

   $-5y = -2x - 4$

   $y = \dfrac{-2x - 4}{-5}$

   $y = \dfrac{2}{5}x + \dfrac{4}{5}$

Applying the intersection of graphs method

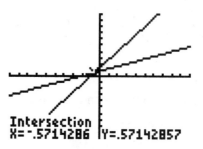

[–10, 10] by [–10, 10]

The solution is $(-0.5714, 0.5714)$ or

$\left(-\dfrac{4}{7}, \dfrac{4}{7}\right)$.

7. $\begin{cases} 2x + 5y = 6 & (Eq\,1) \\ x + 2.5y = 3 & (Eq\,2) \end{cases}$

   $\begin{cases} 2x + 5y = 6 & (Eq\,1) \\ -2x - 5y = -6 & -2 \times (Eq\,2) \end{cases}$

   $0 = 0$

   There are infinitely many solutions to the system. The graphs of both equations represent the same line.

9. $\begin{cases} x - 5y = 12 \\ 3x + 4y = -2 \end{cases}$

Solving the first equation for x

$x - 5y = 12$

$x = 5y + 12$

Substituting into the second equation

$3(5y + 12) + 4y = -2$

$15y + 36 + 4y = -2$

$19y + 36 = -2$

$19y = -38$

$y = -2$

Substituting to find x

$x - 5(-2) = 12$

$x + 10 = 12$

$x = 2$

The solution is $(2, -2)$.

11. $\begin{cases} 2x - 3y = 5 \\ 5x + 4y = 1 \end{cases}$

Solving the first equation for x

$2x - 3y = 5$

$2x = 3y + 5$

$x = \dfrac{3y + 5}{2}$

Substituting into the second equation

$5\left(\dfrac{3y + 5}{2}\right) + 4y = 1$

$2\left[5\left(\dfrac{3y + 5}{2}\right) + 4y\right] = 2[1]$

$15y + 25 + 8y = 2$

$23y + 25 = 2$

$23y = -23$

$y = -1$

Substituting to find x

$2x - 3(-1) = 5$

$2x + 3 = 5$

$2x = 2$

$x = 1$

The solution is $(1, -1)$.

13. $\begin{cases} x + 3y = 5 & (Eq\,1) \\ 2x + 4y = 8 & (Eq\,2) \end{cases}$

$\begin{cases} -2x - 6y = -10 & -2 \times (Eq\,1) \\ 2x + 4y = 8 & (Eq\,2) \end{cases}$

$-2y = -2$

$y = 1$

Substituting to find x

$x + 3(1) = 5$

$x + 3 = 5$

$x = 2$

The solution is $(2, 1)$.

15. $\begin{cases} 5x + 3y = 8 & (Eq1) \\ 2x + 4y = 8 & (Eq2) \end{cases}$

$\begin{cases} -10x - 6y = -16 & -2 \times (Eq1) \\ 10x + 20y = 40 & 5 \times (Eq2) \end{cases}$

$14y = 24$

$y = \dfrac{24}{14} = \dfrac{12}{7}$

Substituting to find x

$2x + 4\left(\dfrac{12}{7}\right) = 8$

$7\left[2x + \left(\dfrac{48}{7}\right)\right] = 7[8]$

$14x + 48 = 56$

$14x = 8$

$x = \dfrac{8}{14} = \dfrac{4}{7}$

The solution is $\left(\dfrac{4}{7}, \dfrac{12}{7}\right)$.

17. $\begin{cases} 0.3x + 0.4y = 2.4 & (Eq1) \\ 5x - 3y = 11 & (Eq2) \end{cases}$

$\begin{cases} 9x + 12y = 72 & 30 \times (Eq1) \\ 20x - 12y = 44 & 4 \times (Eq2) \end{cases}$

$29x = 116$

$x = \dfrac{116}{29} = 4$

Substituting to find x

$5(4) - 3y = 11$

$20 - 3y = 11$

$-3y = -9$

$y = 3$

The solution is $(4, 3)$.

19. $\begin{cases} 3x + 6y = 12 & (Eq1) \\ 2x + 4y = 8 & (Eq2) \end{cases}$

$\begin{cases} -6x - 12y = -24 & -2 \times (Eq1) \\ 6x + 12y = 24 & 3 \times (Eq2) \end{cases}$

$0 = 0$

Infinitely many solutions. The lines are the same. This is a dependent system.

21. $\begin{cases} 6x - 9y = 12 & (Eq1) \\ 3x - 4.5y = -6 & (Eq2) \end{cases}$

$\begin{cases} 6x - 9y = 12 & (Eq1) \\ -6x + 9y = 12 & -2 \times (Eq2) \end{cases}$

$0 = 24$

No solution. Lines are parallel.

23. $\begin{cases} y = 3x - 2 \\ y = 5x - 6 \end{cases}$

Substituting the first equation into the second equation

$3x - 2 = 5x - 6$

$-2x - 2 = -6$

$-2x = -4$

$x = 2$

Substituting to find y

$y = 3(2) - 2 = 6 - 2 = 4$

The solution is $(2, 4)$.

25. $\begin{cases} 4x+6y=4 \\ x=4y+8 \end{cases}$

Substituting the second equationinto the first equation

$4(4y+8)+6y=4$

$16y+32+6y=4$

$22y+32=4$

$22y=-28$

$y=\dfrac{-28}{22}=-\dfrac{14}{11}$

Substituting to find x

$x=4\left(-\dfrac{14}{11}\right)+8$

$x=-\dfrac{56}{11}+\dfrac{88}{11}=\dfrac{32}{11}$

The solution is $\left(\dfrac{32}{11},-\dfrac{14}{11}\right)$.

27. $\begin{cases} 2x-5y=16 & (Eq1) \\ 6x-8y=34 & (Eq2) \end{cases}$

$\begin{cases} -6x+15y=-48 & -3\times(Eq1) \\ 6x-8y=34 & (Eq2) \end{cases}$

$7y=-14$

$y=-2$

Substituting to find x

$2x-5(-2)=16$

$2x+10=16$

$2x=6$

$x=3$

The solution is $(3,-2)$.

29. $\begin{cases} 3x-7y=-1 & (Eq1) \\ 4x+3y=11 & (Eq2) \end{cases}$

$\begin{cases} -12x+28y=4 & -4\times(Eq1) \\ 12x+9y=33 & 3\times(Eq2) \end{cases}$

$37y=37$

$y=\dfrac{37}{37}=1$

Substituting to find x

$3x-7(1)=-1$

$3x-7=-1$

$3x=6$

$x=2$

The solution is $(2,1)$.

31. $\begin{cases} 4x-3y=9 & (Eq1) \\ 8x-6y=16 & (Eq2) \end{cases}$

$\begin{cases} -8x+6y=-18 & -2\times(Eq1) \\ 8x-6y=16 & (Eq2) \end{cases}$

$0=-2$

No solution. Lines are parallel.

Section 1.7 Exercises

33.
$$R = C$$
$$76.50x = 2970 + 27x$$
$$49.50x = 2970$$
$$x = \frac{2970}{49.50}$$
$$x = 60$$

Applying the intersections of graphs method yields $x = 60$.

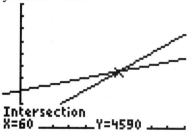

Intersection
X=60   Y=4590

[–10, 100] by [–10, 10,000]

Break-even occurs when the number of units produced and sold is 60.

35. a. Let $p = 60$ and solve for q.
Supply function
$$60 = 5q + 20$$
$$5q = 40$$
$$q = 8$$
Demand function
$$60 = 128 - 4q$$
$$-4q = -68$$
$$q = \frac{-68}{-4} = 17$$

When the price is $60, the quantity supplied is 8, while the quantity demanded is 17.

b. Equilibrium occurs when the demand equals the supply,

$$5q + 20 = 128 - 4q$$
$$9q + 20 = 128$$
$$9q = 108$$
$$q = \frac{108}{9} = 12$$

Substituting to calculate p
$$p = 5(12) + 20 = 80$$

When the price is $80, 12 units are produced and sold. This level of production and price represents equilibrium.

37. a. Applying the intersection of graphs method

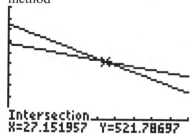

Intersection
X=27.151957 Y=521.78697

[0, 50] by [–100, 1000]

The solution is $(27.152, 521.787)$. The number of active duty Navy personnel equals the number of active duty Air Force personnel in 1987.

b. Considering the solution in part a, approximately 521,787 people will be on active duty in each service branch.

39. a. $\begin{cases} y = 24.5x + 93.5 \\ y = -0.2x + 1007 \end{cases}$

Substituting the first equation

into the second equation

$24.5x + 93.5 = -0.2x + 1007$

$10(24.5x + 93.5) = 10(-0.2x + 1007)$

$245x + 935 = -2x + 10{,}070$

$247x + 935 = 10{,}070$

$247x = 9135$

$x = \dfrac{9135}{247} = 36.9838 \approx 37$

b. In $1990 + 37 = 2027$, mint sales and gum sales are equal.

c. The graphs are misleading. Notice that the scales are different. Mint sales are measured between $0 and $300 million, while gum sales are measured between $0 and $1000 million. Also note that the first tick mark on the y-axis for each graph represents inconsistent units when compared with the remainder of the graph.

41. Let l = the low stock price, and let h = the high stock price.

$\begin{cases} h + l = 83.5 & (Eq\,1) \\ h - l = 21.88 & (Eq\,2) \end{cases}$

$2h = 105.38$

$h = \dfrac{105.38}{2} = 52.69$

Substituting to calculate l

$52.69 + l = 83.5$

$l = 30.81$

The high stock price is $52.69, while the low stock price is $30.81.

43. a. $x + y = 2400$

b. $30x$

c. $45y$

d. $30x + 45y = 84{,}000$

e. $\begin{cases} x + y = 2400 & (Eq\,1) \\ 30x + 45y = 84{,}000 & (Eq\,2) \end{cases}$

$\begin{cases} -30x - 30y = -72{,}000 & -30 \times (Eq\,1) \\ 30x + 45y = 84{,}000 & (Eq\,2) \end{cases}$

$15y = 12{,}000$

$y = \dfrac{12{,}000}{15} = 800$

Substituting to calculate x

$x + 800 = 2400$

$x = 1600$

The promoter needs to sell 1600 tickets at $30 per ticket and 800 tickets at $45 per ticket.

45. a. Let x = the amount in the safer account, and let y = the amount in the riskier account.

$$\begin{cases} x + y = 100,000 & (Eq\,1) \\ 0.08x + 0.12y = 9000 & (Eq\,2) \end{cases}$$

$$\begin{cases} -0.08x - 0.08y = -8000 & -0.08 \times (Eq\,1) \\ 0.08x + 0.12y = 9000 & (Eq\,2) \end{cases}$$

$$0.04y = 1000$$

$$y = \frac{1000}{0.04} = 25,000$$

Substituting to calculate x

$$x + 25,000 = 100,000$$

$$x = 75,000$$

$75,000 is invested in the 8% account, and $25,000 is invested in the 12% account.

b. Using two accounts minimizes investment risk.

47. Let x = the number of glasses of milk, and let y = the number of quarter pound servings of meat.

Protein equation:

$$8.5x + 22y = 69.5$$

Iron equation:

$$0.1x + 3.4y = 7.1$$

$$\begin{cases} 8.5x + 22y = 69.5 & (Eq\,1) \\ 0.1x + 3.4y = 7.1 & (Eq\,2) \end{cases}$$

$$\begin{cases} 8.5x + 22y = 69.5 & (Eq\,1) \\ -8.5x - 289y = -603.5 & -85 \times (Eq\,2) \end{cases}$$

$$-267y = -534$$

$$y = \frac{-534}{-267} = 2$$

Substituting to calculate x

$$8.5x + 22(2) = 69.5$$

$$8.5x + 44 = 69.5$$

$$8.5x = 25.5$$

$$x = \frac{25.5}{8.5} = 3$$

The person on the special diet needs to consume 3 glasses of milk and 2 quarter pound portions of meat to reach the required iron and protein content in the diet.

49. Let x = the amount of 10% solution, and let y = the amount of 5% solution.

$x + y = 20$

Medicine concentration:

$10\%x + 5\%y = 8\%(20)$

$$\begin{cases} x + y = 20 & (Eq1) \\ 0.10x + 0.05y = 1.6 & (Eq2) \end{cases}$$

$$\begin{cases} -0.10x - 0.10y = -2 & -0.10\times(Eq1) \\ 0.10x + 0.05y = 1.6 & (Eq2) \end{cases}$$

$-0.05y = -0.4$

$y = \dfrac{-0.4}{-0.05} = 8$

Substituting to calculate x

$x + 8 = 20$

$x = 12$

The nurse needs to mix 12 cc of the 10% solution with 8 cc of the 5% solution to obtain 20 cc of an 8% solution.

51.

L1	L2	L3	1
50	210	0	
60	190	40	
70	170	80	
80	150	120	
100	110	200	
------	------	------	

L1(1)=50

a. Demand function: Finding a linear model using L₂ as input and L₁ as output yields

$p = -\dfrac{1}{2}q + 155$.

```
LinReg
 y=ax+b
 a=-.5
 b=155
```

b. Supply function: Finding a linear model using L₃ as input and L₁ as output yields $p = \dfrac{1}{4}q + 50$.

```
LinReg
 y=ax+b
 a=.25
 b=50
```

c.   Applying the intersection of graphs method

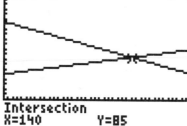

Intersection
X=140   Y=85

[0, 200] by [−50, 200]

When the price is $85, 140 units are both supplied and demanded. Therefore, equilibrium occurs when the price is $85 per unit.

53. Applying the intersection of graphs method

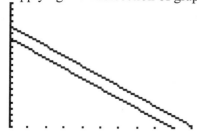

[0, 10] by [55, 75]

Note that the lines do not intersect. The slopes are the same, but the y-intercepts are different. There is no solution to the system. Based on the two models, the percentages are never equal.

55. a. $300x + 200y = 100,000$

  b. $x = 2y$

$300(2y) + 200y = 100,000$

$800y = 100,000$

$y = \dfrac{100,000}{800} = 125$

Substituting to calculate x

$x = 2(125) = 250$

There are 250 clients in the first group and 125 clients in the second group.

57. The slope of the demand function is

$$m = \frac{y_2 - y_1}{x_2 - x_1} = \frac{10 - 60}{900 - 400} = \frac{-50}{500} = \frac{-1}{10}.$$

Calculating the equation:

$$y - y_1 = m(x - x_1)$$

$$y - 10 = \frac{-1}{10}(x - 900)$$

$$y - 10 = \frac{-1}{10}x + 90$$

$$y = \frac{-1}{10}x + 100 \text{ or}$$

$$p = \frac{-1}{10}q + 100$$

Likewise, the slope of the supply function is

$$m = \frac{y_2 - y_1}{x_2 - x_1} = \frac{30 - 50}{700 - 1400} = \frac{-20}{-700} = \frac{2}{70}.$$

Calculating the equation

$$y - y_1 = m(x - x_1)$$

$$y - 30 = \frac{2}{70}(x - 700)$$

$$y - 30 = \frac{2}{70}x - 20$$

$$y = \frac{2}{70}x + 10 \text{ or}$$

$$p = \frac{2}{70}q + 10$$

The quantity, q, that produces market equilibrium is 700.

$$\frac{-1}{10}q + 100 = \frac{2}{70}q + 10$$

$$70\left(\frac{-1}{10}q + 100\right) = 70\left(\frac{2}{70}q + 10\right)$$

$$-7q + 7000 = 2q + 700$$

$$-9q = -6300$$

$$q = \frac{-6300}{-9} = 700$$

The price, p, at market equilibrium is \$30.

$$p = \frac{2}{70}(700) + 10$$

$$p = 2(10) + 10$$

$$p = 30$$

700 units priced at \$30 represents the market equilibrium.

Section 1.8 Skills Check

1. Algebraically:

$3x - 7 \le 5 - x$

$4x - 7 \le 5$

$4x \le 12$

$x \le \dfrac{12}{4}$

$x \le 3$

Graphically:

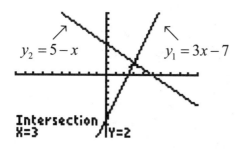

Intersection
X=3   Y=2

[–10, 10] by [–10, 10]

$3x - 7 \le 5 - x$ implies that the solution region is $x \le 3$.

The interval notation is $(-\infty, 3]$.

The graph of the solution is

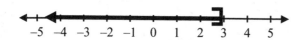

3. Algebraically:

$4(3x - 2) \le 5x - 9$

$12x - 8 \le 5x - 9$

$7x \le -1$

$\dfrac{7x}{7} \le \dfrac{-1}{7}$

$x \le -\dfrac{1}{7}$

Graphically:

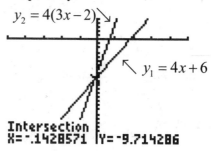

Intersection
X=-.1428571   Y=-9.714286

[–5, 5] by [–25, 5]

$4(3x - 2) \le 5x - 9$ implies that the solution region is $x \le -\dfrac{1}{7}$.

The interval notation is $\left(-\infty, -\dfrac{1}{7}\right]$.

The graph of the solution is

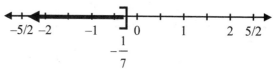

5. Algebraically:

$4x + 1 < -\dfrac{3}{5}x + 5$

$5(4x + 1) < 5\left(-\dfrac{3}{5}x + 5\right)$

$20x + 5 < -3x + 25$

$23x + 5 < 25$

$23x < 20$

$x < \dfrac{20}{23}$

Graphically:

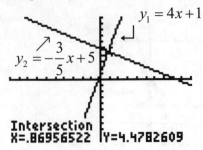

$$y_1 = 4x + 1$$

$$y_2 = -\frac{3}{5}x + 5$$

Intersection
X=.86956522 Y=4.4782609

[–10, 10] by [–10, 10]

$4x + 1 < -\dfrac{3}{5}x + 5$ implies that the solution

region is $x < \dfrac{20}{23}$.

The interval notation is $\left(-\infty, \dfrac{20}{23}\right)$.

The graph of the solution is

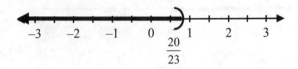

7. Algebraically:

$$\frac{x-5}{2} < \frac{18}{5}$$

$$10\left(\frac{x-5}{2}\right) < 10\left(\frac{18}{5}\right)$$

$$5(x-5) < 2(18)$$

$$5x - 25 < 36$$

$$5x < 61$$

$$x < \frac{61}{5}$$

Graphically:

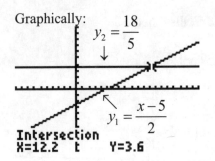

$$y_2 = \frac{18}{5}$$

$$y_1 = \frac{x-5}{2}$$

Intersection
X=12.2 t   Y=3.6

[–10, 20] by [–10, 10]

$\dfrac{x-5}{2} < \dfrac{18}{5}$ implies that the solution

region is $x < \dfrac{61}{5}$.

The interval notation is $\left(-\infty, \dfrac{61}{5}\right)$.

The graph of the solution is

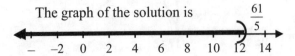

9. Algebraically:

$$\frac{3(x-6)}{2} \geq \frac{2x}{5} - 12$$

$$10\left(\frac{3(x-6)}{2}\right) \geq 10\left(\frac{2x}{5} - 12\right)$$

$$5(3(x-6)) \geq 2(2x) - 120$$

$$15(x-6) \geq 4x - 120$$

$$15x - 90 \geq 4x - 120$$

$$11x - 90 \geq -120$$

$$11x \geq -30$$

$$x \geq -\frac{30}{11}$$

Graphically:

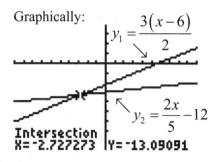

[−10, 10] by [−35, 15]

$\frac{3(x-6)}{2} \geq \frac{2x}{5} - 12$ implies that the solution region is $x \geq -\frac{30}{11}$.

The interval notation is $\left[-\frac{30}{11}, \infty\right)$.

The graph of the solution is

11. Algebraically:

$2.2x - 2.6 \geq 6 - 0.8x$

$3.0x - 2.6 \geq 6$

$3.0x \geq 8.6$

$x \geq \frac{8.6}{3.0}$

$x \geq 2.8\overline{6}$

Graphically:

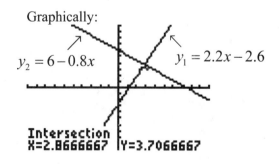

[−10, 10] by [−10, 10]

$2.2x - 2.6 \geq 6 - 0.8x$ implies that the solution region is $x \geq 2.8\overline{6}$.

The interval notation is $\left[2.8\overline{6}, \infty\right)$.

The graph of the solution is

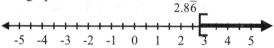

13. Applying the intersection of graphs method yields:

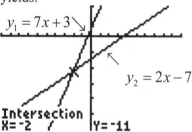

[−10, 10] by [−30, 10]

$7x + 3 < 2x - 7$ implies that the solution region is $x < -2$.

The interval notation is $(-\infty, -2)$.

15. To apply the x-intercept method, first rewrite the inequality so that zero is on one side of the inequality.

$5(2x+4) \geq 6(x-2)$

$10x + 20 \geq 6x - 12$

$4x + 32 \geq 0$

Let $f(x) = 4x + 32$, and determine graphically where $f(x) \geq 0$.

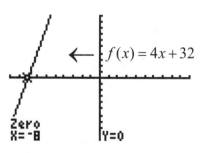

[−10, 10] by [−10, 10]

$f(x) \geq 0$ implies that the solution region is $x \geq -8$.

The interval notation is $[-8, \infty)$.

17. a. The x-coordinate of the intersection point is the solution. $x = -1$.

    b. $(-\infty, -1)$

19.
$$17 \leq 3x - 5 < 31$$
$$17 + 5 \leq 3x - 5 + 5 < 31 + 5$$
$$22 \leq 3x < 36$$
$$\frac{22}{3} \leq \frac{3x}{3} < \frac{36}{3}$$
$$\frac{22}{3} \leq x < 12$$

The interval notation is $\left[\frac{22}{3}, 12\right)$.

21.
$$2x + 1 \geq 6 \text{ and } 2x + 1 \leq 21$$
$$6 \leq 2x + 1 \leq 21$$
$$5 \leq 2x \leq 20$$
$$\frac{5}{2} \leq x \leq 10$$
$$x \geq \frac{5}{2} \text{ and } x \leq 10$$

The interval notation is $\left[\frac{5}{2}, 10\right]$.

23. $3x + 1 < -7$ and $2x - 5 > 6$

Inequality 1

$$3x + 1 < -7$$
$$3x < -8$$
$$x < -\frac{8}{3}$$

Inequality 2

$$2x - 5 > 6$$
$$2x > 11$$
$$x > \frac{11}{2}$$

$$x < -\frac{8}{3} \text{ and } x > \frac{11}{2}$$

25. $\frac{3}{4}x - 2 \geq 6 - 2x$ or $\frac{2}{3}x - 1 \geq 2x - 2$

Inequality 1

$$4\left(\frac{3}{4}x - 2\right) \geq 4\left(6 - 2x\right)$$
$$3x - 8 \geq 24 - 8x$$
$$11x \geq 32$$
$$x \geq \frac{32}{11}$$

Inequality 2

$$3\left(\frac{2}{3}x - 1\right) \geq 3\left(2x - 2\right)$$
$$2x - 3 \geq 6x - 6$$
$$-4x \geq -3$$
$$x \leq \frac{3}{4}$$

$$x \geq \frac{32}{11} \text{ or } x \leq \frac{3}{4}$$

27.
$$37.002 \le 0.554x - 2.886 \le 77.998$$
$$37.002 + 2.886 \le 0.554x - 2.886 + 2.886 \le 77.998 + 2.886$$
$$39.888 \le 0.554x \le 80.884$$
$$\frac{39.888}{0.554} \le \frac{0.554x}{0.554} \le \frac{80.884}{0.554}$$
$$72 \le x \le 146$$

The interval notation is $[72, 146]$.

Section 1.8 Exercises

29. a. $p \geq 0.1$

b. Considering x as a discrete variable representing the number of drinks, then if $x \geq 6$, the 220-lb male is intoxicated.

31. $F \leq 32$

$\frac{9}{5}C + 32 \leq 32$

$\frac{9}{5}C \leq 0$

$C \leq 0$

A Celsius temperature at or below zero degrees is "freezing."

33. Position 1 income $= 3100$

Position 2 income $= 2000 + 0.05x$, where x represents the sales within a given month When does the income from the second position exceed the income from the first position? Consider the inequality

$2000 + 0.05x > 3100$

$0.05x > 1100$

$x > \dfrac{1100}{0.05}$

$x > 22{,}000$

When monthly sales exceed \$22,000, the second position is more profitable than the first position.

35. Let x = Jill's final exam grade.

$80 \leq \dfrac{78 + 69 + 92 + 81 + 2x}{6} \leq 89$

$6(80) \leq 6\left(\dfrac{78 + 69 + 92 + 81 + 2x}{6}\right) \leq 6(89)$

$480 \leq 320 + 2x \leq 534$

$480 - 320 \leq 320 - 320 + 2x \leq 534 - 320$

$160 \leq 2x \leq 214$

$\dfrac{160}{2} \leq \dfrac{2x}{2} \leq \dfrac{214}{2}$

$80 \leq x \leq 107$

If the final exam does not contain any bonus points, Jill needs to score between 80 and 100 to earn a grade of B for the course.

37. Let $x = 6$, and solve for p.

$30p - 19(6) = 1$

$30p - 114 = 1$

$30p = 115$

$p = \dfrac{115}{30} = 3.8\overline{3}$

Let $x = 10$, and solve for p.

$30p - 10(19) = 1$

$30p - 190 = 1$

$30p = 191$

$p = \dfrac{191}{30} = 6.3\overline{6}$

Therefore, between 1996 and 2000, the percentage of marijuana use is between 3.83% and 6.37%. In symbols, $3.83 \leq p \leq 6.37$.

39. $y \geq 1000$

$0.97x + 128.3829 \geq 1000$

$0.97x \geq 871.6171$

$x \geq \dfrac{871.6171}{0.97}$

$x \geq 898.57$ or approximately $x \geq 899$

Old scores greater than or equal to 899 are equivalent to new scores.

41. Let x represent the actual life of the HID headlights.

$$1500 - 10\%(1500) \leq x \leq 1500 + 10\%(1500)$$
$$1500 - 150 \leq x \leq 1500 + 150$$
$$1350 \leq x \leq 1650$$

43. a. Let $y > 50$.

$$-0.763x + 85.284 > 50$$

Applying the intersection of graphs method:

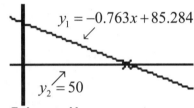

Intersection
X=46.243775 ·Y=50

[−5, 75] by [−5, 100]

When $x < 46.24, y > 50$.
The marriage rate per 1000 women is greater than 50 prior to 1996.

b. Let $y < 45$.

$$-0.763x + 85.284 > 45$$

Applying the intersection of graphs method:

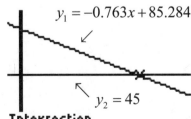

Intersection
X=52.796855 ·Y=45

[−5, 75] by [−5, 100]

When $x > 52.80, y < 45$.
The marriage rate per 1000 women will be less than 45 beyond 2002.

45. a. Since the rate of increase is constant, the equation modeling the value of the home is linear.

$$m = \frac{y_2 - y_1}{x_2 - x_1}$$
$$= \frac{270,000 - 190,000}{4 - 0}$$
$$= \frac{80,000}{4}$$
$$= 20,000$$

Solving for the equation:
$$y - y_1 = m(x - x_1)$$
$$y - 190,000 = 20,000(x - 0)$$
$$y - 190,000 = 20,000x$$
$$y = 20,000x + 190,000$$

b. $y > 400,000$
$$20,000x + 190,000 > 400,000$$
$$20,000x > 210,000$$
$$x > \frac{210,000}{20,000}$$
$$x > 10.5$$

2010 corresponds to
$x = 2010 - 1996 = 14$.
Therefore, $11 \leq x < 14$.
Or, $y > 400,000$
between 2007 and 2010.

The value of the home will be greater than $400,000 between 2007 and 2010.

47.
$$P(x) > 10,900$$
$$6.45x - 2000 > 10,900$$
$$6.45x > 12,900$$
$$x > \frac{12,900}{6.45}$$
$$x > 2000$$

A production level above 2000 units will yield a profit greater than $10,900.

49.
$$P(x) \geq 0$$
$$6.45x - 9675 \geq 0$$
$$6.45x \geq 9675$$
$$x \geq \frac{9675}{6.45}$$
$$x \geq 1500$$

Sales of 1500 feet or more of PVC pipe will avoid a loss for the hardware store.

51. Recall that Profit = Revenue − Cost.
Let $x =$ the number of boards manufactured and sold.
$$P(x) = R(x) - C(x)$$
$$R(x) = 489x$$
$$C(x) = 125x + 345,000$$
$$P(x) = 489x - (125x + 345,000)$$
$$P(x) = 489x - 125x - 345,000$$
$$P(x) = 364x - 345,000$$
To make a profit, $P(x) > 0$.
$$364x - 345,000 > 0$$

53.
$$245 < y < 248$$
$$245 < 0.155x + 244.37 < 248$$
$$245 - 244.37 < 0.155x + 244.37 - 244.37 < 248 - 244.37$$
$$0.63 < 0.155x < 3.63$$
$$\frac{0.63}{0.155} < \frac{0.155}{0.155}x < \frac{3.63}{0.155}$$
$$4.06 < x < 23.42$$

Considering x as a discrete variable yields $4 < x < 23$.

From 1974 until 1993 the reading scores were between 245 and 248.

55. a. $t = 1998 - 1975 = 23$

$p(23) = 75.4509 - 0.706948(23)$

$= 75.4509 - 16.259804$

$= 59.191096 \approx 59.2\%$

In 1998 the percent of high school seniors who have tried cigarettes is estimated to be 59.2%.

b. $0 \le p \le 100$

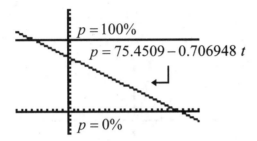

X	Y₁	Y₂
-37	101.61	0
-36	100.9	0
-35	100.19	0
-34	99.487	0
-33	98.78	0
-32	98.073	0
-31	97.366	0

X= -34

X	Y₁	Y₂
104	1.9283	0
105	1.2214	0
106	.51441	0
107	-.1925	0
108	-.8995	0
109	-1.606	0
110	-2.313	0

X=106

$-34 \le t \le 106$

c. Considering part *b* above, the model is valid between $1975 - 34 = 1941$ and $1975 + 106 = 2081$ inclusive. It is not valid before 1941 or after 2081.

Chapter 1 Skills Check

1. The table represents a function because every x matches with exactly one y.

2. Domain: $\{-3,-1,1,3,5,7,9,11,13\}$

   Range: $\{9,6,3,0,-3,-6,-9,-12,-15\}$

3. $f(3)=0$

4. Yes. The first differences are constant. The slope is
$$m=\frac{y_2-y_1}{x_2-x_1}=\frac{6-9}{-1-(-3)}=\frac{-3}{2}=-\frac{3}{2}.$$

   Calculating the equation:
   $$y-y_1=m(x-x_1)$$
   $$y-9=-\frac{3}{2}(x-(-3))$$
   $$y-9=-\frac{3}{2}x-\frac{9}{2}$$
   $$y=-\frac{3}{2}x-\frac{9}{2}+9$$
   $$y=\frac{3}{2}x-\frac{9}{2}+\frac{18}{2}$$
   $$y=\frac{3}{2}x+\frac{9}{2}$$

5. a. $C(3)=16-2(3)^2=16-2(9)$
      $=16-18=-2$

   b. $C(-2)=16-2(-2)^2=16-2(4)$
      $=16-8=8$

   c. $C(-1)=16-2(-1)^2=16-2(1)$
      $=16-2=14$

6. a. $f(-3)=1$

   b. $f(-3)=-10$

7.

   $[-10, 10]$ by $[-10, 10]$

8.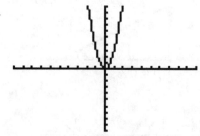

   $[-10, 10]$ by $[-10, 10]$

9.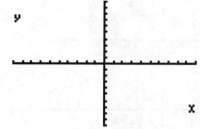

   $[-10, 10]$ by $[-10, 10]$

   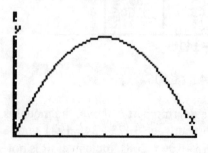

   $[0, 40]$ by $[0, 5000]$

   The second view is better.

10.

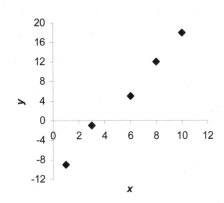

11.

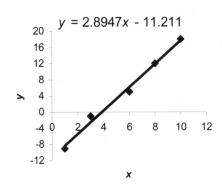

12.

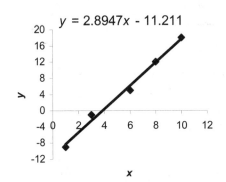

13. No. Data points do not necessarily fit a linear model exactly.

14. $m = \dfrac{y_2 - y_1}{x_2 - x_1} = \dfrac{-16 - 6}{8 - (-4)} = \dfrac{-22}{12} = -\dfrac{11}{6}$

15. a. *x*-intercept: Let $y = 0$ and solve for *x*.

$2x - 3(0) = 12$

$2x = 12$

$x = 6$

y-intercept: Let $x = 0$ and solve for *y*.

$2(0) - 3y = 12$

$-3y = 12$

$y = -4$

x-intercept: $(6, 0)$, *y*-intercept: $(0, -4)$

b. Solving for *y* :

$2x - 3y = 12$

$-3y = -2x + 12$

$\dfrac{-3y}{-3} = \dfrac{-2x + 12}{-3}$

$y = \dfrac{2}{3}x - 4$

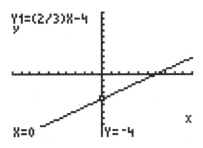

$[-10, 10]$ by $[-10, 10]$

16. Since the model is linear, the rate of change is equal to the slope of the equation. The slope, *m*, is $\dfrac{2}{3}$.

17. $m = -6$.

$(0, 3)$ is the *y*-intercept.

18. Since the function is linear, the rate of change is the slope. $m = -6$.

19. $y = mx + b$

$y = \dfrac{1}{3}x + 3$

20. $y - y_1 = m(x - x_1)$

$y - (-6) = -\dfrac{3}{4}(x - 4)$

$y + 6 = -\dfrac{3}{4}x + 3$

$y = -\dfrac{3}{4}x - 3$

21. The slope is

$m = \dfrac{y_2 - y_1}{x_2 - x_1} = \dfrac{6 - 3}{2 - (-1)} = \dfrac{3}{3} = 1.$

Solving for the equation:

$y - y_1 = m(x - x_1)$

$y - 6 = 1(x - 2)$

$y - 6 = x - 2$

$y = x + 4$

22. $\begin{cases} 3x + 2y = 0 & (Eq\,1) \\ 2x - y = 7 & (Eq\,2) \end{cases}$

$\begin{cases} 3x + 2y = 0 & (Eq\,1) \\ 4x - 2y = 14 & 2 \times (Eq\,2) \end{cases}$

$7x = 14$

$x = 2$

Substituting to find y

$3(2) + 2y = 0$

$6 + 2y = 0$

$2y = -6$

$y = -3$

The solution is $(2, -3)$.

23. $\begin{cases} 3x + 2y = -3 & (Eq\,1) \\ 2x - 3y = 3 & (Eq\,2) \end{cases}$

$\begin{cases} 9x + 6y = -9 & 3 \times (Eq\,1) \\ 4x - 6y = 6 & 2 \times (Eq\,2) \end{cases}$

$13x = -3$

$x = -\dfrac{3}{13}$

Substituting to find y

$3\left(-\dfrac{3}{13}\right) + 2y = -3$

$-\dfrac{9}{13} + 2y = -\dfrac{39}{13}$

$2y = -\dfrac{30}{13}$

$y = -\dfrac{15}{13}$

The solution is $\left(-\dfrac{3}{13}, -\dfrac{15}{13}\right)$.

24. $\begin{cases} -4x + 2y = -14 & (Eq\,1) \\ 2x - y = 7 & (Eq\,2) \end{cases}$

$\begin{cases} -4x + 2y = -14 & (Eq\,1) \\ 4x - 2y = 14 & 2 \times (Eq\,2) \end{cases}$

$0 = 0$

Dependent system. Infinitely many solutions.

25. $\begin{cases} -6x + 4y = 10 & (Eq\,1) \\ 3x - 2y = 5 & (Eq\,2) \end{cases}$

$\begin{cases} -6x + 4y = 10 & (Eq\,1) \\ 6x - 4y = 10 & 2 \times (Eq\,2) \end{cases}$

$0 = 10$

No solution. Lines are parallel.

26. $\begin{cases} 2x+3y=9 & (Eq1) \\ -x-y=-2 & (Eq2) \end{cases}$

$\begin{cases} 2x+3y=9 & (Eq1) \\ -2x-2y=-4 & 2\times(Eq2) \end{cases}$

$y=5$

Substituting to find x

$2x+3(5)=9$

$2x+15=9$

$2x=-6$

$x=-3$

The solution is $(-3,5)$.

27. $\begin{cases} 2x+y=-3 & (Eq1) \\ 4x-2y=10 & (Eq2) \end{cases}$

$\begin{cases} 4x+2y=-6 & 2\times(Eq1) \\ 4x-2y=10 & (Eq2) \end{cases}$

$8x=4$

$x=\dfrac{1}{2}$

Substituting to find y

$2\left(\dfrac{1}{2}\right)+y=-3$

$1+y=-3$

$y=-4$

The solution is $\left(\dfrac{1}{2},-4\right)$.

28. a. $f(x+h)=5-4(x+h)$

$\qquad\qquad =5-4x-4h$

b. $f(x+h)-f(x)$

$=\left[5-4(x+h)\right]-\left[5-4x\right]$

$=5-4x-4h-5+4x$

$=-4h$

c. $\dfrac{f(x+h)-f(x)}{h}$

$=\dfrac{-4h}{h}$

$=-4$

29. a. $f(x+h)$

$=10(x+h)-50$

$=10x+10h-50$

b. $f(x+h)-f(x)$

$=\left[10x+10h-50\right]-\left[10x-50\right]$

$=10x+10h-50-10x+50$

$=10h$

c. $\dfrac{f(x+h)-f(x)}{h}$

$=\dfrac{10h}{h}$

$=10$

30. a. $3x+22=8x-12$

$3x-8x+22=8x-8x-12$

$-5x+22=-12$

$-5x+22-22=-12-22$

$-5x=-34$

$\dfrac{-5x}{-5}=\dfrac{-34}{-5}$

$x=\dfrac{34}{5}$

b. Applying the intersections of graphs method yields $x=6.8$.

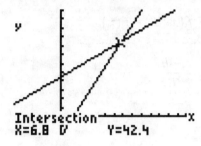

[−5, 15] by [−10, 60]

31. a. $\dfrac{3(x-2)}{5} - x = 8 - \dfrac{x}{3}$

LCM: 15

$15\left(\dfrac{3(x-2)}{5} - x\right) = 15\left(8 - \dfrac{x}{3}\right)$

$3(3(x-2)) - 15x = 120 - 5x$

$3(3x-6) - 15x = 120 - 5x$

$9x - 18 - 15x = 120 - 5x$

$-6x - 18 = 120 - 5x$

$-1x - 18 = 120$

$-1x = 138$

$x = -138$

b. Applying the intersections of graphs method yields $x = -138$.

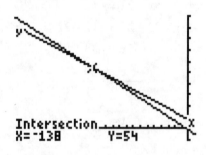

[−250, 10] by [−10, 100]

32. If $x = 0$, then $y = (0)^2 = 0$.

If $x = 3$, then $y = (3)^2 = 9$.

The average rate of change between the

points is $\dfrac{y_2 - y_1}{x_2 - x_1} = \dfrac{9 - 0}{3 - 0} = \dfrac{9}{3} = 3$.

33. Solving for y:

$4x - 3y = 6$

$-3y = -4x + 6$

$y = \dfrac{-4x + 6}{-3}$

$y = \dfrac{4}{3}x - 2$

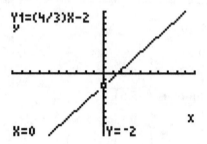

34. Algebraically:

$3x + 8 < 4 - 2x$

$5x + 8 < 4$

$5x < -4$

$x < -\dfrac{4}{5}$

Graphically:

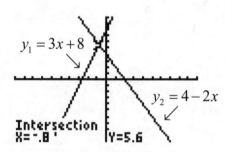

[−10, 10] by [−10, 10]

$3x + 8 < 4 - 2x$ implies that the solution region is $x < -\dfrac{4}{5}$.

The interval notation is $\left(-\infty, -\dfrac{4}{5}\right)$.

35. Algebraically:

$$3x - \frac{1}{2} \leq \frac{x}{5} + 2$$

$$10\left(3x - \frac{1}{2}\right) \leq 10\left(\frac{x}{5} + 2\right)$$

$$30x - 5 \leq 2x + 20$$

$$28x - 5 \leq 20$$

$$28x \leq 25$$

$$x \leq \frac{25}{28}$$

Graphically:

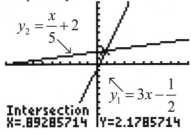

[–10, 10] by [–10, 10]

$3x - \frac{1}{2} \leq \frac{x}{5} + 2$ implies that the solution region is $x \leq \frac{25}{28}$.

The interval notation is $\left(-\infty, \frac{25}{28}\right]$.

Graphically:

$y = 42$

$y = 2x + 6$

$y = 18$

[–5, 25] by [–10, 50]

$18 \leq 2x + 6 < 42$ implies that the solution region is $6 \leq x < 18$.

The interval notation is $[6, 18)$.

19. Algebraically:

$$18 \leq 2x + 6 < 42$$

$$18 - 6 \leq 2x + 6 - 6 < 42 - 6$$

$$12 \leq 2x < 36$$

$$\frac{12}{2} \leq \frac{2x}{2} < \frac{36}{2}$$

$$6 \leq x < 18$$

Chapter 1 Review Exercises

37. a. Yes. Every year matches with exactly one Democratic Party percentage.

b. $f(1992) = 82$. The table indicates that in 19992, 82% of African American voters supported a Democratic candidate for president.

c. When $f(y) = 94, y = 1964$. The table indicates that in 1964, 94% of African American voters supported a Democratic candidate for president.

38. a. The domain is
$$\{1960, 1964, 1968, 1972, 1976, 1980,$$
$$1984, 1992, 1996\}.$$

b. No. 1982 was not a presidential election year.

c. Discrete. The input values are the presidential election years. There are 4-year gaps between the inputs.

39.

40. a. $m = \dfrac{y_2 - y_1}{x_2 - x_1}$

$$= \frac{84 - 85}{1996 - 1968}$$

$$= \frac{-1}{28} \approx -0.357$$

b. $\dfrac{f(b) - f(a)}{b - a}$

$$= \frac{84 - 85}{1996 - 1968}$$

$$= \frac{-1}{28}$$

$$\approx -0.357$$

c. No.

$$\frac{f(b) - f(a)}{b - a}$$

$$= \frac{86 - 85}{1980 - 1968}$$

$$= \frac{1}{28}$$

$$\approx 0.357$$

d. No. Consider the scatter plot in problem 39 above.

41. a. Every amount borrowed matches with exactly one monthly payment. The change in y is fixed at 89.62 for a fixed change in x of 5000.

b. $f(25,000) = 448.11$. Therefore, borrowing $25,000 to buy a car from the dealership results in a monthly payment of $448.11.

c. If $f(A) = 358.49$, then $A = 20,000$.

42. a. Domain: $\{10,000, 15,000, 20,000,$
$$25,000, 30,000\}$$

Range: $\{179.25, 268.87, 358.49$
$$448.11, 537.73\}$$

b. No. $12,000 is not in the domain of the function.

c. Discrete. There are gaps between the possible inputs.

43. a. Yes. As each amount borrowed increases by $5000, the monthly payment increases by $89.62.

b. Yes. Since the first differences are constant, a linear model will fit the data exactly.

44. a.

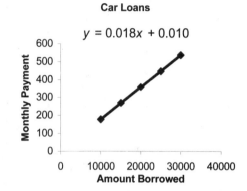

$$P = f(A) = 0.018A + .010$$

b. $f(28,000) = 0.018(28,000) + 0.010$
$$= 504.01$$
The predicted monthly payment on a car loan of $28,000 is $504.01

c. Yes. Any input could be used for A.

d.
$$f(A) \leq 500$$
$$0.018A + 0.010 \leq 500$$
$$0.018A \leq 499.99$$
$$A \leq \frac{499.99}{0.018}$$
$$A \leq 27,777.\overline{2}$$
The loan amount needs to be less than or equal to approximately $27,777.22.

45. a. $f(1960) = 15.9$. A 65-year old woman in 1960 is expected to live 15.9 more

years. Her overall life expectancy is 80.9 years.

b. $f(2010) = 19.4$. A 65-year old woman in 2010 has a life expectancy of 84.4 years.

c. $f(1990) = 19$

46. a. $g(2020) = 16.9$.
A 65-year old man in 2020 is expected to live 16.9 more years. His overall life expectancy is 81.9 years.

b. $g(1950) = 12.8$. A 65-year old man in 1950 has a life expectancy of 77.8 years.

c. $g(1990) = 15$

47. a. $t = 2000 - 1990 = 10$
$$f(10) = 982.06(10) + 32,903.77$$
$$f(10) = 42,724.37$$

b. $t = 15$
$$f(15) = 982.06(15) + 32,903.77$$
$$f(15) = 47,634.67$$
Based on the model in 2005 average teacher salaries will be $47,634.67.

c. Increasing

48. a. Y1=982.06X+32903.77

X=7.5 . . . . Y=40269.22 .

[0, 15] by [10,000, 60,000]

b. From 1990 through 2005

49. a.
$$m = \frac{y_2 - y_1}{x_2 - x_1}$$
$$= \frac{14.5 - 12.0}{1999 - 1992}$$
$$= \frac{2.5}{7} \approx 0.357$$

b. Assuming that drug use follows a linear model, the annual rate of change is equal to the slope calculated in part a). Each year, the number of people using illicit drugs increases by 0.357 million or 357,000.

50. $f(x) = 4500$

51. a. Let x = the number of months past December 1997, and $f(x)$ = average weekly hours worked. Then $f(x) = 34.6$.

b. Yes. The average rate of change is zero.

52. a. Let x = monthly sales.
$$2100 = 1000 + 5\% x$$
$$2100 = 1000 + 0.05x$$
$$1100 = 0.05x$$
$$x = \frac{1100}{0.05} = 22,000$$
If monthly sales are $22,000, both positions will yield the same monthly income.

b. Considering the solution from part a), if sales exceed $22,000 per month, the 2nd position will yield a greater salary.

53. Profit $= 10\%(24,000 \cdot 12) = 28,800$
Cost $= 24,000 \cdot 12 = 288,000$
Revenue $= 8(24,000 + 12\% \cdot 24,000) + 4x$,
where x is the selling price of the remaining four cars.
Profit = Revenue − Cost
$$28,800 = (215,040 + 4x) - 288,000$$
$$28,800 = 4x - 72,960$$
$$4x = 101,760$$
$$x = \frac{101,760}{4} = 25,440$$
The remaining four cars should be sold for $25,440 each.

54. Let x = amount invested in the safe account, and let $420,000 - x$ = amount invested in the risky account.
$$6\% x + 10\%(420,000 - x) = 30,000$$
$$0.06x + 42,000 - 0.10x = 30,000$$
$$-0.04x = -12,000$$
$$x = \frac{-12,000}{-0.04}$$
$$x = 300,000$$
The couple invests $300,000 in the safe account and $120,000 in the risky account.

55. Let $y = 285$, and solve for x.
$$285 = -0.629x + 293.871$$
$$285 - 293.871 =$$
$$\qquad -0.629x + 293.871 - 293.871$$
$$-8.871 = -0.629x$$
$$\frac{-0.629x}{-0.629} = \frac{-8.871}{-0.629}$$
$$x \approx 14.1$$
Therefore, the writing score is 285 in $1980 + 14 = 1994$.

56. a. $R(120) = 564(120) = 67,680$

b. $C(120) = 40,000 + 64(120) = 47,680$

c. Marginal Cost $= \overline{MC} = 64$

Marginal Revenue $= \overline{MR} = 564$

d. $m = 64$

e.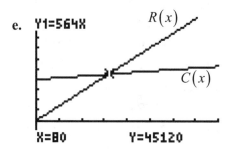

[0, 200] by [0, 200,000]

57. a. $P(x) = 564x - (40,000 + 64x)$

$= 564x - 40,000 - 64x$

$= 500x - 40,000$

b. $P(120) = 500(120) - 40,000$

$= 60,000 - 40,000$

$= 20,000$

c. Break-even occurs when $R(x) = C(x)$
or alternately $P(x) = 0$.

$500x - 40,000 = 0$

$500x = 40,000$

$x = \dfrac{40,000}{500} = 80$

80 units represents break-even for
the company.

d. $\overline{MP} =$ the slope of $P(x) = 500$

e. $\overline{MP} = \overline{MR} - \overline{MC}$

58. a. Let $x = 0$, and solve for y.

$y + 3000(0) = 300,000$

$y = 300,000$

The initial value of the property is
$300,000.

b. Let $y = 0$, and solve for x.

$0 + 3000x = 300,000$

$3000x = 300,000$

$x = 100$

The value of the property after 100 years
is zero dollars.

59. a. $m = \dfrac{y_2 - y_1}{x_2 - x_1} = \dfrac{895 - 455}{250 - 150} = \dfrac{440}{100} = 4.4$

The average rate of change is
$4.40 per unit.

b. For a linear function, the slope is the
average rate of change. Referring to
part a), the slope is 4.4.

c. $y - y_1 = m(x - x_1)$

$y - 455 = 4.4(x - 150)$

$y - 455 = 4.4x - 660$

$y = 4.4x - 205$

$P(x) = 4.4x - 205$

d. $\overline{MP} =$ the slope of $P(x) = 4.4$ or
$4.40 per unit.

e. Break-even occurs when $R(x) = C(x)$
or alternately $P(x) = 0$.

$4.4x - 205 = 0$

$4.4x = 205$

$x = \dfrac{205}{4.4} = 46.59\overline{09} \approx 47$

The company will break even selling
approximately 47 units.

60. a.

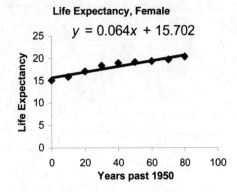

b. See part a) above.

c. $f(104) = 0.064(104) + 15.702$
$= 22.358$

In 2054 the average woman is expected to live 22.36 years beyond age 65. Her life expectancy is 87.36 years.

d. $y \geq 84 - 65$

$0.064x + 15.702 \geq 19$

$0.064x \geq 3.298$

$x \geq \dfrac{3.298}{0.064}$

$x \geq 51.53$

For years 2002 and beyond, the average woman is expected to live at least 84 years.

c. $g(130) = 0.065(130) + 12.324$
$= 8.45 + 12.324$
$= 20.774 \approx 20.8$

In 2080 (1950 + 130), a 65-year old male is expected to live 20.8 more years. The overall life expectancy is 85.8 years.

d. A life expectancy of 90 years translates into $90 - 65 = 25$ years beyond age 65. Therefore, let $g(x) = 25$.

$0.065x + 12.324 = 25$

$0.065x = 25 - 12.324$

$0.065x = 12.676$

$x = \dfrac{12.676}{0.065} = 195.0153846 \approx 195$

In approximately the year 2145 (1950 +195), male life expectancy will be 90 years.

e. A life expectancy of 81 years translates into $81 - 65 = 16$ years beyond age 65. $g(x) \leq 16$

$0.065x + 12.324 \leq 16$

$0.065x \leq 3.676$

$x \leq \dfrac{3.676}{0.065}$

$x \leq 56.6$

61. a.

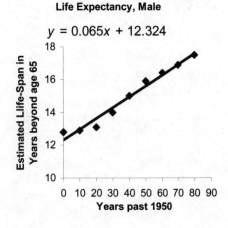

b. See part a) above.

62. a.

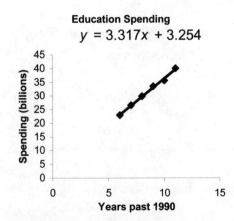

A linear model is reasonable.

b. See part a) above.

c. $y = 3.317x + 3.254$

$y = 3.317(2002 - 1990) + 3.254$

$y = 3.317(12) + 3.254$

$y = 43.058$

Approximately $43.1 billion

Using the unrounded model:

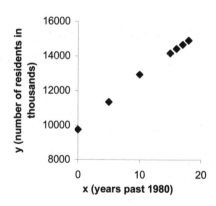

The unrounded model predicts that education spending in 2002 will be $43.06 billion.

63. a.

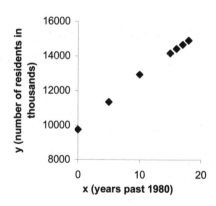

Yes. A linear model would fit the data well. The data points lie approximately along a line.

b.

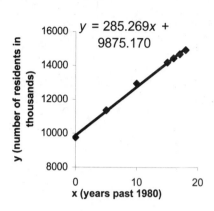

c. In 2002, $x = 2002 - 1980 = 22$.

$y = 285.269(22) + 9875.170$

$= 6275.918 + 9875.170$

$= 16,151.088$

In 2002, the predicted population is 16,151,088.

Using the unrounded model:

$Y_1 = 16151.0938134$

The unrounded model predicts that the population in 2002 will be approximately 16,151,094.

64. a.

$y = 0.301x + 0.477$

A linear model is reasonable.

b. See part a) above.

c. See part a) above. The line seems to fit the data points very well.

65. a.

$$y > 48.66$$
$$-0.763x + 85.284 > 48.66$$
$$-0.763x > -36.624$$
$$x < \frac{-36.624}{-0.763} \qquad \text{(Note the inequality sign switch.)}$$
$$x < 48$$

For years less than $1950 + 48 = 1998$, the marriage rate is less than 48.66 per 1000 women.

b.

$$y < 41.03$$
$$-0.763x + 85.284 < 41.03$$
$$-0.763x < -44.254$$
$$x > \frac{-44.254}{-0.763} \qquad \text{(Note the inequality sign switch.)}$$
$$x > 58$$

For years beyond $1950 + 58 = 2008$, the marriage rate per 1000 women will be less than 41.03.

66. $30p - 19x = 1$

Let $p = 3.2$.

$30(3.2) - 19x = 1$

$96 - 19x = 1$

$-19x = -95$

$x = \dfrac{-95}{-19} = 5$

The year is $1990 + 5 = 1995$.

Let $p = 7$.

$30(7) - 19x = 1$

$210 - 19x = 1$

$-19x = -209$

$x = \dfrac{-209}{-19} = 11$

The year is $1990 + 11 = 2001$

From 1995 until 2001, Marijuana use is in the range of 3.2%-7%.

67. Let $3(12) \leq x \leq 5(12)$ or

$36 \leq x \leq 60$.

Then

$0.554(36) - 2.886 \leq y \leq 0.554(60) - 2.886$.

Therefore, $17.058 \leq y \leq 30.354$. Or,

rounding to the nearest month,

$17 \leq y \leq 30$.

The criminal is expected to serve between 17 and 30 months inclusive.

68. Let x = the amount in the safer fund, and y = the amount in the riskier fund.

$\begin{cases} x + y = 240,000 & (Eq\,1) \\ 0.08x + 0.12y = 23,200 & (Eq\,2) \end{cases}$

$\begin{cases} -0.08x - 0.08y = -19,200 & -0.08 \times (Eq\,1) \\ 0.08x + 0.12y = 23,200 & (Eq\,2) \end{cases}$

$0.04y = 4000$

$y = \dfrac{4000}{0.04} = 100,000$

Substituting to calculate x

$x + 100,000 = 240,000$

$x = 140,000$

The safer fund contains \$140,000, while the riskier fund contains \$100,000.

69. Let x = number of units.

$R = C$

$565x = 6000 + 325x$

$240x = 6000$

$x = 25$

The number of units that produced to create a break even point is 25.

70. Let x = dosage of Medication A, and let y = dosage of Medication B.

$$\begin{cases} 6x + 2y = 25.2 & (Eq\,1) \\ \dfrac{x}{y} = \dfrac{2}{3} & (Eq\,2) \end{cases}$$

Solving $(Eq\,2)$ for x yields

$$3x = 2y$$

$$x = \frac{2}{3}y$$

Substituting

$$6\left(\frac{2}{3}y\right) + 2y = 25.2$$

$$4y + 2y = 25.2$$

$$6y = 25.2$$

$$y = 4.2$$

Substituting to calculate x

$$x = \frac{2}{3}(4.2)$$

$$x = 2.8$$

Medication A dosage is 2.8 mg while Medication B dosage is 4.2 mg.

71. Let p = price and q = quantity.

$$\begin{cases} 3q + p = 340 & (Eq\,1) \\ -4q + p = -220 & (Eq\,2) \end{cases}$$

$$\begin{cases} -3q - 1p = -340 & -1 \times (Eq\,1) \\ -4q + p = -220 & (Eq\,2) \end{cases}$$

$$-7q = -560$$

$$q = \frac{-560}{-7} = 80$$

Substituting to calculate p

$$3(80) + p = 340$$

$$240 + p = 340$$

$$p = 100$$

Equilibrium occurs when the price is $100, and the quantity is 80 pairs.

72. Let p = price and q = quantity.

$$\begin{cases} p = \dfrac{q}{10} + 8 & (Eq\,1) \\ 10p + q = 1500 & (Eq\,2) \end{cases}$$

Substituting

$$10\left(\frac{q}{10} + 8\right) + q = 1500$$

$$q + 80 + q = 1500$$

$$2q = 1420$$

$$q = 710$$

Substituting to calculate p

$$p = \frac{710}{10} + 8$$

$$p = 79$$

Equilibrium occurs when the price is $79, and the quantity is 710 units.

73. a. $x + y = 2600$

b. $40x$

c. $60y$

d. $40x + 60y = 120,000$

e.
$$\begin{cases} x + y = 2600 & (Eq\,1) \\ 40x + 60y = 120,000 & (Eq\,2) \end{cases}$$

$$\begin{cases} -40x - 40y = -104,000 & -40 \times (Eq\,1) \\ 40x + 60y = 120,000 & (Eq\,2) \end{cases}$$

$$20y = 16,000$$

$$y = \frac{16,000}{20} = 800$$

Substituting to calculate x

$$x + 800 = 2600$$

$$x = 1800$$

The promoter needs to sell 1800 tickets at $40 per ticket and 800 tickets at $60 per ticket.

74. a. $x + y = 500,000$

b. $0.12x$

c. $0.15y$

d. $0.12x + 0.15y = 64,500$

e.
$$\begin{cases} x + y = 500,000 & (Eq1) \\ 0.12x + 0.15y = 64,500 & (Eq2) \end{cases}$$

$$\begin{cases} -0.12x - 0.12y = -60,000 & -0.12 \times (Eq1) \\ 0.12x + 0.15y = 64,500 & (Eq2) \end{cases}$$

$$0.03y = 4500$$

$$y = \frac{4500}{0.03} = 150,000$$

Substituting to calculate x

$$x + 150,000 = 500,000$$

$$x = 350,000$$

Devote \$350,000 in the 12% investment and \$150,000 in the 15% investment.

Extended Application I

1. **a.** A person uses the table to determine his or her BMI by locating the entry in the table that corresponds to the person's height and weight. The entry in the table is the person's BMI.

b. If a person's BMI is 30 or higher, the person is considered obese and at risk for health problems.

c. **1.** Determine the heights and weights that produce a BMI of exactly 30 based on the table.

Height (inches)	Weight (pounds)
61	160
63	170
65	180
67	190
68	200
69	200
72	220
73	230

2.

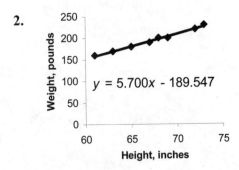

$y = 5.700x - 189.547$

A linear model is reasonable, but not exact.

3. See part *2* above.

4. See part *2* above. The scatter plot fits the data points well, but not perfectly.

5. Any data point that lies exactly along the line generated from the model will yield a BMI of 30. If a height is substituted into the model, the output weight would generate a BMI of 30. That weight or any higher weight for the given height would place a person at risk for health problems.

**Chapter 2
Quadratic and Other Nonlinear Functions**

Algebra Toolbox

1. a. $x^4 \cdot x^3 = x^{4+3} = x^7$

   b. $\dfrac{x^{12}}{x^7} = x^{12-7} = x^5$

   c. $(4ay)^4 = (4)^4 a^4 y^4 = 256 a^4 y^4$

   d. $\left(\dfrac{3}{z}\right)^4 = \dfrac{3^4}{z^4} = \dfrac{81}{z^4}$

   e. $2^3 \cdot 2^2 = 2^{3+2} = 2^5 = 32$

   f. $\left(x^4\right)^2 = x^{4 \cdot 2} = x^8$

2. a. $y^5 \cdot y = y^5 \cdot y^1 = y^{5+1} = y^6$

   b. $\dfrac{w^{10}}{w^4} = w^{10-4} = w^6$

   c. $(6bx)^3 = (6)^3 b^3 x^3 = 216 b^3 x^3$

   d. $\left(\dfrac{5z}{2}\right)^3 = \dfrac{5^3 z^3}{2^3} = \dfrac{125 z^3}{8}$

   e. $3^2 \cdot 3^3 = 3^{2+3} = 3^5 = 243$

   f. $\left(2y^3\right)^4 = 2^4 y^{4 \cdot 3} = 16 y^{12}$

3. a. $\sqrt{16} = \sqrt{4 \cdot 4} = 4$

   b. $-\sqrt{16} = -\sqrt{4 \cdot 4} = -4$

   c. $\sqrt{-16}$ has no solution in the real number system, since there is a negative under the even radical.

d. $\sqrt[3]{27} = \sqrt[3]{3^3} = 3$

e. $-\sqrt[5]{-32} = -\sqrt[5]{(-2)^5} = -(-2) = 2$

4. a. $\sqrt{18} = \sqrt{2 \cdot 3^2} = 3\sqrt{2}$

   b. $\sqrt{216} = \sqrt{2^2 \cdot 2 \cdot 3^2 \cdot 3} =$
   $2 \cdot 3\sqrt{2 \cdot 3} = 6\sqrt{6}$

   c. $\sqrt{(-12)^2 - 4(2)(-3)}$
   $= \sqrt{144 + 24}$
   $= \sqrt{168}$
   $= \sqrt{2^2 \cdot 2 \cdot 3 \cdot 7}$
   $= 2\sqrt{2 \cdot 3 \cdot 7}$
   $= 2\sqrt{42}$

   d. $\sqrt[3]{243 a^7 b^4}$
   $= \sqrt[3]{3^5 a^7 b^4}$
   $= \sqrt[3]{3^3 \cdot 3^2 \cdot a^6 \cdot a \cdot b^3 \cdot b}$
   $= 3 a^2 b \sqrt[3]{9ab}$

5. a. $\sqrt{x^3} = \sqrt[2]{x^3} = x^{\frac{3}{2}}$

   b. $\sqrt[4]{x^3} = x^{\frac{3}{4}}$

   c. $\sqrt[5]{x^3} = x^{\frac{3}{5}}$

   d. $\sqrt[6]{27 y^9}$
   $= \left(27 y^9\right)^{\frac{1}{6}}$
   $= \left(3^3 y^9\right)^{\frac{1}{6}}$
   $= 3^{\frac{3}{6}} y^{\frac{9}{6}}$
   $= 3^{\frac{1}{2}} y^{\frac{3}{2}}$

e. $27\sqrt[6]{y^9} = 27y^{\frac{9}{6}} = 27y^{\frac{3}{2}}$

6. a. $a^{\frac{3}{4}} = \sqrt[4]{a^3}$

   b. $-15x^{\frac{5}{8}} = -15\sqrt[8]{x^5}$

   c. $(-15x)^{\frac{5}{8}} = \sqrt[8]{(-15x)^5}$

7. The degree of the polynomial is 2, the highest exponent in the expression. The leading coefficient is –3, the coefficient of the highest degree term. The constant term, the term without a variable, is 8. The polynomial contains only one variable, x.

8. The degree of the polynomial is 4, the highest exponent in the expression. The leading coefficient is 5, the coefficient of the highest degree term. The constant term, the term without a variable, is –3. The polynomial contains only one variable, x.

9. The degree of the polynomial is 7, the greatest sum of exponents in any one term. The leading coefficient is –2, the coefficient of the highest degree term. The constant term, the term without a variable, is –119. The polynomial contains two variables, x and y.

10. $(4x^2 y^3)(-3a^2 x^3)$
    $= -12a^2 x^{2+3} y^3$
    $= -12a^2 x^5 y^3$

11. $2xy^3 (2x^2 y + 4xz - 3z^2)$
    $= (2xy^3)(2x^2 y) + (2xy^3)(4xz)$
    $\quad - (2xy^3)(3z^2)$
    $= 4x^3 y^4 + 8x^2 y^3 z - 6xy^3 z^2$

12. $(x-7)(2x+3)$
    $x \cdot 2x + x \cdot 3 - 7 \cdot 2x - 7 \cdot 3$
    $2x^2 + 3x - 14x - 21$
    $2x^2 - 11x - 21$

13. $(k-3)^2$
    $(k-3)(k-3)$
    $k^2 - 3k - 3k + 9$
    $k^2 - 6k + 9$
    or
    $(k-3)^2$
    $(k)^2 - 2(k)(3) + (3)^2$
    $k^2 - 6k + 9$

14. $(4x - 7y)(4x + 7y)$
    $16x^2 + 28xy - 28xy - 49y^2$
    $16x^2 - 49y^2$
    or
    $(4x - 7y)(4x + 7y)$
    $(4x)^2 - (7y)^2$
    $16x^2 - 49y^2$

15. $3x^2 - 12x = 3x(x - 4)$

16. $12x^5 - 24x^3 = 12x^3(x^2 - 2)$

17. Difference of two squares

$$9x^2 - 25m^2 = (3x + 5m)(3x - 5m)$$

18. Find two numbers whose product is 15 and whose sum is -8.

$$x^2 - 8x + 15 = (x - 5)(x - 3)$$

19. Find two numbers whose product is -35 and whose sum is -2.

$$x^2 - 2x - 35 = (x - 7)(x + 5)$$

20. To factor by grouping, first multiply the 2nd degree term by the constant term:

$3x^2 \bullet - 2 = -6x^2$.

Then, find two terms whose product is $-6x^2$ and whose sum is $-5x$, the middle term.

$$3x^2 - 5x - 2$$
$$= 3x^2 - 6x + 1x - 2$$
$$= (3x^2 - 6x) + (1x - 2)$$
$$= 3x(x - 2) + 1(x - 2)$$
$$= (3x + 1)(x - 2)$$

21. To factor by grouping, first multiply the 2nd degree term by the constant term:

$8x^2 \bullet 5 = 40x^2$.

Then, find two terms whose product is $40x^2$ and whose sum is $-22x$, the middle term.

$$8x^2 - 22x + 5$$
$$= 8x^2 - 2x - 20x + 5$$
$$= (8x^2 - 2x) + (-20x + 5)$$
$$= 2x(4x - 1) + (-5)(4x - 1)$$
$$= (2x - 5)(4x - 1)$$

22. $6n^2 + 18 + 39n$
$$= 6n^2 + 39n + 18$$
$$= 3(2n^2 + 13n + 6)$$

To factor by grouping, first multiply the 2nd degree term by the constant term:

$2n^2 \bullet 6 = 12n^2$.

Then, find two terms whose product is $12n^2$ and whose sum is $13n$, the middle term.

$$3(2n^2 + 13n + 6)$$
$$= 3(2n^2 + 1n + 12n + 6)$$
$$= 3\left[(2n^2 + 1n) + (12n + 6)\right]$$
$$= 3\left[n(2n + 1) + 6(2n + 1)\right]$$
$$= 3(n + 6)(2n + 1)$$

23. $3y^4 + 9y^2 - 12y^2 - 36$

$= 3\left[y^4 + 3y^2 - 4y^2 - 12 \right]$

$= 3\left[\left(y^4 + 3y^2 \right) + \left(-4y^2 - 12 \right) \right]$

$= 3\left[y^2 \left(y^2 + 3 \right) + \left(-4 \right)\left(y^2 + 3 \right) \right]$

$= 3\left(y^2 - 4 \right)\left(y^2 + 3 \right)$

$= 3\left(y - 2 \right)\left(y + 2 \right)\left(y^2 + 3 \right)$

24. $18p^2 + 12p - 3p - 2$

$= \left(18p^2 + 12p \right) + \left(-3p - 2 \right)$

$= 6p\left(3p + 2 \right) + \left(-1 \right)\left(3p + 2 \right)$

$= \left(6p - 1 \right)\left(3p + 2 \right)$

25. $5x - 4xy = 2y$

$5x = 2y + 4xy$

$5x = y\left(2 + 4x \right)$

$\dfrac{5x}{\left(2 + 4x \right)} = \dfrac{y\left(2 + 4x \right)}{\left(2 + 4x \right)}$

$y = \dfrac{5x}{\left(2 + 4x \right)}$

26. $2xy = \dfrac{6}{y} + \dfrac{x}{3}$

The LCM is $3y$.

$3y\left(2xy \right) = 3y\left(\dfrac{6}{y} + \dfrac{x}{3} \right)$

$6xy^2 = 18 + xy$

$6xy^2 - xy = 18$

$x\left(6y^2 - y \right) = 18$

$\dfrac{x\left(6y^2 - y \right)}{\left(6y^2 - y \right)} = \dfrac{18}{\left(6y^2 - y \right)}$

$x = \dfrac{18}{\left(6y^2 - y \right)}$

27. a. Imaginary. The number has a non-zero real part and pure imaginary part.

  b. Pure imaginary. The real part is zero.

  c. Real. The imaginary part is zero.

  d. Real. $2 - 5i^2 = 2 - 5\left(-1 \right) = 7$

28. a. Imaginary. The number has a non-zero real part and pure imaginary part.

  b. Real. The imaginary part is zero.

  c. Pure imaginary. The real part is zero.

  d. Imaginary. The number has a non-zero real part and pure imaginary part.
  $2i^2 - i = 2\left(-1 \right) - i = -2 - i$

29. $a + bi = 4 + 0i$. Therefore,
  $a = 4, b = 0.$

30. $a + 3i = 15 - bi$
  Therefore, $a = 15$ and
  $-b = 3$ or $b = -3$.

31. $a + bi = 2 + 4i$
  Therefore, $a = 2$ and $b = 4$.

32. $\left(3 + 7i \right) + \left(2i - 4 \right)$
  $= \left(3 - 4 \right) + \left(7 + 2 \right)i$
  $= -1 + 9i$

33. $\left(4 - 2i \right) - \left(2 - 3i \right)$
  $= 4 - 2i - 2 + 3i$
  $= 2 + 1i$
  $= 2 + i$

34. $2(3-i)-(7-2i)$

$\quad = 6-2i-7+2i$

$\quad = -1$

35. $2(2i+3)-(4i+5)$

$\quad = 4i+6-4i-5$

$\quad = 1$

36. $(2-i)(2+i)$

$\quad = 4+2i-2i-i^2$

$\quad = 4-(-1)$

$\quad = 5$

or

$(2-i)(2+i)$

$\quad = (2)^2-(i)^2$

$\quad = 4-(-1)$

$\quad = 5$

37. $(6-3i)(4+5i)$

$\quad = 24+30i-12i-15i^2$

$\quad = 24+18i-15(-1)$

$\quad = 39+18i$

38. $(2i-1)^2$

$\quad = (2i-1)(2i-1)$

$\quad = 4i^2-2i-2i+1$

$\quad = 4i^2-4i+1$

$\quad = 4(-1)+1-4i$

$\quad = -3-4i$

or

$(2i-1)^2$

$(2i)^2-2(2i)(1)+(1)^2$

$\quad = 4i^2-4i+1$

$\quad = 4(-1)+1-4i$

$\quad = -3-4i$

39. $\dfrac{4-3i}{3-3i} \cdot \dfrac{3+3i}{3+3i}$

$\quad = \dfrac{12+12i-9i-9i^2}{9+9i-9i-9i^2}$

$\quad = \dfrac{12+3i+9}{9+9}$

$\quad = \dfrac{21+3i}{18}$

$\quad = \dfrac{21}{18}+\dfrac{3}{18}i$

$\quad = \dfrac{7}{6}+\dfrac{1}{6}i$

40. $\dfrac{\sqrt{3}-2i}{\sqrt{5}+i} \cdot \dfrac{\sqrt{5}-i}{\sqrt{5}-i}$

$= \dfrac{\sqrt{15}-i\sqrt{3}-2i\sqrt{5}+2i^2}{\sqrt{25}-i\sqrt{5}+i\sqrt{5}-i^2}$

$= \dfrac{\sqrt{15}-i\sqrt{3}-2i\sqrt{5}-2}{5+1}$

$= \dfrac{\sqrt{15}-i\sqrt{3}-2i\sqrt{5}-2}{6}$

$= \dfrac{\left(\sqrt{15}-2\right)-i\left(\sqrt{3}+2\sqrt{5}\right)}{6}$

$= \dfrac{\sqrt{15}-2}{6}-\dfrac{\sqrt{3}+2\sqrt{5}}{6}i$

41. $\sqrt{-36}=\sqrt{36}\cdot\sqrt{-1}=6i$

42. $\sqrt{-98}=\sqrt{98}\cdot\sqrt{-1}=i\sqrt{2\cdot49}=7i\sqrt{2}$

43. $\left(\sqrt{-4}\right)^2$

$=\left(\sqrt{-4}\right)\left(\sqrt{-4}\right)$

$=\left(i\sqrt{4}\right)\left(i\sqrt{4}\right)$

$=i^2\sqrt{16}$

$=-4$

44. $\sqrt{-6}\sqrt{-15}$

$=\left(i\sqrt{6}\right)\left(i\sqrt{15}\right)$

$=i^2\sqrt{90}$

$=i^2\sqrt{9\cdot10}$

$=-3\sqrt{10}$

45. $\left(3+\sqrt{-4}\right)+\left(2-\sqrt{9}\right)$

$=\left(3+2\sqrt{-1}\right)+\left(2-3\sqrt{-1}\right)$

$=\left(3+2i\right)+\left(2-3i\right)$

$=5-i$

46. $\left(2+\sqrt{-2}\right)-\left(3-\sqrt{-8}\right)$

$=\left(2+i\sqrt{2}\right)-\left(3-i\sqrt{8}\right)$

$=\left(2+i\sqrt{2}\right)-\left(3-i\sqrt{4\cdot2}\right)$

$=2+i\sqrt{2}-3+2i\sqrt{2}$

$=-1+3i\sqrt{2}$

47. $\left(1+\sqrt{-4}\right)\left(2-\sqrt{-9}\right)$

$=\left(1+\sqrt{-1\cdot4}\right)\left(2-\sqrt{-1\cdot9}\right)$

$=\left(1+2i\right)\left(2-3i\right)$

$=2-3i+4i-6i^2$

$=2+i+6$

$=8+i$

48. $\left(2-3\sqrt{-1}\right)^2$

$=\left(2-3i\right)^2$

$=\left(2-3i\right)\left(2-3i\right)$

$=4-6i-6i+9i^2$

$=4-12i-9$

$=-5-12i$

49. $\dfrac{1+\sqrt{-1}}{\sqrt{-1}}$

$=\dfrac{1+i}{i}$

$=\dfrac{1+i}{i}\cdot\dfrac{-i}{-i}$

$=\dfrac{-i-i^2}{-i^2}$

$=\dfrac{-i+1}{1}$

$=-i+1$

$=1-i$

Section 2.1 Skills Check

1. **a.** Yes. The equation fits the form
    $$f(x) = ax^2 + bx + c, a \neq 0.$$

    b. Since $a = 2 > 0$, the graph opens up and is therefore concave up.

    c. Since the graph is concave up, the vertex point is a minimum.

3. Not quadratic. The equation does not fit the form $f(x) = ax^2 + bx + c, a \neq 0$. The highest exponent is 3.

5. **a.** Yes. The equation fits the form
    $$f(x) = ax^2 + bx + c, a \neq 0.$$

    b. Since $a = -5 < 0$, the graph opens down and is therefore concave down.

    c. Since the graph is concave down, the vertex point is a maximum.

7. **a.**

    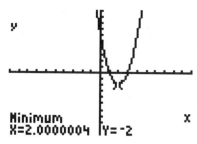

    b. Yes.

9. **a.**

    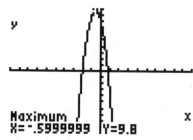

    b. Yes.

11. **a.**

    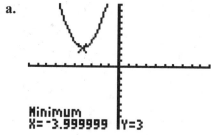

    b. Yes.

13. **a.**

    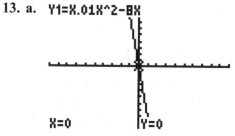

    b. No. The complete graph will be a parabola.

15. **a.** $(1, -3)$

    b.

    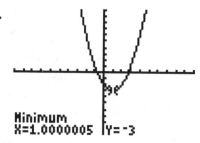

    $[-10, 10]$ by $[-10, 10]$

17. a. $(-8,8)$

b.

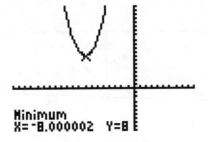

[–20, 10] by [–10, 20]

19. a. $y = 12x - 3x^2$

$y = -3x^2 + 12x$

$a = -3, b = 12, c = 0$

$h = \dfrac{-b}{2a} = \dfrac{-12}{2(-3)} = \dfrac{-12}{-6} = 2$

$k = f(h) = f(2) = 12(2) - 3(2)^2 =$

$24 - 3(4) = 24 - 12 = 12$

The vertex is $(2, 12)$.

b.

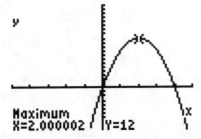

[–5, 5] by [–10, 20]

21. a. $y = 3x^2 + 18x - 3$

$a = 3, b = 18, c = -3$

$h = \dfrac{-b}{2a} = \dfrac{-18}{2(3)} = \dfrac{-18}{6} = -3$

$k = f(h)$

$= f(-3)$

$= 3(-3)^2 + 18(-3) - 3$

$= 3(9) - 54 - 3$

$= 27 - 54 - 3 = -30$

The vertex is $(-3, -30)$.

b.

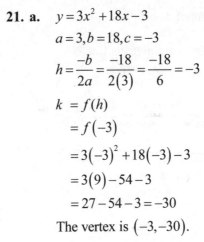

[–10, 10] by [–40, 10]

23. a. $y = 2x^2 - 40x + 10$

$a = 2, b = -40, c = 10$

$h = \dfrac{-b}{2a} = \dfrac{-(-40)}{2(2)} = \dfrac{40}{4} = 10$

b.

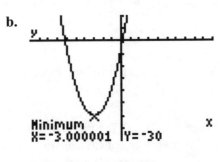

[–10, 30] by [–250, 30]

25. a. $y = -0.2x^2 - 32x + 2$

$a = -0.2, b = -32, c = 2$

$h = \dfrac{-b}{2a} = \dfrac{-(-32)}{2(-0.2)} = \dfrac{32}{-0.4} = -80$

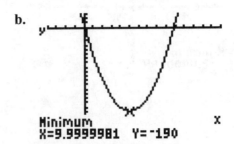

b.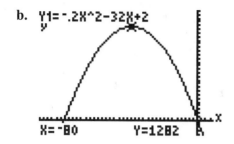

[–190, 30] by [–200, 1500]

27.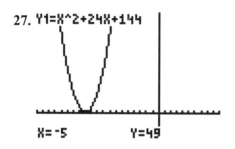

[–20, 10] by [–5, 20]

29.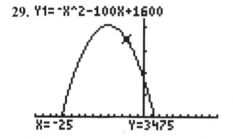

[–150, 100] by [–500, 5000]

31.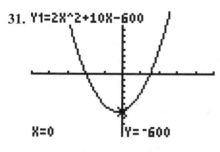

[–50, 50] by [–1000, 1000]

33. $(1,0), (3,0)$

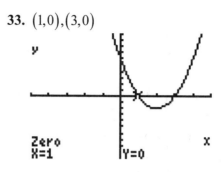

[–5, 5] by [–10, 10]

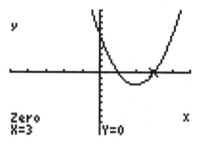

[–5, 5] by [–10, 10]

35. $(11,0), (-10,0)$

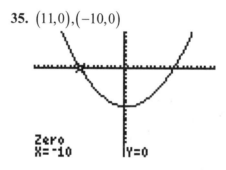

[–20, 20] by [–250, 100]

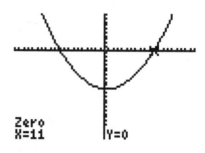

[–20, 20] by [–250, 100]

37. $(-2,0),(0.8,0)$

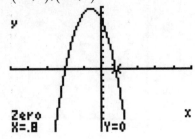

$[-5, 5]$ by $[-10, 10]$

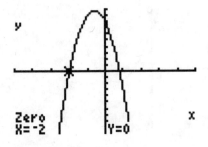

$[-5, 5]$ by $[-10, 10]$

Section 2.1 Exercises

39. a.

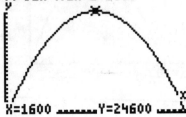

$[0, 3200]$ by $[0, 25,000]$

b. For x between 1 and 1600, the profit is increasing.

c. For x greater than 1600, the profit decreases.

41. a.

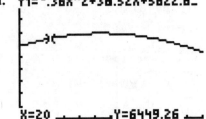

$[0, 120]$ by $[0, 10,000]$

b. In 2010, $x = 2010-1990 = 20$. Based on the graph of the model in part *a*, the population is 6449.3 million or 6.4493 billion people.

43. a.

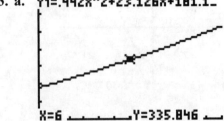

$[0, 12]$ by $[0, 700]$

b. In 2002, $x = 2002-1987 = 15$.

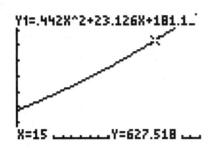

[0, 20] by [0, 800]

The projected tourism spending is $627.52 billion.

c. The calculation part b) is an extrapolation because $x = 15$ is beyond the scope of the original x-values. The model is applicable only in years between 1987 and 1999.

45. a. The model is a quadratic function. The graph of the model is a concave down parabola.

b. $h = \dfrac{-b}{2a} = \dfrac{-96}{2(-16)} = \dfrac{-96}{-32} = 3$

$k = f(h) = f(3)$

$= 100 + 96(3) - 16(3)^2$

$= 100 + 288 - 144$

$= 244$

The vertex is $(3, 244)$.

c. Three seconds into the flight of the ball, the ball is 244 feet above the ground. The ball reaches its maximum height of 244 feet in 3 seconds.

47. a. `Y1=270X-90X^2`

X=2.5   Y=112.5

[0, 5] by [−50, 250]

b. $h = \dfrac{-b}{2a} = \dfrac{-270}{2(-90)} = \dfrac{-270}{-180} = 1.5$

1.5 lumens yields the maximum rate of photosynthesis.

49. Note that the maximum profit occurs at the vertex of the quadratic function, since the function is concave down.

a. $h = \dfrac{-b}{2a} = \dfrac{-40}{2(-0.01)} = \dfrac{-40}{-0.02} = 2000$

b. $k = P(h) = P(2000)$

$= 40(2000) - 3000 - 0.01(2000)^2$

$= 80,000 - 3000 - 0.01(4,000,000)$

$= 80,000 - 3000 - 40,000$

$= 37,000$

Producing 2000 units yields a maximum profit of $37,000.

51. Note that the maximum profit occurs at the vertex of the quadratic function, since the function is concave down.

a. $h = \dfrac{-b}{2a} = \dfrac{-1500}{2(-0.02)} = \dfrac{-1500}{-0.04} = 37,500$

b. $k = R(h) = R(37,500)$

$= 1500(37,500) - 0.02(37,500)^2$

$= 56,250,000 - 28,125,000$

$= 28,125,000$

Selling 37,500 units yields a maximum revenue of $28,125,000.

53. a. Yes. $A = x(100 - x) = 100x - x^2$. Note that A fits the form

$f(x) = ax^2 + bx + c, a \neq 0$.

b. The maximum area will occur at the vertex of graph of function A.

$$h = \frac{-b}{2a} = \frac{-100}{2(-1)} = \frac{-100}{-2} = 50$$

$$k = A(h) = A(50)$$

$$= 100(50) - (50)^2$$

$$= 5000 - 2500$$

$$= 2500$$

The maximum area of the pen is 2500 square feet.

55.

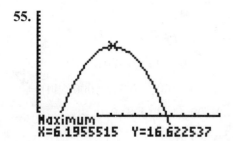

[0, 15] by [−5, 25]

The number of schools with satellite dishes reached a maximum between 1992 and 1997. An x-value of approximately 6.2 corresponds with the year 1996. The maximum number of dishes is approximately 16,623.

57. a. Since $a = 0.003 > 0$, the graph is concave up. The vertex is a minimum.

b. Using a graphing utility yields,

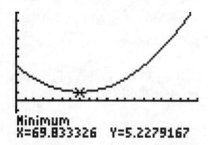

[0, 200] by [−20, 50]

The vertex is approximately $(69.8, 5.2)$.

c. In 1970 (when $x = 69.8$), the U.S. population reached a minimum percentage for foreign born people of 5.2%.

d. Consider part a above.

59. a.

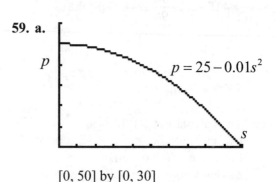

[0, 50] by [0, 30]

b. Decreasing. The graph is falling as s increases.

c. $(0, 25)$.

d. When the wind speed is zero, the amount of particulate pollution is 25 ounces per cubic yard.

61.

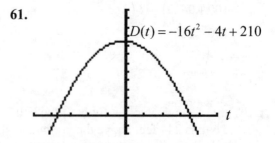

[−5, 5] by [−50, 300]

The t-intercepts are approximately $(-3.75, 0)$ and $(3.5, 0)$. The $(3.5, 0)$ makes sense because time is understood to be positive. In the context of the question, the tennis ball will hit the swimming pool in 3.5 seconds.

63. a.

Rent	Number of Apartments Rented	Total Revenue
$1200	100	$120,000
$1240	98	$121,520
$1280	96	$122,880
$1320	94	$124,080

b. Yes. Revenue = Rent multiplied by Number of Apartments Rented. $(1200 + 40x)$ represents the rent amount, while $(100 - 2x)$ represents the number of apartments rented.

c.

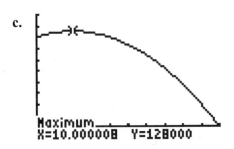

[0, 50] by [−15,000, 150,000]

The maximum occurs when $x = 10$. Therefore the most profitable rent to charge is $1200 + 40(10) = \$1600$.

65. a.

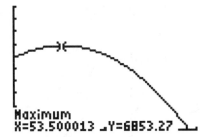

[0, 200] by [0, 10,000]

The vertex is $(53.5, 6853)$.

b. The model predicts that in 2044 the population of the world will by 6853 million.

c. The model predicts a population increase from 1990 until 2044. After 2004, based on the model, the population of the world will decrease.

Section 2.2 Skills Check

1. $x^2 - 3x - 10 = 0$
    $(x-5)(x+2) = 0$
    $x-5 = 0, x+2 = 0$
    $x = 5, x = -2$

3. $x^2 - 11x + 24 = 0$
    $(x-8)(x-3) = 0$
    $x-8 = 0, x-3 = 0$
    $x = 8, x = 3$

5. $2x^2 + 2x - 12 = 0$
    $2(x^2 + x - 6) = 0$
    $2(x+3)(x-2)$
    $x+3 = 0, x-2 = 0$
    $x = -3, x = 2$

7. $0 = 2t^2 - 11t + 12$
    $2t^2 - 11t + 12 = 0$
    Note that $2t^2 \cdot 12 = 24t^2$. Look for two terms whose product is $24t^2$ and whose sum is the middle term, $-11t$.
    $2t^2 - 8t - 3t + 12 = 0$
    $(2t^2 - 8t) + (-3t + 12) = 0$
    $2t(t-4) - 3(t-4) = 0$
    $(2t-3)(t-4) = 0$
    $2t-3 = 0, t-4 = 0$
    $x = \dfrac{3}{2}, x = 4$

9. $6x^2 + 10x = 4$
    $6x^2 + 10x - 4 = 0$
    $2(3x^2 + 5x - 2) = 0$
    Note that $3x^2 \cdot -2 = -6x^2$. Look for two terms whose product is $-6x^2$ and whose sum is the middle term, $5x$.
    $2(3x^2 + 6x - 1x - 2) = 0$
    $2\left[(3x^2 + 6x) + (-1x - 2)\right] = 0$
    $2\left[3x(x+2) + (-1)(x+2)\right] = 0$
    $2(3x-1)(x+2) = 0$
    $3x-1 = 0, x+2 = 0$
    $x = \dfrac{1}{3}, x = -2$

11. $y = x^2 - 3x - 10$

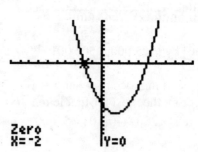

[−10, 10] by [−20, 10]

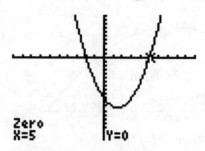

[−10, 10] by [−20, 10]

The x-intercepts are $(-2, 0)$ and $(5, 0)$.

13. $y = 3x^2 - 8x + 4$

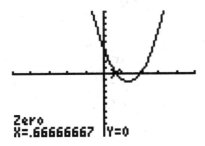

Zero
X=.66666667   Y=0

$[-5, 5]$ by $[-10, 10]$

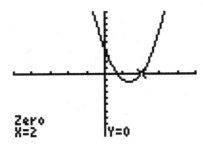

Zero
X=2   Y=0

$[-5, 5]$ by $[-10, 10]$

The x-intercepts are $\left(\dfrac{2}{3}, 0\right)$ and $(2, 0)$.

15. $y = 2x^2 + 7x - 4$

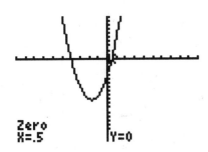

Zero
X=.5   Y=0

$[-10, 10]$ by $[-20, 10]$

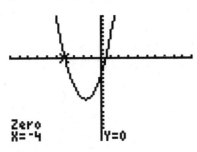

Zero
X=-4   Y=0

$[-10, 10]$ by $[-20, 10]$

The x-intercepts are $(-4, 0)$ and $(0.5, 0)$.

17.

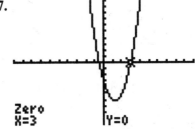

Zero
X=3   Y=0

$[-10, 10]$ by $[-10, 10]$

Since $w = 3$ is an x-intercept, then
$w - 3$ is a factor.
$2w^2 - 5w - 3 = 0$
$(w - 3)(2w + 1) = 0$
$w - 3 = 0, 2w + 1 = 0$
$w = 3, w = -\dfrac{1}{2}$

19.

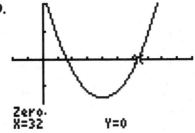

Zero.
X=32   Y=0

$[-10, 50]$ by $[-250, 200]$

Since $x = 32$ is an x-intercept, then
$x - 32$ is a factor.
$x^2 - 40x + 256 = 0$
$(x - 32)(x - 8) = 0$
$x - 32 = 0, x - 8 = 0$
$x = 32, x = 8$

21.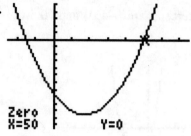

$[-25, 75]$ by $[-2500, 1000]$

Since $x = 50$ is an x-intercept for
$2s^2 - 70s - 1500 = 0$, then $s - 50$
is a factor.

$(s - 50)(2s + 30) = 0$

$2(s - 50)(s + 15) = 0$

$s - 50 = 0, s + 15 = 0$

$s = 50, s = -15$

23. $4x^2 - 9 = 0$

$4x^2 = 9$

$\sqrt{4x^2} = \pm\sqrt{9}$

$2x = \pm 3$

$x = \pm\dfrac{3}{2}$

25. $x^2 - 32 = 0$

$x^2 = 32$

$\sqrt{x^2} = \pm\sqrt{32}$

$x = \pm\sqrt{32} = \pm\sqrt{16 \cdot 2} = \pm 4\sqrt{2}$

27. $x^2 - 4x - 9 = 0$

$\left(\dfrac{-4}{2}\right)^2 = (-2)^2 = 4$

$x^2 - 4x + 4 - 4 - 9 = 0$

$(x^2 - 4x + 4) + (-4 - 9) = 0$

$(x - 2)^2 - 13 = 0$

$(x - 2)^2 = 13$

$\sqrt{(x - 2)^2} = \pm\sqrt{13}$

$x - 2 = \pm\sqrt{13}$

$x = 2 \pm \sqrt{13}$

29. $x^2 - 3x + 2 = 0$

$\left(\dfrac{-3}{2}\right)^2 = \dfrac{9}{4}$

$x^2 - 3x + \dfrac{9}{4} - \dfrac{9}{4} + 2 = 0$

$\left(x^2 - 3x + \dfrac{9}{4}\right) + \left(-\dfrac{9}{4} + 2\right) = 0$

$\left(x - \dfrac{3}{2}\right)^2 + \left(-\dfrac{9}{4} + \dfrac{8}{4}\right) = 0$

$\left(x - \dfrac{3}{2}\right)^2 - \dfrac{1}{4} = 0$

$\left(x - \dfrac{3}{2}\right)^2 = \dfrac{1}{4}$

$\sqrt{\left(x - \dfrac{3}{2}\right)^2} = \pm\sqrt{\dfrac{1}{4}}$

$x - \dfrac{3}{2} = \pm\dfrac{1}{2}$

$x = \dfrac{3}{2} \pm \dfrac{1}{2}$

$x = \dfrac{3}{2} + \dfrac{1}{2}, x = \dfrac{3}{2} - \dfrac{1}{2}$

$x = 2, x = 1$

31. $x^2 - 5x + 2 = 0$

$a = 1, b = -5, c = 2$

$$x = \frac{-b \pm \sqrt{b^2 - 4ac}}{2a}$$

$$x = \frac{-(-5) \pm \sqrt{(-5)^2 - 4(1)(2)}}{2(1)}$$

$$x = \frac{5 \pm \sqrt{25 - 8}}{2}$$

$$x = \frac{5 \pm \sqrt{17}}{2}$$

33. $5x + 3x^2 = 8$

$3x^2 + 5x - 8 = 0$

$a = 3, b = 5, c = -8$

$$x = \frac{-b \pm \sqrt{b^2 - 4ac}}{2a}$$

$$x = \frac{-(5) \pm \sqrt{(5)^2 - 4(3)(-8)}}{2(3)}$$

$$x = \frac{-5 \pm \sqrt{25 + 96}}{6}$$

$$x = \frac{-5 \pm \sqrt{121}}{6}$$

$$x = \frac{-5 \pm 11}{6}$$

$$x = \frac{-5 + 11}{6}, x = \frac{-5 - 11}{6}$$

$$x = 1, x = -\frac{16}{6} = -\frac{8}{3}$$

35.

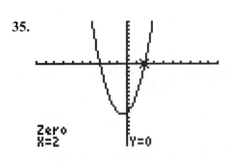

[–10, 10] by [–20, 10]

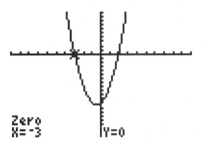

[–10, 10] by [–20, 10]

The solutions are $x = 2$ and $x = -3$.

37.

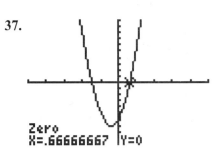

[–5, 5] by [–10, 10]

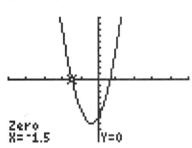

[–5, 5] by [–10, 10]

The solutions are $x = \frac{2}{3}$ and $x = -\frac{3}{2}$.

39. $4x + 2 = 6x^2 + 3x$

$0 = 6x^2 + 3x - 4x - 2$

$6x^2 - x - 2 = 0$

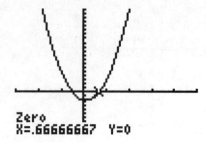

[−3, 5] by [−10, 20]

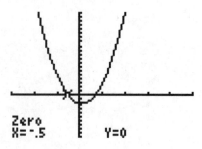

[−3, 5] by [−10, 20]

The solutions are $x = \dfrac{2}{3}$ and $x = -\dfrac{1}{2}$.

41. $x^2 + 25 = 0$

$x^2 = -25$

$\sqrt{x^2} = \pm\sqrt{-25}$

$x = \pm 5i$

Graphical check

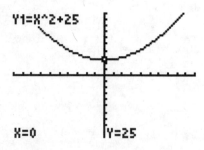

[−10, 10] by [−100, 100]

Note that the graph has no x-intercepts.

43. $(x-1)^2 = -4$

$\sqrt{(x-1)^2} = \pm\sqrt{-4}$

$x - 1 = \pm 2i$

$x = 1 \pm 2i$

Graphical check

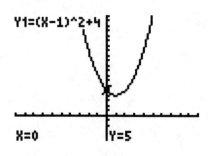

[−10, 10] by [−5, 20]

Note that the graph has no x-intercepts.

45. $x^2 + 4x + 8 = 0$

$x = \dfrac{-b \pm \sqrt{b^2 - 4ac}}{2a}$

$x = \dfrac{-4 \pm \sqrt{4^2 - 4(1)(8)}}{2(1)}$

$x = \dfrac{-4 \pm \sqrt{-16}}{2}$

$x = \dfrac{-4 \pm 4i}{2} = -2 \pm 2i$

Graphical check

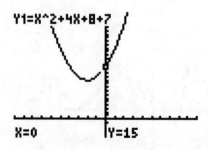

[−10, 10] by [−5, 30]

Note that the graph has no x-intercepts.

47. $2x^2 - 8x + 9 = 0$

$$x = \frac{-b \pm \sqrt{b^2 - 4ac}}{2a}$$

$$x = \frac{-(-8) \pm \sqrt{(-8)^2 - 4(2)(9)}}{2(2)}$$

$$x = \frac{8 \pm \sqrt{-8}}{4}$$

$$x = \frac{8 \pm 2i\sqrt{2}}{4} = \frac{4 \pm i\sqrt{2}}{2}$$

Graphical check

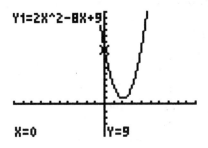

Y1=2X^2-8X+9

X=0   Y=9

[−10, 10] by [−5, 15]

Note that the graph has no x-intercepts.

Section 2.2 Exercises

49. Let $S = 228$ and solve for t.

$$228 = 100 + 96t - 16t^2$$

$$-16t^2 + 96t - 128 = 0$$

$$-16(t^2 - 6t + 8) = 0$$

$$-16(t - 4)(t - 2) = 0$$

$$t - 4 = 0, t - 2 = 0$$

$$t = 4, t = 2$$

The ball is 228-feet high after 2 seconds and 4 seconds.

51. Let $P(x) = 0$ and solve for x.

$$-12x^2 + 1320x - 21,600 = 0$$

$$-12(x^2 - 110x + 1800) = 0$$

$$-12(x - 90)(x - 20) = 0$$

$$x - 90 = 0, x - 20 = 0$$

$$x = 90, x = 20$$

Producing and selling either 20 units or 90 units produces a profit of zero dollars. Therefore 20 units or 90 units represent the break-even point for manufacturing and selling this product.

53. a. $P(x) = R(x) - C(x)$

$$P(x) = 550x - (10,000 + 30x + x^2)$$

$$P(x) = 550x - 10,000 - 30x - x^2$$

$$P(x) = -x^2 + 520x - 10,000$$

b. Let $x = 18$.

$$P(18) = -(18)^2 + 520(18) - 10,000$$

$$P(18) = -324 + 9360 - 10,000$$

$$p(18) = -964$$

When 18 units are produced, there is a loss of $964.

c. Let $x = 32$.

$P(32) = -(32)^2 + 520(32) - 10,000$

$P(32) = -1024 + 16,640 - 10,000$

$p(32) = 5616$

When 32 units are produced, there is a profit of \$5616.

d. Let $P(x) = 0$ and solve for x.

$0 = -x^2 + 520x - 10,000$

$-1(x^2 - 520x + 10,000) = 0$

$-1(x - 500)(x - 20) = 0$

$x - 500 = 0, x - 20 = 0$

$x = 500, x = 20$

To break even on this product the company needs to manufacture and sell either 20 units or 500 units.

55. a. Let $p = 0$ and solve for s.

$0 = 25 - 0.01s^2$

$-25 = -0.01s^2$

$\dfrac{-25}{-0.01} = \dfrac{-0.01}{-0.01} s^2$

$2500 = s^2$

$s = \pm\sqrt{2500} = \pm 50$

When s is 50 or –50, $p = 0$. Since a wind speed of –50 does not make physical sense, the only solution is 50. Therefore, when the wind speed is 50 mph the pollution in the air above the power plant is zero.

b. When $p = 0$, the pollution in the air above the power plant is zero.

c. Only the positive solution, $s = 50$, makes sense because wind speed must be positive.

57. a. Let $v = 0.02$ and solve for r.

$0.02 = 2(0.01 - r^2)$

$\dfrac{0.02}{2} = \dfrac{2(0.01 - r^2)}{2}$

$0.01 = 0.01 - r^2$

$-r^2 = 0$

$r = 0$

A distance of 0 cm produces a velocity of 0.02 cm/sec.

b. Let $v = 0.015$ and solve for r.

$0.015 = 2(0.01 - r^2)$

$\dfrac{0.015}{2} = \dfrac{2(0.01 - r^2)}{2}$

$0.0075 = 0.01 - r^2$

$-r^2 = 0.0075 - 0.01$

$-r^2 = -0.0025$

$r^2 = 0.0025$

$r = \pm\sqrt{0.0025} = \pm 0.05$

A distance of 0.05 cm produces a velocity of 0.02 cm/sec. Since r is a distance, only $r = 0.05$ makes sense in the physical context of the question.

c. Let $v = 0$ and solve for r.

$0 = 2(0.01 - r^2)$

$\dfrac{0}{2} = \dfrac{2(0.01 - r^2)}{2}$

$0 = 0.01 - r^2$

$r^2 = 0.01$

$r = \pm\sqrt{0.01} = \pm 0.1$

A distance of 0.01 cm produces a velocity of 0 cm/sec. Since r is a distance, only $r = 0.1$ makes sense in the physical context of the question.

59. Equilibrium occurs when demand is equal to supply. Solve

$$109.70 - 0.10q = 0.01q^2 + 5.91$$

$$0.01q^2 + 0.10q - 103.79 = 0$$

$$a = 0.01, b = 0.10, c = -103.79$$

$$q = \frac{-b \pm \sqrt{b^2 - 4ac}}{2a}$$

$$q = \frac{-0.10 \pm \sqrt{(0.10)^2 - 4(0.01)(-103.79)}}{2(0.01)}$$

$$q = \frac{-0.10 \pm \sqrt{0.01 + 4.1516}}{0.02}$$

$$q = \frac{-0.10 \pm \sqrt{4.1616}}{0.02}$$

$$q = \frac{-0.10 \pm 2.04}{0.02}$$

$$q = \frac{-0.10 + 2.04}{0.02}, q = \frac{-0.10 - 2.04}{0.02}$$

$$q = 97, q = -107$$

Since q represents the quantity of trees at equilibrium, q must be positive. A q-value of -107 does not make sense in the physical context of the question. Producing 9700 trees, when $q = 97$, creates an equilibrium price. The equilibrium price is given by

$$p = 109.70 - 0.10q$$

$$p = 109.70 - 0.10(97)$$

$$p = 109.70 - 9.70$$

$$p = 100.00 \text{ or } \$100.00 \text{ per tree.}$$

61. a.

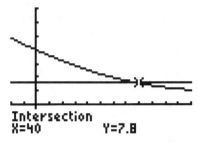

[–10, 60] by [–10, 35]

b. Let $y = 7.8$ and solve for x.

$$7.8 = 0.003x^2 - 0.42x + 19.8$$

$$0.003x^2 - 0.42x + 12 = 0$$

$$0.003(x^2 - 140x + 4000) = 0$$

$$0.003(x - 40)(x - 100) = 0$$

$$x = 40, x = 100$$

7.8% of the U.S. population is foreign born in 1940 and 2000.

63. a. Begin by setting the x-range to [0, 10] in order to capture the years between 1989 and 1997. Then, note that when $x = 0$, $C(0) = 1662.14$. Adjust the y-range accordingly.

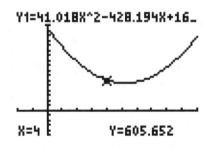

[–2, 10] by [–500, 2000]

b.

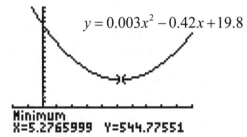

$$y = 0.003x^2 - 0.42x + 19.8$$

[–2, 10] by [–500, 2000]

The minimum number of complaints occurred in approximately $1989 + 5 = 1994$

c.

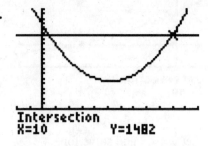

Intersection
X=10   Y=1482

[–2, 12] by [0, 1800]
Baggage complaints are 1482 in $1989 + 10 = 1999$.

d. The result in part c) is an extrapolation, since 1999 is beyond the scope of the original data, 1989–1997.

65.

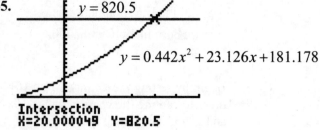

Intersection
X=20.000049 Y=820.5

[–10, 30] by [–300, 1000]

In approximately $1987 + 20 = 2007$ tourism spending will be \$820.5 billion.

67. a. To determine the change in the number of hospital admissions between 1990 and 1997, calculate
$A(1997 - 1980) - A(1990 - 1980) = A(17) - A(10)$.

$$A(17) = 38.228(17)^2 - 1069.60(17) + 40{,}698.547$$
$$= 11{,}047.892 - 18{,}183.2 + 40{,}698.547$$
$$= 33{,}563.239$$

$$A(10) = 38.228(10)^2 - 1069.60(10) + 40{,}698.547$$
$$= 3822.8 - 10{,}696 + 40{,}698.547$$
$$= 33{,}825.347$$

$$A(17) - A(10) = 33{,}563.239 - 33{,}825.347$$
$$= -262.108 \approx 262$$

There is a decrease of approximately
262 thousand people.

b.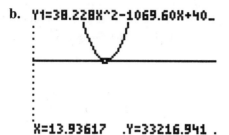

X=13.93617   .Y=33216.941 .

[10, 20] by [33,100, 33,300]

In about 1994 (1980 + 14) hospital admissions were 33,217,000.

c. $A(25) = 38.228(25)^2 - 1069.60(25) + 40,698.547$

$A(25) = 23,892.5 - 26,740 + 40,698.547$

$A(25) = 37,851.047$

Assuming the model continues to be valid beyond 1997, hospital admissions would be about 37,851 thousand or 37,851,000 in 2005.

69. a. The funds available in 1990 correspond to

$f(10) = 9.032(10)^2 + 99.970(10) + 3645.90$

$= 903.2 + 999.70 + 3645.90$

$= 5548.8$ or \$5548.8 million

The funds available in 2000 correspond to

$f(20) = 9.032(20)^2 + 99.970(20) + 3645.90$

$= 3612.8 + 1999.4 + 3645.90$

$= 9258.10$ or \$9258.10 million

Calculating the change between 1990 and 2000

$f(20) - f(10)$

$= 9258.10 - 5548.8$

$= 3709.3$ or \$3709.3 million

b.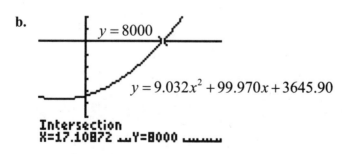

Intersection
X=17.10872 ⌐Y=8000 ⌐⌐⌐⌐⌐

[−10, 30] by [0, 10,000]

Federal aid exceeds \$8000 million in 1998, 17.1 years after 1980.

71. a.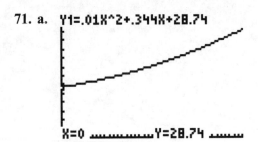

[0, 42] by [0, 70]

b.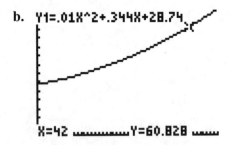

[0, 50] by [0, 70]

In 2002 about 60.8% of the people who smoked after age nineteen will have quit.

c.

X	Y1
28	46.212
29	47.126
30	48.06
31	49.014
32	49.988
33	50.982
34	51.996

X=32

[0, 42] by [0, 70]

In 1992 (when $x = 32$), about 50% of people who smoked after age nineteen will have quit.

73.

X	Y1
30	6654.5
31	6671
32	6686.9
33	6702
34	6716.4
35	6730.1
36	6743

X=33

When $x = 33$, the population is predicted to be 6702 million. The year is 2023.

Section 2.3 Skills Check

1. Y1=X^3

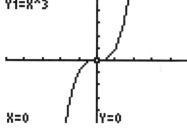

[−5, 5] by [−5, 5]

3. Y1=X^(1/3)

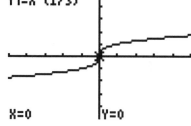

[−5, 5] by [−5, 5]

5. Y1=√(X)+2

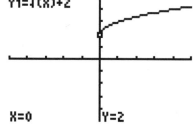

[−5, 5] by [−5, 5]

7. Y1=(1/X)−3

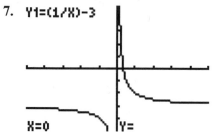

[−5, 5] by [−5, 5]

9. Y1=−1(X<0)

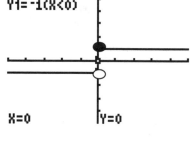

[−5, 5] by [−5, 5]

11.a.

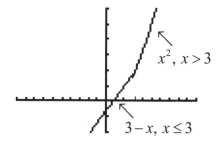

[−10, 10] by [−10, 30]

b. $f(2) = 4(2) - 3 = 8 - 3 = 5$
$f(4) = (4)^2 = 16$

c. Domain: $(-\infty, \infty)$

13. a. Y1=abs(X)

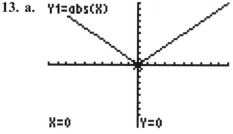

[−10, 10] by [−10, 10]

b. $f(-2) = |-2| = 2$
$f(5) = |5| = 5$

c. Domain: $(-\infty, \infty)$

15. a. $f(-1) = (-1-2)^3 = (-3)^3 = -27$

b. $f(3) = (3-2)^3 = (1)^3 = 1$

17. a. $f(-1) = \sqrt{-1-2} = \sqrt{-3}$
No real number solution

b. $f(3) = \sqrt{3-2} = \sqrt{1} = 1$

19. a. $f(-1) = \dfrac{1}{-1-2} = \dfrac{1}{-3} = -\dfrac{1}{3}$

b. $f(3) = \dfrac{1}{3-2} = \dfrac{1}{1} = 1$

21. a. $f(-1) = \dfrac{1}{-1-4} - 9$

$= -\dfrac{1}{5} - 9$

$= -9.2$

b. $f(3) = \dfrac{1}{3-4} - 9$

$= -1 - 9$

$= -10$

23. a. $f(-1) = 5$, since $x \le 1$.

b. $f(3) = 6$, since $x > 1$.

25. a. $f(-1) = (-1)^2 - 1 = 0$, since $x \le 0$.

b. $f(3) = (3)^3 + 2 = 29$, since $x > 0$.

27. The function is increasing for all values of x.

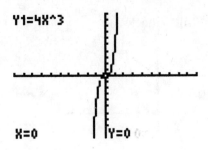

[−10, 10] by [−10, 10]

a.   increasing

b.   increasing

29. Concave down.

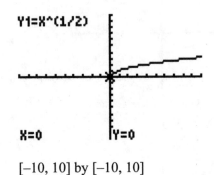

[−10, 10] by [−10, 10]

31. Concave up.

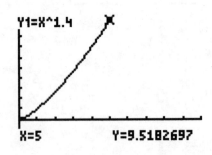

[0, 10] by [−2, 10]

33. Y1=X(X≥0)

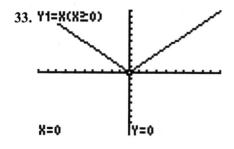

X=0   Y=0

[–10, 10] by [–10, 10]

Section 2.3 Exercises

35. **a.** The given equation is a power function. It fits the form
$f(x) = ax^b$, where b is positive.

b. $f(5) = y = 20,000(5)^{0.11} = 23,873.53$
In 1965 (1960 + 5) the model predicts that there were approximately 23,874 suicides.

c. In 1982, the x-value is 1982–1960 = 22.
$f(22) = 20,000(22)^{0.11} = 28,099.36$.
The model predicts approximately 28,099 suicides for 1982.

37. **a.** Y1=489X^.6

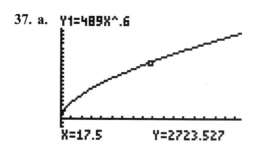

X=17.5   Y=2723.527

[0, 35] by [–1000, 5000]

b. Y1=489X^.6

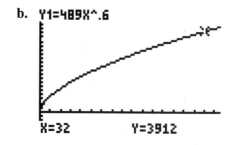

X=32   Y=3912

[0, 35] by [–1000, 5000]

If the company employs 32 taxi drivers then 3912 miles are driven per day.

c. The function is increasing. As the number of drivers increases, it is reasonable to expect the number of miles to increase.

39. a.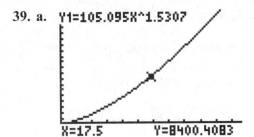

[0, 35] by [–2000, 20,000]

The function is increasing.

b. Concave up.

c.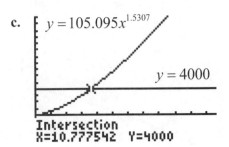

[0, 35] by [–4000, 15,000]

The model predicts that assets will reach $4000 billion when $x = 10.777542$, which corresponds to the year 2001.

41. a.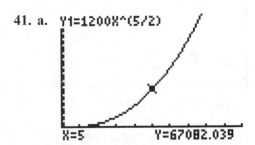

[0, 10] by [0, 200,000]

b. Considering the picture in part *a*, the graph is concave up.

43. a.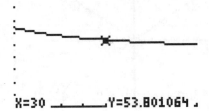

[10, 50] by [40, 70]

b. Voter turnout is decreasing, since the graph is falling as the number of years increases.

c. Since 1996 corresponds to $x = 46$, the voter turnout in 1996 is predicted to be 50.277%

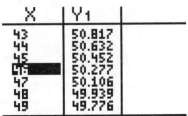

[10, 50] by [40, 70]

d. Since 2000 corresponds to $x = 50$, the voter turnout in 2000 is predicted to be 49.617%

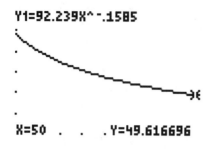

X=50 . . .Y=49.616696

[10, 50] by [40, 70]

e. The actual voter turnout in 2000 was higher than the value predicted by the model.

45. Let $x = 2000$ and calculate $C(x)$.

$$C(2000) = 105 + \frac{50,000}{2000}$$
$$= 105 + 25$$
$$= 130 \text{ per unit}$$

47. a. $P(w) = \begin{cases} 0.37 & 0 < w \le 1 \\ 0.60 & 1 < w \le 2 \\ 0.83 & 2 < w \le 3 \\ 1.06 & 3 < w \le 4 \end{cases}$

b. $p(1.2) = 0.60$. The cost of mailing a 1.2 ounce first class letter is $0.60.

c. Domain: $(0, 4]$

d. $P(2) = 0.60$
$p(2.1) = 0.83$

e. Considering the solution to part d, a 2 ounce letter costs $0.60 to mail first class, while a 2.1 ounce letter costs $0.83 to mail first class.

49. a. $T(x) = \begin{cases} 0.15x & 0 \le x \le 43,850 \\ 6577.50 + 0.28(x - 43,850) & 43,850 < x \le 105,950 \end{cases}$

b. $T(42,000) = 0.15(42,000) = 6300$

c. $T(55,000) = 6577.50 + 0.28(55,000 - 43,850)$
$$= 6577.50 + 0.28(11,150)$$
$$= 9699.50$$

The total tax on $55,000 is $9699.50.

d. The statement is incorrect.

$T(43,850) = 0.15(43,850)$
$$= 6577.50$$

The total tax on $43,850 is $6577.50.

$T(43,850 + 1) = 6577.50 + 0.28(43,851 - 43,850)$
$$= 6577.50 + 0.28(1)$$
$$= 6577.78$$

The total tax on $43,851 is $6577.78.

The difference in the tax bills is $0.28. Only the extra dollar is taxed at the 28% rate.

51. a.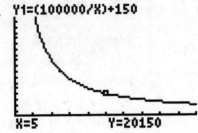

[–0, 10] by [–10,000, 101,000]

b. The function, and therefore the average cost, is decreasing.

53. a. $C = \dfrac{1000}{V}$

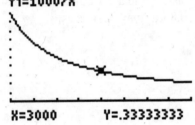

[1000, 5000] by [–0.2, 1.2]

b. Decrease.

Section 2.4 Skills Check

1. Y1=X^3

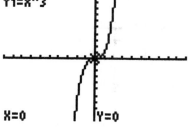

X=0   Y=0

[–10, 10] by [–10, 10]

Y1=X^3+5

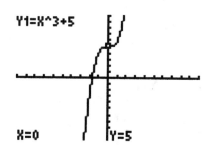

X=0   Y=5

[–10, 10] by [–10, 10]

3. Y1=ƒ(X)

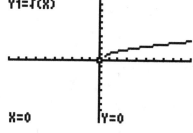

X=0   Y=0

[–10, 10] by [–10, 10]

Y1=ƒ(X–4)

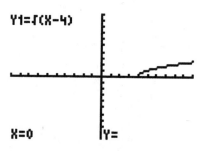

X=0   Y=

[–10, 10] by [–10, 10]

5. Y1=X^(1/3)

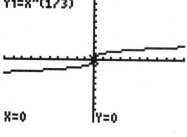

X=0   Y=0

[–10, 10] by [–10, 10]

Y1=(X+2)^(1/3)–1

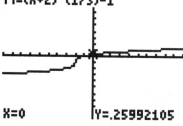

X=0   Y=.25992105

[–10, 10] by [–10, 10]

7. Y1=abs(X)

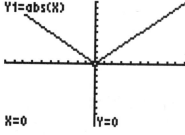

X=0   Y=0

[–10, 10] by [–10, 10]

Y1=abs(X–2)+1

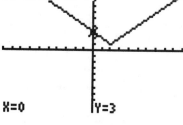

X=0   Y=3

[–10, 10] by [–10, 10]

9. Y1=X^2

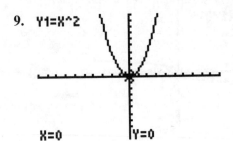

X=0   Y=0

[–10, 10] by [–10, 10]

Y1=-X^2+5

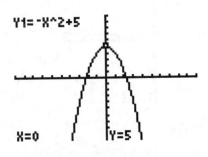

X=0   Y=5

[–10, 10] by [–10, 10]

11. Y1=1/X

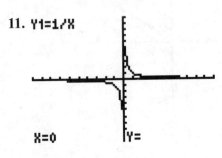

X=0   Y=

[–10, 10] by [–10, 10]

Y1=(1/X)-3

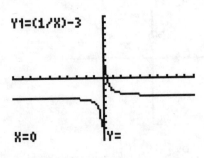

X=0   Y=

[–10, 10] by [–10, 10]

13. Y1=X^2

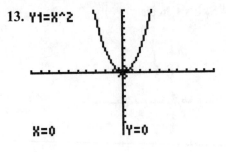

X=0   Y=0

[–10, 10] by [–10, 10]

Y1=(1/3)X^2

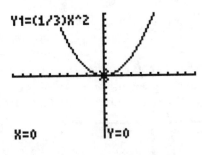

X=0   Y=0

[–10, 10] by [–10, 10]

15. Y1=abs(X)

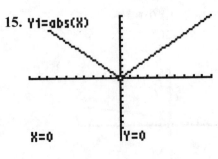

X=0   Y=0

[–10, 10] by [–10, 10]

Y1=3abs(X)

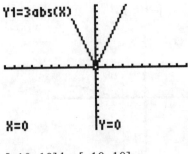

X=0   Y=0

[–10, 10] by [–10, 10]

17. The graph of the function is shifted 2 units right and 3 units up.

19. $y = (x+4)^{\frac{3}{2}}$

21. $y = 3x^{\frac{3}{2}} + 5$

23. x-axis symmetry.

Let $x = -x$

$y = 2(-x)^2 - 3 = 2x^2 - 3$

Since the result matches the original equation, the graph of the equation is symmetric with respect to the x-axis.

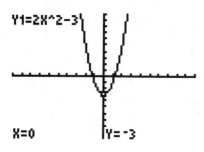

[–10, 10] by [–10, 10]

25. Origin symmetry.

Let $x = -x, y = -y$

$-y = (-x)^3 - (-x)$

$-y = -x^3 + x$

$y = x^3 - x$

Since the result matches the original equation, the graph of the equation is symmetric with respect to the origin.

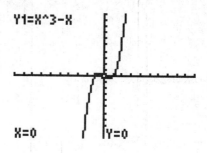

[–10, 10] by [–10, 10]

27. Origin symmetry.

Let $x = -x, y = -y$

$-y = \dfrac{6}{-x}$

$y = \dfrac{6}{x}$

Since the result matches the original equation, the graph of the equation is symmetric with respect to the origin.

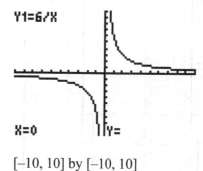

[–10, 10] by [–10, 10]

29. x-axis symmetry.

Let $y = -y$

$x^2 + (-y)^2 = 25$

$x^2 + y^2 = 25$

Since the result matches the original equation, the graph of the equation is symmetric with respect to the x-axis.

y-axis symmetry.

Let $x = -x$

$(-x)^2 + y^2 = 25$

$x^2 + y^2 = 25$

Since the result matches the original equation, the graph of the equation is symmetric with respect to the *y*-axis.

Origin symmetry.

Let $x = -x, y = -y$

$(-x)^2 + (-y)^2 = 25$

$x^2 + y^2 = 25$

Since the result matches the original equation, the graph of the equation is symmetric with respect to the origin.

Since the given equation is not a function, it can not be easily graphed using the graphing calculator. The equation must be rewritten as $y = \pm\sqrt{25 - x^2}$

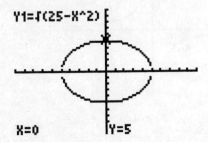

Y1=√(25-X^2)

X=0   Y=5

[–10, 10] by [–10, 10]

31. $f(-x) = |-x| - 5$

$= |-1 \cdot x| - 5$

$= |-1||x| - 5$

$= |x| - 5$

$= f(x)$

Since $f(-x) = f(x)$, the function is even.

33. $f(-x) = \sqrt{(-x)^2 + 3}$

$= \sqrt{x^2 + 3}$

Since $f(-x) = f(x)$, the function is even.

35. $f(-x) = \dfrac{5}{-x}$

$= -\dfrac{5}{x}$

$= -\left(\dfrac{5}{x}\right)$

Since $f(-x) \neq f(x)$ and $f(-x) \neq -f(x)$, the function is neither even nor odd.

Section 2.4 Exercises

37. a. $s = t^3 \; \left(y = x^3\right)$

b.

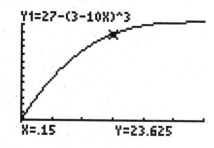

[0, 0.3] by [0, 30]

c. $f(0.3) = 27 - \left(3 - 10(0.3)\right)^3$

$= 27 - (3-3)^3$

$= 27 - 0$

$= 27$

After 0.3 seconds the bullet has traveled 27 inches.

39. a. Since q is in the numerator with an exponent of one, $p = \dfrac{180 + q}{6}$ is a linear function.

b. Since there is a variable in the denominator, $p = \dfrac{30,000}{q} - 20$ is a shifted reciprocal function.

41. a. $y = \dfrac{1}{x}$. Shifted one unit left.

b.

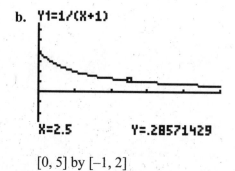

[0, 5] by [−1, 2]

c. The function is decreasing. Therefore, the amount of self-attentiveness of person decreases as the size of the crowd increases.

43. a. Shifted 10 units left and one unit down. Reflected over the x-axis and stretched by a factor of 1000.

b.

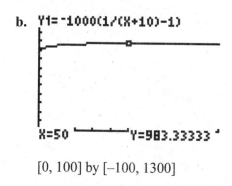

[0, 100] by [−100, 1300]

45. Since in the given function x represents the years since 1990, $x + 5$ represents the years since 1995. Therefore, the new function would be $f(x) = 105.095(x+5)^{1.5307}$.

47. Since in the given function x represents the years since 1986, $x + 4$ represents the years since 1990. Therefore, the new function would be

$f(x) = 0.084(x+4)^2 + 1.124(x+4)$

$+4.028.$

49. a. $t = 2005 - 1980 = 25$

$S(25) = 0.0003488(25)^{4.2077}$

≈ 265.882

The number of subscribers in 2005 is approximately 266 million.

b. Since in the given function x represents the years since 1980, $x + 5$ represents the years since 1985. Therefore, the new function would be

$S(t) = 0.0003488(t+5)^{4.2077}$.

c. $t = 2005 - 1985 = 20$

$$S(20) = 0.0003488(20 + 5)^{4.2077}$$

$$= 0.0003488(25)^{4.2077}$$

Using the shifted model yields the same result as part a) above.

Section 2.5 Skills Check

1.

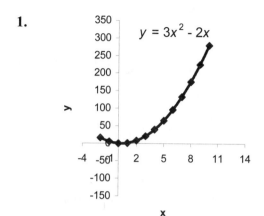

$y = 3x^2 - 2x$

3. The *x*-values are not equally spaced.

5.

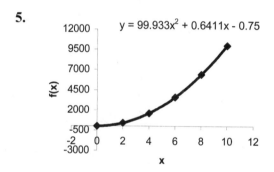

$y = 99.933x^2 + 0.6411x - 0.75$

7. a.

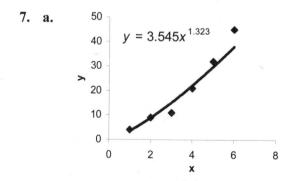

$y = 3.545x^{1.323}$

b.

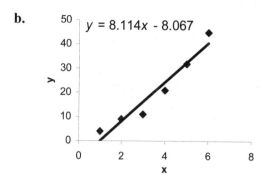

$y = 8.114x - 8.067$

c. The power function is a better fit.

9. a.

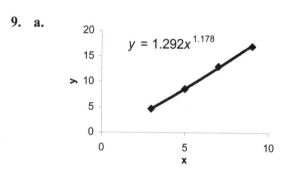

$y = 1.292x^{1.178}$

b.

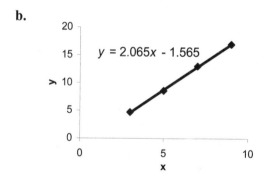

$y = 2.065x - 1.565$

c. Both models appear to be good fits.

Section 2.5 Exercises

11.

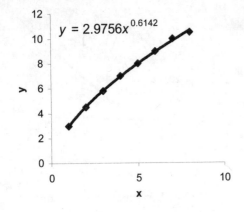

$y = 2.9756x^{0.6142}$

13. a.

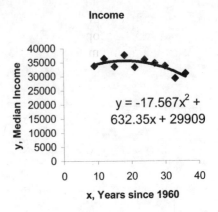

Income

$y = -17.567x^2 + 632.35x + 29909$

y, Median Income

x, Years since 1960

b. In 1997, $x = 1997 - 1960 = 37$

$$y = -17.567(37)^2 + 632.35(37)$$
$$+29,909$$

Using the TI-83 and substituting into the unrounded model

```
37→X
              37
-17.56734006734X
^2+632.346464646
46X+29909.145454
545
       29256.27609
```

$y \approx 29,256$

In 1997 the median annual income is approximately \$29,256.

In 2002, $x = 2002 - 1960 = 42$

$$y = -17.567(42)^2 + 632.35(42) + 29,909$$

Using the TI-83 and substituting into the unrounded model

```
42→X
              42
-17.56734006734X
^2+632.346464646
46X+29909.145454
545
       25478.90909
```

$y \approx 29,256$

In 2002 the median annual income is approximately $25,479.

c. No. Since both 1997 and 2002 are beyond the scope of the base data, the extrapolation may not be valid.

15. a.

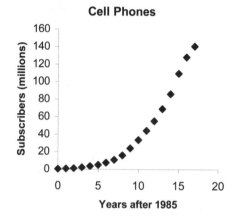
Cell Phones

Yes. It appears a quadratic function will fit the data well.

b.

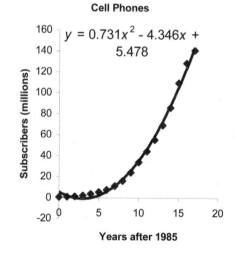
Cell Phones
$y = 0.731x^2 - 4.346x + 5.478$

c. In 2005, $x = 2005 - 1985 = 20$
Using the TI-83 and substituting into the unrounded model

```
20→X
                20
.73121633126935X
^2+ -4.3461162280
702X+5.477701754
386
        211.0419097
```

$y \approx 211$
The number of cell phone subscribers in 2005 is approximately 211 million.

d. There are about 300 million people in the U.S. Therefore, approximately $\frac{211}{300}$ or $\frac{2}{3}$ have cell phones

17. a.

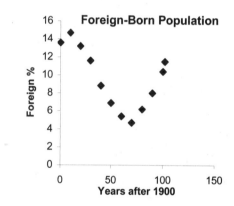
Foreign-Born Population

b. Yes.

c.

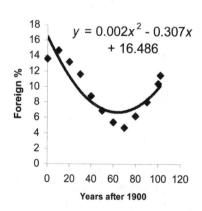
Foreign-Born Population
$y = 0.002x^2 - 0.307x + 16.486$

d. In 2005, $x = 2005 - 1900 = 105$
Using the TI-83 and substituting into the unrounded model:

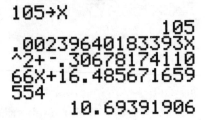

$y \approx 10.69\%$
The percentage of the U.S. population that is foreign born in 2005 is approximately 10.69%.

When $x = 28$, federal funding for education exceeds \$10 billion. An x-value of 28 corresponds to year 1988.

d. The results would be the same. The models are equivalent. The function in part b) is shifted five units left in comparison to the function on part a).

19. a.

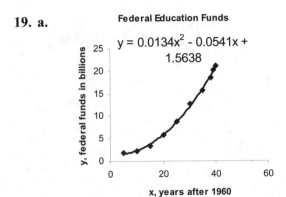

b.

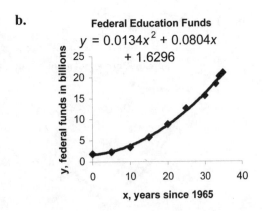

c. Using the TI-83 table feature in conjunction with the unrounded model:

21. a.

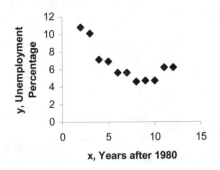

b. Yes. It appears a quadratic model fits the data.

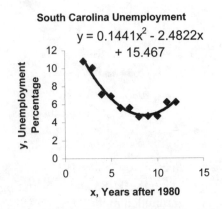

c. Yes. The *y*-intercept, (0, 15.467) represents the unemployment percentage in 1980.

23. a. The second differences are not constant. The situation is not modeled by a quadratic function.

x	*y*	first differences	second differences
1	1/3		
2	2 2/3	2 1/3	
3	9	6 1/3	4
4	21 1/3	12 1/3	6
5	41 2/3	20 1/3	8
6	72	30 1/3	10

b. Modeling with a power function yields:

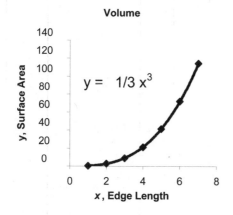

Volume

$$y = \tfrac{1}{3} x^3$$

y, Surface Area vs *x*, Edge Length

25. a.

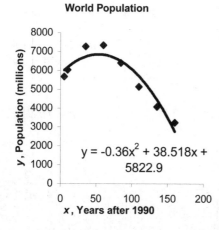

World Population

$$y = -0.36x^2 + 38.518x + 5822.9$$

y, Population (millions) vs *x*, Years after 1990

b. The *x*-intercept is approximately 191. The corresponding year is 2181.

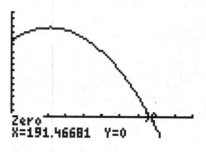

Zero
X=191.46681   Y=0

[0, 250] by [−1500, 8000]

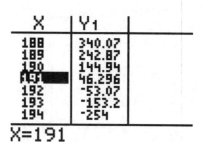

X	Y₁	
188	340.07	
189	242.87	
190	144.94	
191	46.296	
192	-53.07	
193	-153.2	
194	-254	

X=191

c. The model is not valid beyond the year 2181. The model predicts that the population will become negative beginning in 2182!

27. a.

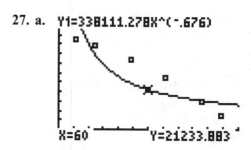

Y1=338111.278X^(-.676)

X=60   Y=21233.883

[0, 120] by [−5000, 60,000]

b.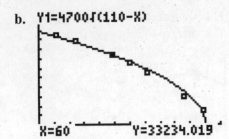

[0, 120] by [–5000, 60,000]

Equation b) fits the data much better.

29. a.

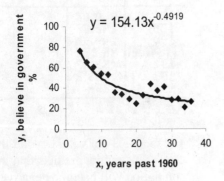

b. In the year 2000,

$x = 2000 - 1960 = 40.$

$y = 154.13x^{-0.4919}$

$y = 154.13(40)^{-0.4919}$

Using the TI-83 and substituting into the unrounded model:

```
40→X
              40
154.13120343008X
^-.49190852626305
      25.10866371
```

$y \approx 25.1\%$

In 2000 approximately 25.1% of people said they trust the government always or most of the time.

31. a.

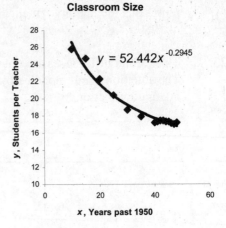

b. The year 2000 corresponds to an x-value of 50. Therefore,

$y = 52.442x^{-0.2945}$

$y = 52.442(50)^{-0.2945}$

Using the TI-83 and substituting into the unrounded model

```
50→X
              50
52.44243249087X^
-.29446156335305
      16.57301274
```

$y \approx 16.6$

The model predicts that in the year 2000, there were 16.6 students per teacher.

c. The function is decreasing.

d. Based on the model, the number of students per teacher will approach zero but will never reach zero.

33. a.

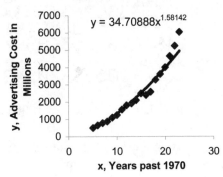

b. The year 2000 corresponds to $x = 30$.

$y = 34.70888x^{1.58142}$

$y = 34.70888(30)^{1.58142}$

Using the TI-83 and substituting into the unrounded model

```
30→X
                    30
34.708877276433X
^1.581419660237I
              7522.967032
```

$y \approx 7523$

Cigarette advertising in 2000 is predicted to be $7523 million.

c.

X	Y1
18	3353.9
19	3653.3
20	3962
21	4279.8
22	4606.5
23	4942
24	5286.1

X=20

When $x = 20$, the year is 1990.

d. Cigarette advertising expenditures are modeled by an increasing function.

35. a.

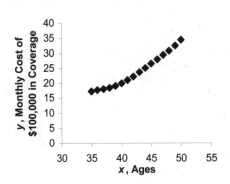

b. Yes. It appears that a quadratic function will fit the data.

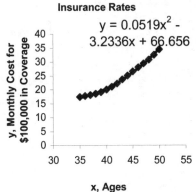

c.

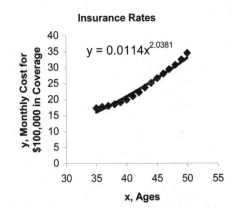

d. It appears that the quadratic model, based on the scatter plots, is the best fit.

37. a. Quadratic model

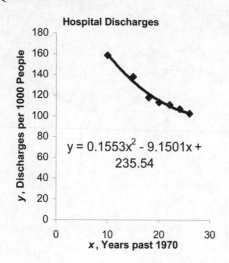

Power model

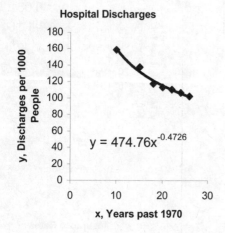

Based on the graphs, both models fit the data reasonably well. The power model might be slightly better.

b. Since 1988 the discharge rate is decreasing by approximately 2 per thousand people each year. Therefore, 84.3 per thousand people is a reasonable guess for the discharge rate in 2005. The quick estimate for the discharge rate in 2005 is 18 per thousand people less than the discharge rate in 1996.

c. In 2005, $x = 2005 - 1970 = 35$.
Using the TI-83 and substituting into the unrounded quadratic model

```
35→X
                    35
.15533309635333X
^2+-9.1500859401
148X+235.5369897
9592
         105.5670249
```

$y \approx 105.6$ discharges per 1000 people

Using the TI-83 and substituting into the unrounded power model

```
35→X
                    35
474.75818816637X
^-.4726347539819
         88.44886076
```

$y \approx 88.4$ discharges per 1000 people

Based on the solution to part *b*, the power function is a better fit to the data. The scatter plots in part *a* also support the conclusion that the power function is a better fit.

Section 2.6 Skills Check

1. **a.** $(f+g)(x)$
$= f(x)+g(x)$
$= (3x-5)+(4-x)$
$= 2x-1$

   b. $(f-g)(x)$
$= f(x)-g(x)$
$= (3x-5)-(4-x)$
$= 3x-5-4+x$
$= 4x-9$

   c. $(f \cdot g)(x)$
$= f(x) \cdot g(x)$
$= (3x-5)(4-x)$
$= -3x^2 + 17x - 20$

   d. $\left(\dfrac{f}{g}\right)(x)$
$= \dfrac{f(x)}{g(x)}$
$= \dfrac{3x-5}{4-x}$

   e. $g(x) \neq 0$
$4-x = 0$
$x = 4$
Domain: $(-\infty,4)\cup(4,\infty)$

3. **a.** $(f+g)(x)$
$= f(x)+g(x)$
$= (x^2-2x)+(1+x)$
$= x^2 - x + 1$

   b. $(f-g)(x)$
$= f(x)-g(x)$
$= (x^2-2x)-(1+x)$
$= x^2 - 2x - 1 - x$
$= x^2 - 3x - 1$

   c. $(f \cdot g)(x)$
$= f(x) \cdot g(x)$
$= (x^2-2x)(1+x)$
$= x^3 - x^2 - 2x$

   d. $\left(\dfrac{f}{g}\right)(x)$
$= \dfrac{f(x)}{g(x)}$
$= \dfrac{x^2-x}{1+x}$

   e. $g(x) \neq 0$
$1+x = 0$
$x = -1$
Domain: $(-\infty,-1)\cup(-1,\infty)$

5. **a.** $(f+g)(x)$
$= f(x)+g(x)$
$= \left(\dfrac{1}{x}\right)+\left(\dfrac{x+1}{5}\right)$
$LCM : 5x$
$= \dfrac{5}{5}\left(\dfrac{1}{x}\right)+\dfrac{x}{x}\left(\dfrac{x+1}{5}\right)$
$= \left(\dfrac{5}{5x}\right)+\left(\dfrac{x^2+x}{5x}\right)$
$= \dfrac{x^2+x+5}{5x}$

b. $(f-g)(x)$

$= f(x) - g(x)$

$= \left(\dfrac{1}{x}\right) - \left(\dfrac{x+1}{5}\right)$

$LCM : 5x$

$= \dfrac{5}{5}\left(\dfrac{1}{x}\right) - \dfrac{x}{x}\left(\dfrac{x+1}{5}\right)$

$= \left(\dfrac{5}{5x}\right) - \left(\dfrac{x^2 + x}{5x}\right)$

$= \dfrac{-x^2 - x + 5}{5x}$

c. $(f \cdot g)(x)$

$= f(x) \cdot g(x)$

$= \left(\dfrac{1}{x}\right)\left(\dfrac{x+1}{5}\right)$

$= \dfrac{x+1}{5x}$

d. $\left(\dfrac{f}{g}\right)(x)$

$= \dfrac{f(x)}{g(x)}$

$= \dfrac{\dfrac{1}{x}}{\dfrac{x+1}{5}}$

$= \dfrac{1}{x} \cdot \dfrac{5}{x+1}$

$= \dfrac{5}{x(x+1)}$

e. $g(x) \neq 0$

$x(x+1) = 0$

$x = 0, -1$

Domain: $(-\infty, -1) \cup (-1, 0) \cup (0, \infty)$

7. a. $(f+g)(x)$

$= f(x) + g(x)$

$= \left(\sqrt{x}\right) + \left(1 - x^2\right)$

$= \sqrt{x} + 1 - x^2$

b. $(f-g)(x)$

$= f(x) - g(x)$

$= \left(\sqrt{x}\right) - \left(1 - x^2\right)$

$= \sqrt{x} - 1 + x^2$

c. $(f \cdot g)(x)$

$= f(x) \cdot g(x)$

$= \left(\sqrt{x}\right)\left(1 - x^2\right)$

d. $\left(\dfrac{f}{g}\right)(x)$

$= \dfrac{f(x)}{g(x)}$

$= \dfrac{\sqrt{x}}{1 - x^2}$

e. $g(x) \neq 0$

$1 - x^2 = 0$

$x = -1, 1$

Domain: $(-\infty, -1) \cup (-1, 1) \cup (1, \infty)$

9. a. $(f+g)(2)$

$= f(2) + g(2)$

$= \left(2^2 - 5(2)\right) + \left(6 - (2)^3\right)$

$= -6 - 2$

$= -8$

b. $(g-f)(-1)$

$\qquad = g(-1) - f(-1)$

$\qquad = \left(6 - (-1)^3\right) - \left((-1)^2 - 5(-1)\right)$

$\qquad = 7 - 6$

$\qquad = 1$

c. $(f \cdot g)(-2)$

$\qquad = f(-2) \cdot g(-2)$

$\qquad = \left((-2)^2 - 5(-2)\right) \cdot \left(6 - (-2)^3\right)$

$\qquad = (14)(14)$

$\qquad = 196$

d. $\left(\dfrac{g}{f}\right)(3)$

$\qquad = \dfrac{g(3)}{f(3)}$

$\qquad = \dfrac{\left(6 - (3)^3\right)}{\left(3^2 - 5(3)\right)}$

$\qquad = \dfrac{-21}{-6}$

$\qquad = 3.5$

11. a. $(f \circ g)(x)$

$\qquad = f(g(x))$

$\qquad = 2(3x - 1) - 6$

$\qquad = 6x - 8$

b. $(g \circ f)(x)$

$\qquad = g(f(x))$

$\qquad = 3(2x - 6) - 1$

$\qquad = 6x - 19$

13. a. $(f \circ g)(x)$

$\qquad = f(g(x))$

$\qquad = \left(\dfrac{1}{x}\right)^2$

$\qquad = \dfrac{1}{x^2}$

b. $(g \circ f)(x)$

$\qquad = g(f(x))$

$\qquad = \dfrac{1}{x^2}$

15. a. $(f \circ g)(x)$

$\qquad = f(g(x))$

$\qquad = \sqrt{(2x - 7) - 1}$

$\qquad = \sqrt{2x - 8}$

b. $(g \circ f)(x)$

$\qquad = g(f(x))$

$\qquad = 2\left(\sqrt{x - 1}\right) - 7$

$\qquad = 2\sqrt{x - 1} - 7$

17. a. $(f \circ g)(x)$

$\qquad = f(g(x))$

$\qquad = |(4x) - 3|$

$\qquad = |4x - 3|$

b. $(g \circ f)(x)$

$\qquad = g(f(x))$

$\qquad = 4(|x - 3|)$

$\qquad = 4|x - 3|$

19. a. $(f \circ g)(x)$

$= f\big(g(x)\big)$

$= \dfrac{3\left(\dfrac{2x-1}{3}\right)+1}{2}$

$= \dfrac{2x-1+1}{2}$

$= \dfrac{2x}{2}$

$= x$

b. $(g \circ f)(x)$

$= g\big(f(x)\big)$

$= \dfrac{2\left(\dfrac{3x+1}{2}\right)-1}{3}$

$= \dfrac{3x+1-1}{3}$

$= \dfrac{3x}{3}$

$= x$

21. a. $f\big(g(2)\big) = 2\left(\dfrac{2-5}{3}\right)^2 = 2(-1)^2 = 2$

b. $g\big(f(-2)\big) = \dfrac{\left[2(-2)^2\right]-5}{3}$

$= \dfrac{8-5}{3}$

$= \dfrac{3}{3}$

$= 1$

23. a. $(f+g)(2) = f(2)+g(2) = 1+(-3) = -2$

b. $(fg)(-1) = f(-1) \cdot g(-1) = -2 \cdot 0 = 0$

c. $\left(\dfrac{f}{g}\right)(4) = \dfrac{f(4)}{g(4)} = \dfrac{3}{-1} = -3$

d. $(f \circ g)(1) = f\big(g(1)\big) = f(-2) = -3$

e. $(g \circ f)(-2) = g\big(f(-2)\big) = g(-3) = 2$

25. a. $P(x) = R(x) - C(x)$

$= (89x) - (23x + 3420)$

$= 89x - 23x - 3420$

$= 66x - 3420$

b. $P(150) = 66(150) - 3420 = 6480$

The profit on the production and sale of 150 bicycles is $6480.

27. a. The revenue function is linear, while the cost function is quadratic. Note that $C(x)$ fits the form $f(x) = ax^2 + bx + c, a \neq 0$.

b. $P(x) = R(x) - C(x)$

$P(x) = 1050x - \left(10{,}000 + 30x + x^2\right)$

$P(x) = 1050x - 10{,}000 - 30x - x^2$

$P(x) = -x^2 + 1020x - 10{,}000$

c. Quadratic. Note that $P(x)$ fits the form $f(x) = ax^2 + bx + c, a \neq 0$.

29. a. $P(x) = R(x) - C(x)$

$P(x) = 550x - \left(10{,}000 + 30x + x^2\right)$

$P(x) = 550x - 10{,}000 - 30x - x^2$

$P(x) = -x^2 + 520x - 10{,}000$

b. Note that the maximum profit occurs at the vertex of the quadratic function, since the function is concave down.

$h = \dfrac{-b}{2a} = \dfrac{-520}{2(-1)} = \dfrac{-520}{-2} = 260$

c. $k = P(h)$

$= P(260)$

$= -(260)^2 + 520(260) - 10,000$

$= -67,600 - 135,200 - 10,000$

$= 57,600$

Producing 260 units yields a maximum profit of $57,600.

31. a. $\overline{C}(x)$ fits the form $\left(\dfrac{f}{g}\right)(x)$ where

$f(x) = C(x)$ and $g(x) = x$. Note that

$\overline{C}(x) = \dfrac{C(x)}{x}$.

b. Let $x = 3000$ and calculate $\overline{C}(x)$.

$\overline{C}(3000) = \dfrac{50,000 + 105(3000)}{3000}$

$= \dfrac{365,000}{3000}$

$= 121.\overline{6} \approx \$121.67$ per unit

33. a. $\overline{C}(x) = \dfrac{C(x)}{x} = \dfrac{3000 + 72x}{x}$

b. $\overline{C}(100) = \dfrac{C(100)}{100}$

$= \dfrac{3000 + 72(100)}{100}$

$= \dfrac{3000 + 7200}{100}$

$= \dfrac{10,200}{100}$

$= 102$ or $102 per printer

35. a. Let $T(p)$ represent the total number of tickets for a home football game.

$T(p)$

$= S(p) + N(p)$

$= (62p + 8500) + (0.5p^2 + 16p + 4400)$

$= 0.5p^2 + 78p + 12,900$

b. Since p represents the winning percentage for the football team, $0 \le p \le 100$. Therefore the domain of the function is $[0,100]$.

c. $T(90) = 0.5(90)^2 + 78(90) + 12,900$

$= 4050 + 7020 + 12,900$

$= 23,970$

The stadium holds 23,970 people.

37. a. $B(8) = 6(8+1)^{\frac{3}{2}} = 6(27) = 162$

On May 8th the number of bushels of tomatoes harvested is 162.

b. $P(8) = 8.5 - 0.12(8) = 7.54$

On May 8th the price per bushel of tomatoes is $7.54.

c. $(B \cdot P)(x)$ represents the worth of the tomatoes on the xth day of May.

$(B \cdot P)(8) = B(8) \cdot P(8)$

$= 162 \cdot 7.54$

$= 1221.48$

On May 8th the worth is $1221.48.

d. $W(x) = (B \cdot P)(x)$

$= B(x) \cdot P(x)$

$= \left[6(x+1)^{\frac{3}{2}}\right] \cdot (8.5 - 0.12x)$

$= 6(x+1)^{\frac{3}{2}}(8.5 - 0.12x)$

39. a.
$$P(x) = R(x) - C(x)$$
$$= (592x) - (32,000 + 432x)$$
$$= 592x - 32,000 - 432x$$
$$= 160x - 32,000$$

b. $P(600) = 160(600) - 32,000 = 64,000$

The profit for producing and selling 600 satellite systems is $64,000.

c. Since the function is linear, the rate of change is constant. For every one unit increase in production, the profit increases by $160.

41. a. Let $P(t)$ represent the total U.S. population under age 5 in millions. Then, $P(t) = B(t) + G(t)$

$$P(t) = (0.0076t^2 - 0.1752t + 10.705) +$$
$$(0.0064t^2 - 0.1448t + 10.12)$$
$$P(t) = 0.014t^2 - 0.32t + 20.825$$

b. $P(2003 - 1990)$
$$= P(13)$$
$$= 0.014(13)^2 - 0.32(13) + 20.825$$
$$= 2.366 - 4.16 + 20.825$$
$$= 19.031$$

The model predicts that in 2003 there are 19.031 million children under age 5.

43. For a)–d), consider the output from the function.

a. Meat in a Styrofoam container

b. Ground meat.

c. Meat ground and then ground again.

d. Ground meat placed in a Styrofoam container.

e. Meat placed in a Styrofoam container and then ground.

f. Only part d), unless the reader enjoys ground Styrofoam!

45. Let $B(x)$ convert a Japanese shoe size, x, into a British shoe size, $B(x)$.

Japanese $\rightarrow$ U.S. $\rightarrow$ British
$$B(x) = (p \circ s)(x)$$
$$= p(s(x))$$
$$= (x - 17) - 1.5$$
$$= x - 18.5$$

47. Chilean pesos $\rightarrow$
Austrian schillings $\rightarrow$ Russian rubles

Let $R(x) = 1.987376x$, where x is schillings and $R(x)$ is rubles.
Let $S(x) = 0.025202x$, where x is pesos and $S(x)$ is schillings.
Let $V(x) = (R \circ S)(x)$, where x is pesos and $V(x)$ is rubles.

$$V(x) = R(S(x))$$
$$= 1.987376(0.025202x)$$
$$= 0.05008585x$$
$$V(1000) = 0.05008585(1000)$$
$$= 50.08585 \approx 50.09$$

49. The function is $100 \cdot \left(\dfrac{f}{g}\right)(x)$. Note that

$$\left(\frac{f}{g}\right)(x) = \frac{f(x)}{g(x)} = \text{the ratio of AOL}$$

customers to total Internet users.
Multiplying by 100 creates a percentage.

51. The function is $(f+g)(x) = f(x) + g(x)$.

53. The normal price is $0.50x$ where x represents retail price. Since the sale price is 20% off the normal price, the sale price is
$0.50x - (0.20)(0.50x) = 0.50x - 0.10x = 0.40x.$
Therefore the books are on sale for 40% of retail price.

Section 2.7 Skills Check

1. Yes, the function is one-to-one and has an inverse.

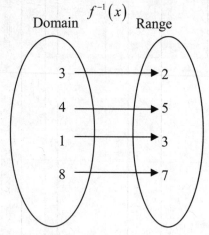

$f^{-1}(x)$

Domain   Range

3. No, the function is not on-to-one. Inputs of 5 and 6 correspond with an output of 2.

5. a. $f\big(g(x)\big)=(f\circ g)(x)=3\left(\dfrac{x}{3}\right)=x$

    $g\big(f(x)\big)=(g\circ f)(x)=\dfrac{(3x)}{3}=x$

    b.   Yes, since $(f\circ g)(x)=(g\circ f)(x)=x$.

7. Is $(f\circ g)(x)=(g\circ f)(x)=x$?

    $(f\circ g)(x)=f\big(g(x)\big)$

    $\qquad=\left(\sqrt[3]{x-1}\right)^{3}+1$

    $\qquad=x-1+1$

    $\qquad=x$

    $(g\circ f)(x)=g\big(f(x)\big)$

    $\qquad=\sqrt[3]{x^{3}+1-1}$

    $\qquad=\sqrt[3]{x^{3}}$

    $\qquad=x$

    Yes, f and g are inverse functions.

9. See the completed table below.

x	$f(x)$	x	$f^{-1}(x)$
-1	-7	-7	-1
0	-4	-4	0
1	-1	-1	1
2	2	2	2
3	5	5	3

Note that values for x in the second table are the values for $f(x)$ in the first table.

11. a. $f(x)=3x-4$

    $\qquad y=3x-4$

    $\qquad x=3y-4$

    $\qquad 3y=x+4$

    $\qquad y=\dfrac{x+4}{3}$

    $f^{-1}(x)=\dfrac{x+4}{3}$

    b. Yes. Substituting the x-values from the table into $f^{-1}(x)$ generates the $f^{-1}(x)$ outputs found in the table.

13. Since x and y are interchanged to create the inverse function, if $(a,\ b)$ is on point on $f(x)$ then $(b,\ a)$ is a point on $f^{-1}(x)$.

15. $f(x) = \dfrac{1}{x}$

$y = \dfrac{1}{x}$

$x = \dfrac{1}{y}$

$xy = 1$

$y = \dfrac{1}{x}$

$f^{-1}(x) = \dfrac{1}{x}$

Note that in this example the function and its inverse are the same function!

17. $f(x) = 4x^2$

$y = 4x^2$

$x = 4y^2$

$y^2 = \dfrac{x}{4}$

$y = \pm\sqrt{\dfrac{x}{4}}$

Since $x \geq 0$, the positive solution represents the inverse function.

$f^{-1}(x) = \sqrt{\dfrac{x}{4}} = \dfrac{\sqrt{x}}{2}$

19.

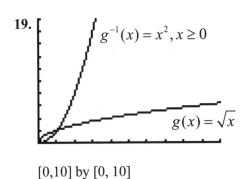

$g^{-1}(x) = x^2, x \geq 0$

$g(x) = \sqrt{x}$

[0,10] by [0, 10]

21.

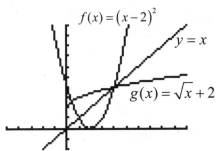

$f(x) = (x-2)^2$

$y = x$

$g(x) = \sqrt{x} + 2$

[–5, 10] by [–2, 10]

Restricting the domain of f to $[2,\infty)$ and the domain of g to $[0,\infty)$ forces f and g to become inverse functions.

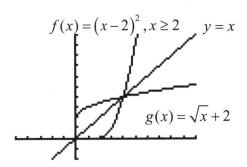

$f(x) = (x-2)^2, x \geq 2$ $y = x$

$g(x) = \sqrt{x} + 2$

[–5, 10] by [–2, 10]

23. Y1=2X^3+1

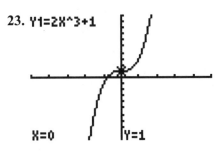

X=0   Y=1

[–5, 5] by [–10, 10]

The function passes the horizontal line test, is one-to-one, and has an inverse function.

25. Yes. Every x matches with exactly one y, and every y matches with exactly one x.

27. The graph fails the horizontal line test. It is not a one-to-one function.

29.

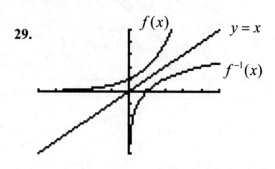

[–5, 5] by [–5, 5]

Section 2.7 Exercises

31. a.
$$d(x) = x + 0.5$$
$$y = x + 0.5$$
$$x = y + 0.5$$
$$y = x - 0.5$$
$$d^{-1}(x) = x - 0.5$$

b. $d^{-1}(8.5) = 8.5 - 0.5 = 8$
The British shoe size is 8.

33.
$$f(t) = 75.451 - 0.707t$$
$$y = 75.451 - 0.707t$$
$$t = 75.451 - 0.707y$$
$$t - 75.451 = -0.707y$$
$$y = \frac{t - 75.451}{-0.707}$$
$$y = \frac{75.451 - t}{0.707}$$
$$f^{-1}(t) = \frac{75.451 - t}{0.707}$$
$$f^{-1}(65) = \frac{75.451 - 65}{0.707}$$
$$= \frac{10.451}{0.707}$$
$$= 14.78218 \approx 15$$

The percentage dropped below 65% in 1990.

35. a.
$$f(x) = 225.304x + 493.432$$
$$y = 225.304x + 493.432$$
$$x = 225.304y + 493.432$$
$$x - 493.432 = 225.304y$$
$$y = \frac{x - 493.432}{225.304}$$
$$f^{-1}(x) = \frac{x - 493.432}{225.304}$$

The inverse function will calculate the number of years beyond 1990 in which

Ritalin consumption in grams per 100,000 people reaches a given level.

b. $f^{-1}(1170) = \dfrac{1170 - 493.432}{225.304} = 3.00$

Ritalin consumption equals 1170 grams in 1993.

37. a. $f(x) = 4\sqrt{4x+1}$

To determine the domain, solve $4x+1 \geq 0$.

$4x \geq -1$

$x \geq -\dfrac{1}{4}$

Domain: $\left[-\dfrac{1}{4}, \infty\right)$

Consequently, the range is $[0, \infty)$.

b. $f(x) = 4\sqrt{4x+1}$

$y = 4\sqrt{4x+1}$

$x = 4\sqrt{4y+1}$

$\dfrac{x}{4} = \sqrt{4y+1}$

$\left(\dfrac{x}{4}\right)^2 = \left(\sqrt{4y+1}\right)^2$

$\dfrac{x^2}{16} = 4y+1$

$4y = \dfrac{x^2}{16} - 1$

$y = \dfrac{\dfrac{x^2 - 16}{16}}{4}$

$y = \dfrac{x^2 - 16}{64}$

$f^{-1}(x) = \dfrac{x^2 - 16}{64}$

c. The domain of the inverse function is the range of the original function, and the range of the inverse function is the

domain of the original function. Therefore,

Domain: $[0, \infty)$

Range: $\left[-\dfrac{1}{4}, \infty\right)$

d. Both the domain and range would be $[0, \infty)$.

39. $C(x) = x+3$

$y = x+3$

$x = y+3$

$x-3 = y$

$y = x-3$

$C^{-1}(x) = x-3$

The decoded numerical sequence is {20 8 5 27 18 5 1 12 27 20 8 9 14 7}, which translates into "The_real_thing."

41. Yes. Each person has a unique social security number. Since no two people have the same number, the function is one-to-one.

43. a. Yes. The function is one-to-one. It passes the horizontal line test.

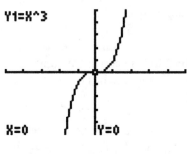

$[-5, 5]$ by $[-5, 5]$

b. $f(x) = x^3$

$$y = x^3$$
$$x = y^3$$
$$\sqrt[3]{x} = \sqrt[3]{y^3}$$
$$y = \sqrt[3]{x}$$
$$f^{-1}(x) = \sqrt[3]{x}$$

c. Both domain and the range are $[0, \infty)$ based on the physical context of the question.

d. The inverse function is used to convert from the volume of the cube into the side length of the cube.

45. a. $f(x) = 0.66832x$

$$y = 0.66832x$$
$$x = 0.66832y$$
$$y = \frac{x}{0.66832}$$
$$f^{-1}(x) = \frac{x}{0.66832}$$

The inverse function converts U.S. dollars into Canadian dollars.

b. If you convert $500 from U.S. to Canadian dollars and then convert the money back to U.S. dollars, you will still have $500 U.S currency. (Note: This assumes there are no transaction fees for the conversion and that the exchange rate has not changed.)

47. a. No. The function is not one-to-one. Note that $I(-1) = I(1) = 300,000$.

b. In the given physical context since x represents distance, the domain is $(0, \infty)$.

c. Yes. Based on the restricted domain $(0, \infty)$, the function is one-to-one.

d. $I(x) = \dfrac{300,000}{x^2}, x > 0$

$$y = \frac{300,000}{x^2}$$
$$x = \frac{300,000}{y^2}$$
$$xy^2 = 300,000$$
$$y^2 = \frac{300,000}{x}$$
$$y = \pm\sqrt{\frac{300,000}{x}}$$

Based on the physical context,

$$I^{-1}(x) = \sqrt{\frac{300,000}{x}}$$
$$I^{-1}(75,000) = \sqrt{\frac{300,000}{75,000}}$$
$$= \sqrt{4}$$
$$= 2$$

When the distance is 2 feet, the intensity of light is 75,000 candlepower.

49. a. $f(x) = 1.7655x$

$$y = 1.7655x$$
$$x = 1.7655y$$
$$y = \frac{x}{1.7655}$$
$$f^{-1}(x) = \frac{x}{1.7655}$$

The inverse function converts U.S. dollars into British pounds.

b. If you convert $1000 from U.S. to British currency and then convert the money back to U.S. dollars, you will still have $1000 U.S currency. (Note: This assumes there are no transaction fees for the conversion and that the exchange rate has not changed.)

51. a. The ball reaches the ground when its height equals zero. Solve $f(x) = 0$.

$$256 + 96x - 16x^2 = 0$$

$$-16(x^2 - 6x - 16) = 0$$

$$-16(x - 8)(x + 2) = 0$$

$$x - 8 = 0, x + 2 = 0$$

$$x = 8, x = -2$$

Since x represents time, $x \geq 0$.

$$x = 8$$

The ball remains in the air for 8 seconds.

b. 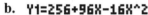 Y1=256+96X-16X^2

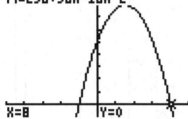

X=8   Y=0

[–10, 10] by [–50, 450]

The function is not one-to-one on the interval $[0,8]$. The graph does not pass the horizontal line test.

c. The function is one-to-one on the interval $[3,\infty)$.

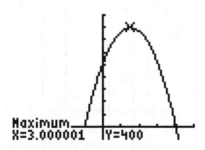

Maximum
X=3.000001   Y=400

[–10, 10] by [–50, 450]

d. $f(x) = 256 + 96x - 16x^2$

$$y = -16(x^2 - 6x - 16)$$

$$x = -16(y^2 - 6y - 16)$$

Completing the square yields,

$$x = -16\left[\left(y^2 - 6y + 9\right) + \left(-9 - 16\right)\right]$$

$$x = -16\left[\left(y - 3\right)^2 - 25\right]$$

$$x = -16(y - 3)^2 + 400$$

$$x - 400 = -16(y - 3)^2$$

$$(y - 3)^2 = \frac{x - 400}{-16}$$

$$\sqrt{(y-3)^2} = \pm\sqrt{\frac{x - 400}{-16}}$$

$$y - 3 = \pm\sqrt{\frac{x - 400}{-16}}$$

$$y = 3 \pm \sqrt{\frac{x - 400}{-16}}$$

For the interval $[0,3]$,

$$f^{-1}(x) = 3 - \sqrt{\frac{400 - x}{16}}$$

The inverse function calculates the time the ball is in air between 0 and 3 seconds, given the height of the ball.

Section 2.8 Skills Check

1. $\sqrt{2x^2 - 1} - x = 0$

$\sqrt{2x^2 - 1} = x$

$\left(\sqrt{2x^2 - 1}\right)^2 = (x)^2$

$2x^2 - 1 = x^2$

$x^2 - 1 = 0$

$(x+1)(x-1) = 0$

$x = 1, x = -1$

-1 does not check

$\sqrt{2(-1)^2 - 1} - (-1) =$

$\sqrt{2(1) - 1} + 1 = \sqrt{1} + 1 = 2 \neq 0$

Applying the intersection of graphs method to check graphically:

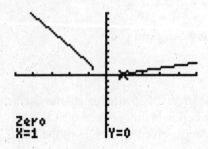

$[-5, 5]$ by $[-10, 10]$

 The only solution that checks is $x = 1$.

3. $\sqrt[3]{x - 1} = -2$

$\left(\sqrt[3]{x - 1}\right)^3 = (-2)^3$

$x - 1 = -8$

$x = -7$

Applying the intersection of graphs method to check graphically:

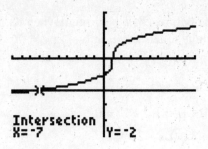

$[-10, 10]$ by $[-5, 3]$

 The solution is $x = -7$.

5. $\sqrt{3x - 2} + 2 = x$

$\sqrt{3x - 2} = x - 2$

$\left(\sqrt{3x - 2}\right)^2 = (x - 2)^2$

$3x - 2 = x^2 - 4x + 4$

$x^2 - 7x + 6 = 0$

$(x - 6)(x - 1) = 0$

$x = 6, x = 1$

Applying the intersection of graphs method to check graphically:

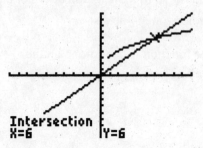

$[-10, 10]$ by $[-10, 10]$

There is only one solution. The x-value of 1 does not check. The solution is $x = 6$.

7. $\sqrt[3]{4x+5} = \sqrt[3]{x^2-7}$

$\left(\sqrt[3]{4x+5}\right)^3 = \left(\sqrt[3]{x^2-7}\right)^3$

$4x+5 = x^2-7$

$x^2-4x-12 = 0$

$(x-6)(x+2) = 0$

$x = 6, x = -2$

Applying the intersection of graphs method to check graphically:

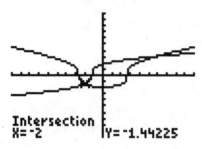

Intersection
X=-2   Y=-1.44225

$[-10, 10]$ by $[-10, 10]$

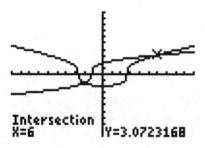

Intersection
X=6   Y=3.0723168

$[-10, 10]$ by $[-10, 10]$

Both solutions check.

9. $\sqrt{x}-1 = \sqrt{x-5}$

$\left(\sqrt{x}-1\right)^2 = \left(\sqrt{x-5}\right)^2$

$x - 2\sqrt{x} + 1 = x - 5$

$-2\sqrt{x} = -6$

$\sqrt{x} = 3$

$\left(\sqrt{x}\right)^2 = (3)^2$

$x = 9$

Checking by substitution

$x = 9$

$\sqrt{x}-1 \overset{?}{=} \sqrt{x-5}$

$\sqrt{9}-1 \overset{?}{=} \sqrt{9-5}$

$2 = 2$

The solution is $x = 9$.

11. $(x+4)^{\frac{2}{3}} = 9$

$\sqrt[3]{(x+4)^2} = 9$

$\left[\sqrt[3]{(x+4)^2}\right]^3 = [9]^3$

$(x+4)^2 = 729$

$\sqrt{(x+4)^2} = \pm\sqrt{729}$

$x+4 = \pm 27$

$x = -4 \pm 27$

$x = 23, x = -31$

Checking by substitution

$x = 23$

$(23+4)^{\frac{2}{3}} = 9$

$(27)^{\frac{2}{3}} = 9$

$9 = 9$

$x = -31$

$(-31+4)^{\frac{2}{3}} = 9$

$(-27)^{\frac{2}{3}} = 9$

$(-3)^2 = 9$

$9 = 9$

The solutions are $x = 23, x = -31$.

13. $|2x-5| = 3$

$2x-5 = 3 \quad | \quad 2x-5 = -3$

$2x = 8 \quad | \quad 2x = 2$

$x = 4 \quad | \quad x = 1$

Using the intersections of graphs method to check the solutions graphically yields,

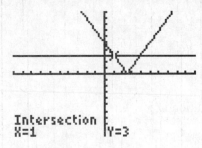

Intersection
X=1   Y=3

[–10, 10] by [–10, 10]

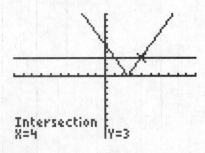

Intersection
X=4   Y=3

[–10, 10] by [–10, 10]

15. $|x| = x^2 + 4x$

$$x = x^2 + 4x \quad \Bigg| \quad x = -\left(x^2 + 4x\right)$$

$$x^2 + 3x = 0 \quad \Bigg| \quad x = -x^2 - 4x$$

$$x(x+3) = 0 \quad \Bigg| \quad x^2 + 5x = 0$$

$$\qquad\qquad\qquad \Bigg| \quad x(x+5) = 0$$

$$x = 0, x = -3 \quad \Bigg| \quad x = 0, x = -5$$

Note that -3 does not check. The solutions are $x = 0, x = -5$.

Using the intersections of graphs method to check the solutions graphically yields,

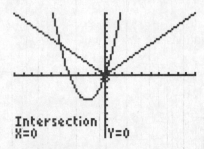

Intersection
X=0   Y=0

[–10, 10] by [–10, 10]

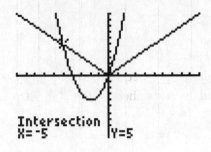

Intersection
X=-5   Y=5

[–10, 10] by [–10, 10]

17. $|3x - 1| = 4x$

$$3x - 1 = 4x \quad \Bigg| \quad 3x - 1 = -4x$$

$$-x = 1 \quad \Bigg| \quad 7x = 1$$

$$x = -1 \quad \Bigg| \quad x = \frac{1}{7}$$

Note that only –1 does not check.
$x = -1$

$$\left|3(-1) - 1\right| = 4(-1)$$

$$\left|-3 - 1\right| = -4$$

$$\left|-4\right| = -4$$

$$4 \neq -4$$

The only solution is $x = \dfrac{1}{7}$.

Using the intersections of graphs method to check the solution graphically yields

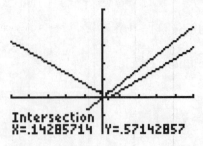

Intersection
X=.14285714   Y=.57142857

[–5, 5] by [–10, 25]

19. $x^2 + 4x < 0$

$x(x+4) < 0$

$x(x+4) = 0$

$x = 0, x = -4$

sign of x $\leftarrow ---_{-4}---_0 +++\rightarrow$

sign of $(x+4)$ $\leftarrow ---_{-4}+++_0 +++\rightarrow$

sign of $x(x+4)$ $\leftarrow +++_{-4}---_0 +++\rightarrow$

Considering the inequality symbol in the original question, the solution is $(-4, 0)$.

21. $9 - x^2 \geq 0$

$-1(x^2 - 9) \geq 0$

$\dfrac{-1(x^2 - 9)}{-1} \leq \dfrac{0}{-1}$

$\boxed{x^2 - 9 \leq 0}$ $\leftarrow$ Solve the simplified question

$(x+3)(x-3) \leq 0$

$(x+3)(x-3) = 0$

$x = -3, x = 3$

sign of $(x+3)$ $\leftarrow ---_{-3}+++_3 +++\rightarrow$

sign of $(x-3)$ $\leftarrow ---_{-3}---_3 +++\rightarrow$

sign of $(x+3)(x-3)$ $\leftarrow +++_{-3}---_3 +++\rightarrow$

Considering the inequality symbol in the simplified question, the solution is $[-3, 3]$.

23. $-x^2 + 9x - 20 > 0$

$-1(x^2 - 9x + 20) > 0$

$\dfrac{-1(x^2 - 9x + 20)}{-1} < \dfrac{0}{-1}$

$\boxed{x^2 - 9x + 20 < 0}$ $\leftarrow$ Solve the simplified question

$(x-5)(x-4) < 0$

$(x-5)(x-4) = 0$

$x = 5, x = 4$

sign of $(x-5)$ $\leftarrow ---_4---_5 +++\rightarrow$

sign of $(x-4)$ $\leftarrow ---_4+++_5 +++\rightarrow$

sign of $(x-5)(x-4)$ $\leftarrow +++_4---_5 +++\rightarrow$

Considering the inequality symbol in the original question, the solution is $(4, 5)$.

25. $2x^2 - 8x \geq 24$

$2x^2 - 8x - 24 \geq 0$

$2(x^2 - 4x - 12) \geq 0$

$2(x-6)(x+2) \geq 0$

$2(x-6)(x+2) = 0$

$2 \neq 0, x = 6, x = -2$

sign of $(x-6)$ $\leftarrow ---_{-2}---_6 +++\rightarrow$

sign of $(x+2)$ $\leftarrow ---_{-2}+++_6 +++\rightarrow$

sign of $(x-6)(x+2)$ $\leftarrow +++_{-2}---_6 +++\rightarrow$

Considering the inequality symbol in the original question, the solution is $(-\infty, -2] \cup [6, \infty)$.

27. $x^2 - 6x < 7$

$\boxed{x^2 - 6x - 7 < 0}$ $\leftarrow$ Solve the simplified problem

$(x-7)(x+1) < 0$

$(x-7)(x+1) = 0$

$t = 7, t = -1$

sign of $(x-7)$ $\leftarrow ---_{-1}---_7 +++\rightarrow$

sign of $(x+1)$ $\leftarrow ---_{-1}+++_7 +++\rightarrow$

sign of $(x-7)(x+1)$ $\leftarrow +++_{-1}---_7 +++\rightarrow$

Considering the inequality symbol in the original question, the solution is $(-1, 7)$.

29. $2x^2 - 7x + 2 \geq 0$

$2x^2 - 7x + 2 = 0$

$x = \dfrac{-b \pm \sqrt{b^2 - 4ac}}{2a}$

$x = \dfrac{-(-7) \pm \sqrt{(-7)^2 - 4(2)(2)}}{2(2)}$

$x = \dfrac{7 \pm \sqrt{49 - 16}}{4}$

$x = \dfrac{7 \pm \sqrt{33}}{4}$

Applying the x-intercept method

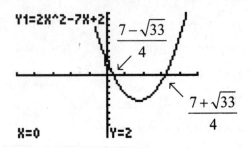

[–5, 5] by [–10, 10]
Considering the graph, the equation is greater than or equal to zero over the interval $\left(-\infty, \dfrac{7-\sqrt{33}}{4}\right] \cup \left[\dfrac{7+\sqrt{33}}{4}, \infty\right)$.

31. $5x^2 \geq 2x + 6$

$\boxed{5x^2 - 2x - 6 \geq 0}$ ← Solve the simplified problem

$5x^2 - 2x - 6 = 0$

$x = \dfrac{-b \pm \sqrt{b^2 - 4ac}}{2a}$

$x = \dfrac{-(-2) \pm \sqrt{(-2)^2 - 4(5)(-6)}}{2(5)}$

$x = \dfrac{2 \pm \sqrt{4 + 120}}{10}$

$x = \dfrac{2 \pm \sqrt{124}}{10}$

$x = \dfrac{2 \pm 2\sqrt{31}}{10}$

$x = \dfrac{1 \pm \sqrt{31}}{5}$

Applying the x-intercept method

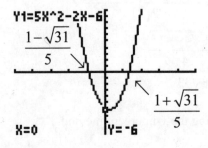

[–5, 5] by [–10, 10]

Considering the graph, the equation is greater than or equal to zero over the interval $\left(-\infty, \dfrac{1-\sqrt{31}}{5}\right] \cup \left[\dfrac{1+\sqrt{31}}{5}, \infty\right)$.

33. $(x+1)^3 < 4$

$\sqrt[3]{(x+1)^3} < \sqrt[3]{4}$

$x + 1 < \sqrt[3]{4}$

$x < \sqrt[3]{4} - 1$

Applying the intersections of graphs method

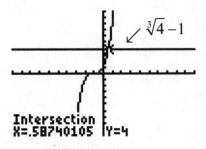

[–10, 10] by [–10, 10]

Considering the graph, the solution is $\left(-\infty, \sqrt[3]{4} - 1\right)$ or approximately $(-\infty, 0.587)$.

35. $(x-3)^5 < 32$

$\sqrt[5]{(x-3)^5} < \sqrt[5]{32}$

$x - 3 < 2$

$x < 5$

Applying the intersections of graphs method

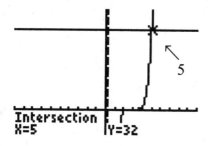

[–10, 10] by [–10, 40]

Considering the graph, the solution is $(-\infty, 5)$.

37. $\qquad |2x-1| < 3$

$\qquad\qquad$ AND

$2x-1 < 3 \quad\bigm|\quad 2x-1 > -3$

$2x < 4 \qquad\bigm|\qquad 2x > -2$

$x < 2 \qquad\bigm|\qquad x > -1$

$\qquad x < 2 \text{ and } x > -1$

$\qquad\qquad (-2, 1)$

Using the intersections of graphs method to check the solution graphically yields,

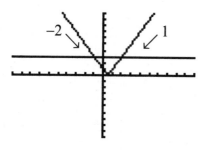

[–10, 10] by [–10, 10]

39. $\qquad |x-6| \ge 2$

$\qquad\qquad$ OR

$x-6 \ge 2 \quad\bigm|\quad x-6 \le -2$

$x \ge 8 \qquad\bigm|\qquad x \le 4$

$\qquad x \le 4 \text{ or } x \ge 8$

$\qquad (-\infty, 4] \cup [8, \infty)$

Using the intersections of graphs method to check the solution graphically yields,

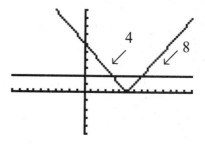

[–10, 15] by [–5, 10]

40. $\qquad |x+8| > 7$

$\qquad\qquad$ OR

$x+8 > 7 \quad\bigm|\quad x+8 < -7$

$x > -1 \qquad\bigm|\qquad x < -15$

$\qquad x < -15 \text{ or } x > -1$

$\qquad (-\infty, -15) \cup (-1, \infty)$

Using the intersections of graphs method to check the solution graphically yields,

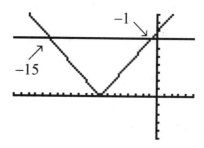

[–20, 5] by [–5, 10]

Section 2.8 Exercises

41. $P(x) > 0$

$-0.3x^2 + 1230x - 120,000 > 0$

Applying the x-intercept method

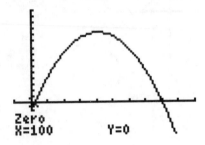

$[-500, 5000]$ by $[-500,000, 1,500,000]$

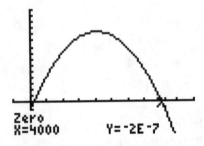

$[-500, 5000]$ by $[-500,000, 1,500,000]$

Considering the graphs, the function is greater than zero over the interval $(100, 4000)$. Producing and selling between 100 and 4000 units, not inclusive, will result in a profit.

43. $P(x)$

$= R(x) - C(x)$

$= (200x - 0.01x^2) - (38x + 0.01x^2 + 16,000)$

$= 200x - 0.01x^2 - 38x - 0.01x^2 - 16,000$

$= -0.02x^2 + 162x - 16,000$

$P(x) > 0$

$-0.02x^2 + 162x - 16,000 > 0$

Applying the x-intercept method

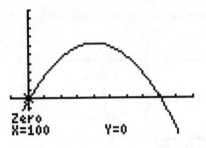

$[-1000, 10,000]$ by $[-200,000, 500,000]$

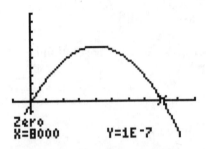

$[-1000, 10,000]$ by $[-200,000, 500,000]$

Considering the graphs, the function is greater than zero over the interval $(100, 8000)$. Producing and selling between 100 and 8000 units, not inclusive, will result in a profit.

45. Applying the intersection of graphs method

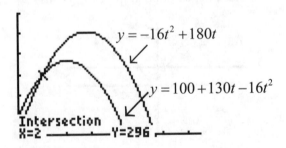

$[0, 15]$ by $[-10, 600]$

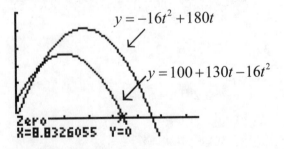

$[0, 15]$ by $[-10, 600]$

Considering the graphs, the second projectile is above the first projectile over the interval $(2, 8.83)$. Between 2 seconds and 8.83 seconds the height of the second projectile exceeds the height of the first projectile.

47. If domestic sales are at least 5,940,000 kg, $y \geq 5.94$. Therefore,

$-0.084x^2 + 1.124x + 4.028 \geq 5.94$.

$$\boxed{-0.084x^2 + 1.124x - 1.912 < 0}$$

Solve the simplified problem above.

$-0.084x^2 + 1.124x - 1.912 = 0$

$$x = \frac{-b \pm \sqrt{b^2 - 4ac}}{2a}$$

$$x = \frac{-(1.124) \pm \sqrt{(1.124)^2 - 4(-0.084)(-1.912)}}{2(-0.084)}$$

$$x = \frac{-1.124 \pm \sqrt{1.263376 - 0.642432}}{-0.168}$$

$$x = \frac{-1.124 \pm \sqrt{0.620944}}{-0.168}$$

$x = 2, x = 11.381$

Applying the x-intercept method

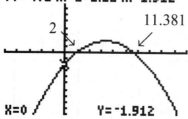

$[-10, 20]$ by $[-10, 10]$

Considering the graph, the equation is greater than or equal to zero over the interval $[2, 11.381]$. Therefore, tobacco sales total at least 5.94 million kg between 1988 and 1990 inclusive.

49. If the number of airplane crashes is below 3000, then $y < 3$. Therefore,

$0.0057x^2 - 0.197x + 3.613 < 3$.

$\boxed{0.0057x^2 - 0.197x + 0.613 < 0}$

Solve the simplified problem above.

$0.0057x^2 - 0.197x + 0.613 = 0$

$x = \dfrac{-b \pm \sqrt{b^2 - 4ac}}{2a}$

$x = \dfrac{-(-0.197) \pm \sqrt{(-0.197)^2 - 4(.0057)(0.613)}}{2(0.0057)}$

$x = \dfrac{0.197 \pm \sqrt{0.038809 - 0.0139764}}{0.0114}$

$x = \dfrac{0.197 \pm \sqrt{0.0248326}}{0.0114}$

$x = 31.104, x = 3.456$

Applying the x-intercept method

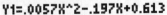

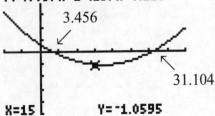

$[-10, 40]$ by $[-5, 5]$

Considering the graph, the equation is less than zero over the interval $(3.456, 31.104)$. Therefore, the number of plane crashes is less than 3000 between 1984 and 2011 inclusive.

51. Applying the intersection of graphs method

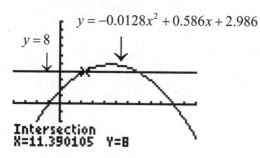

$$y = -0.0128x^2 + 0.586x + 2.986$$
$y = 8$

Intersection
X=11.390105 Y=8

[–20, 60] by [–10, 20]

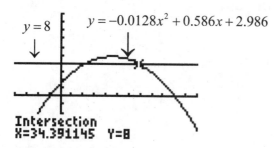

$y = 8$ $y = -0.0128x^2 + 0.586x + 2.986$

Intersection
X=34.391145 Y=8

[–20, 60] by [–10, 20]

Considering the graphs, the homicide rate is above 8 per 100,000 people over the interval $(11.39, 34.39)$. Between 1972 and 1994 inclusive the homicide rate is above 8 per 100,000 people.

53. Applying the intersection of graphs method

$$y = 0.144x^2 - 2.482x + 15.467$$
$y = 6$

Intersection
X=5.6978114 Y=6

[–10, 20] by [–5, 15]

$$y = 0.144x^2 - 2.482x + 15.467$$
$y = 6$

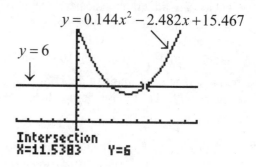

Intersection
X=11.5383 Y=6

[–10, 20] by [–5, 15]

Considering the graphs, the unemployment rate is below 6% over the interval $(5.70, 11.54)$. Between 1986 and 1991 inclusive the unemployment rate is below 6%.

55. Applying the intersection of graphs method

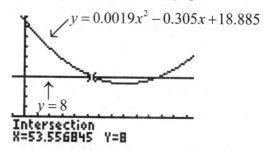

$$y = 0.0019x^2 - 0.305x + 18.885$$
$y = 8$

Intersection
X=53.556845 Y=8

[–10, 135] by [–5, 20]

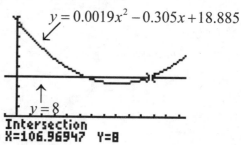

$$y = 0.0019x^2 - 0.305x + 18.885$$
$y = 8$

Intersection
X=106.96947 Y=8

[–10, 135] by [–5, 20]

Considering the graphs, the percentage of
the U.S. population that is foreign born is
below 8% over the interval
$(53.557, 106.969)$. Between 1954 and 2006
inclusive the percentage of the U.S.
population that is foreign born is below 8%.

57. a. Let x equal a person's height in inches.
Then, $|x - 68| > 8$ represents heights
that will be uncomfortable.

b. $|x - 68| > 8$
                            OR
$$x - 68 > 8 \quad | \quad x - 68 < -8$$
$$x > 76 \quad \quad | \quad x < 60$$
$$x < 60 \text{ or } x > 76$$
$$(-\infty, 60) \cup (76, \infty)$$

In the context of the problem, people
with heights below 60 inches or above
76 inches will be uncomfortable

Chapter 2 Skills Check

1. $f(x) = 3x^2 - 6x - 24$

    $h = \dfrac{-b}{2a} = \dfrac{-(-6)}{2(3)} = \dfrac{6}{6} = 1$

    $k = f(h) = f(1)$

    $f(1) = 3(1)^2 - 6(1) - 24$

    $f(1) = 3 - 6 - 24 = -27$

    The vertex is $(1, -27)$.

2.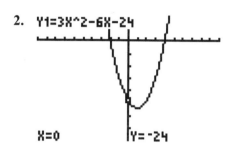

    [–10, 10] by [–40, 10]

3. Algebraically:
    Let $f(x) = 0$

    $3x^2 - 6x - 24 = 0$

    $3(x^2 - 2x - 8) = 0$

    $3(x - 4)(x + 2) = 0$

    $x - 4 = 0, x + 2 = 0$

    $x = 4, x = -2$

    The x-intercepts are $(4, 0)$ and $(-2, 0)$.
    Graphically:

    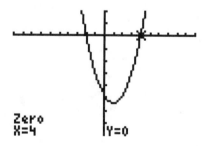

    [–10, 10] by [–40, 10]

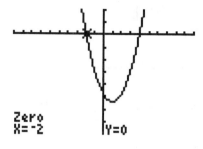

[–10, 10] by [–40, 10]

Again, the x-intercepts are $(4, 0)$ and $(-2, 0)$.

4. Solving $f(x) = 0$ produces the x-intercepts. See problem 3 above. The x-intercepts are $(4, 0)$ and $(-2, 0)$.

5. $x^2 - 5x + 4 = 0$

    $(x - 4)(x - 1) = 0$

    $x = 4, x = 1$

6. $6x^2 + x - 2 = 0$

    Note that $6x^2 \cdot -2 = -12x^2$. Look for two terms whose product is $-12x^2$ and whose sum is the middle term, x.

    $6x^2 + 4x - 3x - 2 = 0$

    $(6x^2 + 4x) + (-3x - 2) = 0$

    $2x(3x + 2) - 1(3x + 2) = 0$

    $(2x - 1)(3x + 2) = 0$

    $2x - 1 = 0, 3x + 2 = 0$

    $x = \dfrac{1}{2}, x = \dfrac{-2}{3}$

7. $5x^2 - x - 4 = 0$

    $(x - 1)(5x + 4) = 0$

    $x = 1, x = -\dfrac{4}{5}$

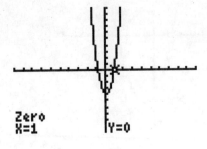

[–10, 10] by [–10, 10]

8. $3x^2 + 4x - 4 = 0$

$(x+2)(3x-2) = 0$

$x = -2, x = \dfrac{2}{3}$

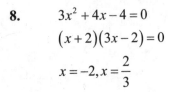

[–10, 10] by [–10, 10]

9. $x^2 - 4x + 3 = 0$

$a = 1, b = -4, c = 3$

$x = \dfrac{-b \pm \sqrt{b^2 - 4ac}}{2a}$

$x = \dfrac{-(-4) \pm \sqrt{(-4)^2 - 4(1)(3)}}{2(1)}$

$x = \dfrac{4 \pm \sqrt{16 - 12}}{2}$

$x = \dfrac{4 \pm \sqrt{4}}{2}$

$x = \dfrac{4 \pm 2}{2}$

$x = 3, x = 1$

10. $4x^2 + 4x - 3 = 0$

$a = 4, b = 4, c = -3$

$x = \dfrac{-b \pm \sqrt{b^2 - 4ac}}{2a}$

$x = \dfrac{-(4) \pm \sqrt{(4)^2 - 4(4)(-3)}}{2(4)}$

$x = \dfrac{-4 \pm \sqrt{16 + 48}}{8}$

$x = \dfrac{-4 \pm \sqrt{64}}{8}$

$x = \dfrac{-4 \pm 8}{8}$

$x = \dfrac{-4 + 8}{8}, x = \dfrac{-4 - 8}{8}$

$x = \dfrac{4}{8} = \dfrac{1}{2}, x = -\dfrac{12}{8} = -\dfrac{3}{2}$

11. a.

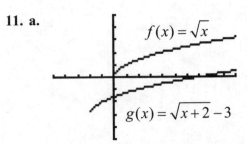

[–5, 10] by [–5, 5]

b. The function *g(x)* is shifted two units left and three units down in comparison to *f(x)*.

12. $g(x) = \sqrt{x+2} - 3$

$x + 2 \geq 0$

$x \geq -2$

Domain: $[-2, \infty)$.

13. Y1= -3X^2

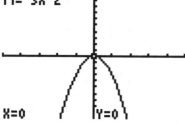

X=0   Y=0

[–5, 5] by [–10, 10]

The function is increasing when $x < 0$ and decreasing when $x > 0$. In interval notation, the function is increasing on the interval $(-\infty, 0)$ and decreasing on the interval $(0, \infty)$.

14. a. Y1=X^(3/2)

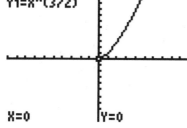

X=0   Y=0

[–10, 10] by [–10, 10]
The graph is concave up.

b. Y1=X^(1/2)

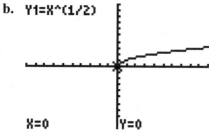

X=0   Y=0

[–10, 10] by [–10, 10]

The graph is concave down.

15.
$$f(x) = 3x - 2$$
$$y = 3x - 2$$
$$x = 3y - 2$$
$$3y = x + 2$$
$$y = \frac{x+2}{3}$$
$$f^{-1}(x) = \frac{x+2}{3}$$

16.
$$g(x) = \sqrt[3]{x-1}$$
$$y = \sqrt[3]{x-1}$$
$$x = \sqrt[3]{y-1}$$
$$x^3 = \left(\sqrt[3]{y-1}\right)^3$$
$$x^3 = y - 1$$
$$y = x^3 + 1$$
$$g^{-1}(x) = x^3 + 1$$

17. Note that $f(x)$ is one-to-one on the given interval [–1, 10]. Therefore,
$$f(x) = (x+1)^2$$
$$y = (x+1)^2$$
$$x = (y+1)^2$$
$$y + 1 = \sqrt{x}, \quad x \geq 0, y \geq -1$$
$$y = -1 + \sqrt{x}$$
$$f^{-1}(x) = -1 + \sqrt{x}$$

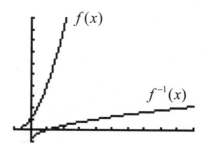

[–1, 10] by [–1, 10]

18.

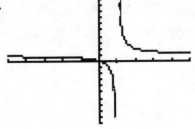

[–5, 5] by [–10, 10]

The function passed the horizontal line test. It is one-to-one.

19.

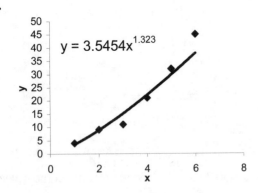

$y = 3.5454x^{1.323}$

20.

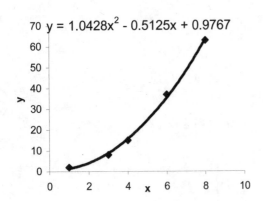

$y = 1.0428x^2 - 0.5125x + 0.9767$

21. f

22. c

23. e

24.

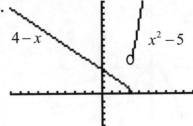

$4 - x$ $x^2 - 5$

[–10, 10] by [–5, 15]

25. $x^2 - 7x \le 18$

$x^2 - 7x - 18 \le 0$

$x^2 - 7x - 18 = 0$

$(x - 9)(x + 2) = 0$

$x = 9, x = -2$

sign of $(x - 9)$ $\leftarrow - - -_{-2} - - -_9 + + + \rightarrow$

sign of $(x + 2)$ $\leftarrow - - -_{-2} + + +_9 + + + \rightarrow$

sign of $(x - 9)(x + 2)$ $\leftarrow + + +_{-2} - - -_9 + + + \rightarrow$

Considering the inequality symbol in the simplified question, the solution is $[-2, 9]$.

26. $2x^2 + 5x \ge 3$

$2x^2 + 5x - 3 \ge 0$

$2x^2 + 5x - 3 = 0$

$(2x - 1)(x + 3) = 0$

$x = \dfrac{1}{2}, x = -3$

sign of $(2x - 1)$ $\leftarrow - - -_{-3} - - -_{\frac{1}{2}} + + + \rightarrow$

sign of $(x + 3)$ $\leftarrow - - -_{-3} + + +_{\frac{1}{2}} + + + \rightarrow$

sign of $(2x - 1)(x + 2)$ $\leftarrow + + +_{-3} - - -_{\frac{1}{2}} + + + \rightarrow$

Considering the inequality symbol in the simplified question, the solution is

$\left(-\infty, -3\right] \cup \left[\dfrac{1}{2}, \infty\right)$.

27. $z^2 - 4z + 6 = 0$

$a = 1, b = -4, c = 6$

$z = \dfrac{-b \pm \sqrt{b^2 - 4ac}}{2a}$

$z = \dfrac{-(-4) \pm \sqrt{(-4)^2 - 4(1)(6)}}{2(1)}$

$z = \dfrac{4 \pm \sqrt{16 - 24}}{2}$

$z = \dfrac{4 \pm \sqrt{-8}}{2}$

$z = \dfrac{4 \pm 2i\sqrt{2}}{2}$

$z = 2 \pm i\sqrt{2}$

28. $w^2 - 4w + 5 = 0$

$a = 1, b = -4, c = 5$

$w = \dfrac{-b \pm \sqrt{b^2 - 4ac}}{2a}$

$w = \dfrac{-(-4) \pm \sqrt{(-4)^2 - 4(1)(5)}}{2(1)}$

$w = \dfrac{4 \pm \sqrt{16 - 20}}{2}$

$w = \dfrac{4 \pm \sqrt{-4}}{2}$

$w = \dfrac{4 \pm 2i}{2}$

$w = 2 \pm i$

29. $4x^2 - 5x + 3 = 0$

$a = 4, b = -5, c = 3$

$x = \dfrac{-b \pm \sqrt{b^2 - 4ac}}{2a}$

$x = \dfrac{-(-5) \pm \sqrt{(-5)^2 - 4(4)(3)}}{2(4)}$

$x = \dfrac{5 \pm \sqrt{25 - 48}}{8}$

$x = \dfrac{5 \pm \sqrt{-23}}{8}$

$x = \dfrac{5 \pm i\sqrt{23}}{8}$

30. $4x^2 + 2x + 1 = 0$

$a = 4, b = 2, c = 1$

$x = \dfrac{-b \pm \sqrt{b^2 - 4ac}}{2a}$

$x = \dfrac{-(2) \pm \sqrt{(2)^2 - 4(4)(1)}}{2(4)}$

$x = \dfrac{-2 \pm \sqrt{4 - 16}}{8}$

$x = \dfrac{-2 \pm \sqrt{-12}}{8}$

$x = \dfrac{-2 \pm \sqrt{-4 \cdot 3}}{8}$

$x = \dfrac{-2 \pm 2i\sqrt{3}}{8}$

$x = \dfrac{-1 \pm i\sqrt{3}}{4}$

31. $(x - 4)^3 < 4096$

$\sqrt[3]{(x - 4)^3} < \sqrt[3]{4096}$

$x - 4 < 16$

$x < 20$

The solution is $(-\infty, 20)$.

32. $(x+2)^2 \geq 512$

$\sqrt{(x+2)^2} = \sqrt{512}$

$x+2 = \pm\sqrt{256 \cdot 2}$

$x+2 = \pm 16\sqrt{2}$

$x = -2 \pm 16\sqrt{2}$

Considering $(x+2)^2 - 512 \geq 0$

$\leftarrow ++++\underset{-2-16\sqrt{2}}{} ----\underset{2+16\sqrt{2}}{} ++++ \rightarrow$

The solution is
$\left(-\infty, -2-16\sqrt{2}\right] \cup \left[-2+16\sqrt{2}, \infty\right)$.

33. $\sqrt{4x^2+1} = 2x+2$

$\left(\sqrt{4x^2+1}\right)^2 = (2x+2)^2$

$4x^2+1 = 4x^2+8x+4$

$8x = -3$

$x = -\dfrac{3}{8}$

34. $\sqrt{3x^2-8} + x = 0$

$\left(\sqrt{3x^2-8}\right)^2 = (-x)^2$

$3x^2-8 = x^2$

$2x^2 = 8$

$x^2 = 4$

$x = \pm 2$

Substituting to check

$x = 2$

$\sqrt{3(2)^2-8} + 2 \overset{?}{=} 0$

$4 \neq 0$

$x = -2$

$\sqrt{3(-2)^2-8} + (-2) \overset{?}{=} 0$

$0 = 0$

The only solution is $x = -2$.

35. $|3x-6| = 24$

$3x-6 = 24$	$3x-6 = -24$
$3x = 30$	$3x = -18$
$x = 10$	$x = -6$

$x = 10, x = -6$

36. $|2x+3| = 13$

$2x+3 = 13$	$2x+3 = -13$
$2x = 10$	$2x = -16$
$x = 5$	$x = -8$

$x = 5, x = -8$

37. $|2x-4| \leq 8$

AND

$2x-4 \leq 8$	$2x-4 \geq -8$
$2x \leq 12$	$2x \geq -4$
$x \leq 6$	$x \geq -2$

$x \geq -2$ and $x \leq 6$

$[-2, 6]$

38. $|4x-3| \geq 15$

OR

$4x-3 \geq 15$	$4x-3 \leq -15$
$4x \geq 18$	$4x \leq -12$
$x \geq \dfrac{9}{2}$	$x \leq -3$

$x \geq \dfrac{9}{2}$ or $x \leq -3$

$(-\infty, -3] \cup \left[\dfrac{9}{2}, \infty\right)$

Chapter 2 Review Exercises

39. a. The maximum profit occurs at the vertex.

$$P(x) = -0.01x^2 + 62x - 12,000$$

$$h = \frac{-b}{2a} = \frac{-62}{2(-0.01)} = \frac{-62}{-0.02} = 3100$$

$$P(3100) = -0.01(3100)^2 + 62(3100)$$
$$-12,000$$

$$P(3100) = -96,100 + 192,200 - 12,000$$

$$p(3100) = 84,100$$

The vertex is $(3100, 84,100)$.

Producing and selling 3100 units produces the maximum profit of $84,100.

b. See part *a*. The maximum profit is $84,100.

40. $P(x)$

$$= R(x) - C(x)$$

$$= \left(200x - 0.01x^2\right) - \left(38x + 0.01x^2 + 16,000\right)$$

$$= 200x - 0.01x^2 - 38x - 0.01x^2 - 16,000$$

$$= -0.02x^2 + 162x - 16,000$$

Let $P(x) = 0$

$$-0.02x^2 + 162x - 16,000 = 0$$

$$x = \frac{-b \pm \sqrt{b^2 - 4ac}}{2a}$$

$$x = \frac{-(162) \pm \sqrt{(162)^2 - 4(-0.02)(-16,000)}}{2(-0.02)}$$

$$x = \frac{-162 \pm \sqrt{26,244 - 1280}}{-0.04}$$

$$x = \frac{-162 \pm \sqrt{24,964}}{-0.04}$$

$$x = \frac{-162 \pm 158}{-0.04}$$

$$x = \frac{-162 + 158}{-0.04}, x = \frac{-162 - 158}{-0.04}$$

$$x = 100, x = 8000$$

The company will break even producing and selling 100 units or 8000 units.

41. a. $h = \dfrac{-b}{2a} = \dfrac{-64}{2(-16)} = 2$. The ball reaches its maximum height in 2 seconds.

b. $S(2) = 192 + 64(2) - 16(2)^2$
$$= 256$$
The ball reaches a maximum height of 256 feet.

42. a. $h = \dfrac{-b}{2a} = \dfrac{-29.4}{2(-9.8)} = 1.5$. The ball reaches its maximum height in 1.5 seconds.

b. $S(1.5) = 60 + 29.4(1.5) - 9.8(1.5)^2$ The
$$= 82.05$$
ball reaches a maximum height of 82.05 meters.

43 a.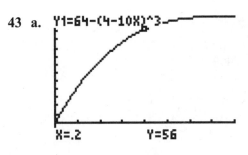

[0, 0.4] by [−10, 65]

b.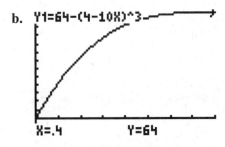

[0, 0.4] by [−10, 65]

The bullet travels 64 inches in 0.4 seconds.

44 a. Note that the function is quadratic written in standard form, $y = a(x-h)^2 + k$. The vertex is $(h,k) = (4,380)$. Therefore the maximum height occurs 4 seconds into the flight of the rocket.

b. Referring to part *a*, the maximum height of the rocket is 380 feet.

c. In comparison to $y = t^2$ the graph is shifted 4 units right, 380 up, stretched by a factor of 16, and reflected across the *x*-axis.

45. a.

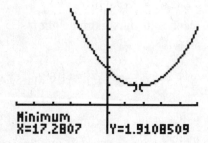

Minimum
X=17.2807   Y=1.9108509

[–50, 50] by [–3, 10]

The model predicts that the minimum number of airplane crashes occurs in 1997.

b.

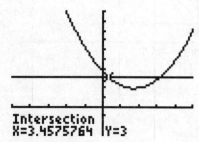

Intersection
X=3.4575764   Y=3

[–50, 50] by [–3, 10]

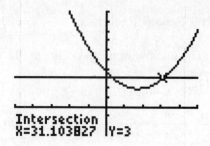

Intersection
X=31.103827   Y=3

[–50, 50] by [–3, 10]

The model predicts that between 1984 and 2011 inclusive the number of plane crashes will be below 3000.

46. a. Using the table feature of the TI-83 calculator along with the given function

X	Y1
5	6.657
6	5.759
7	5.149
8	4.827
9	4.793
10	5.047
11	5.589

X=9

Unemployment reaches a minimum in approximately 1989.

b. In 1989 the unemployment rate is 4.79%.

c. Using the table feature of the TI-83 calculator along with the given function

X	Y1
6	5.759
7	5.149
8	4.827
9	4.793
10	5.047
11	5.589
12	6.419

X=10

Based on the table the unemployment rate is 5.047% in 1990.

47. $3600 - 150x + x^2 = 0$

$x^2 - 150x + 3600 = 0$

$(x - 30)(x - 120) = 0$

$x = 30, x = 120$

Break-even occurs when 30 or 120 units are produced and sold.

48. The ball is on the ground when $s = 0$.

$400 - 16t^2 = 0$

$-16t^2 = -400$

$t^2 = 25$

$t = \pm 5$

In the physical context of the question, $t \geq 0$. The ball reaches the ground in 5 seconds.

49. $-0.3x^2 + 1230x - 120,000 = 324,000$

$-0.3x^2 + 1230x - 444,000 = 0$

$x = \dfrac{-b \pm \sqrt{b^2 - 4ac}}{2a}$

$= \dfrac{-(1230) \pm \sqrt{(1230)^2 - 4(-0.3)(-444,000)}}{2(-0.3)}$

$= \dfrac{-1230 \pm \sqrt{1,512,900 - 532,800}}{-0.6}$

$= \dfrac{-1230 \pm \sqrt{980,100}}{-0.6}$

$= \dfrac{-1230 \pm 990}{-0.6}$

$x = 400, x = 3700$

A profit of \$324,000 occurs when 400 or 3700 units are produced and sold.

50. Applying the intersection of graphs method with the given function and $y = 6.745$

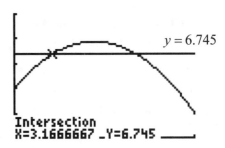

Intersection
X=3.1666667 _Y=6.745

[0, 15] by [0, 10]

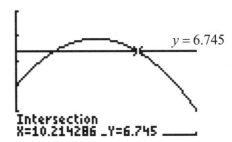

Intersection
X=10.214286 _Y=6.745

[0, 15] by [0, 10]

Tobacco sales equal 6.745 million in 1989 and 1996.

51. Applying the intersection of graphs method with the given function and $y = 8$

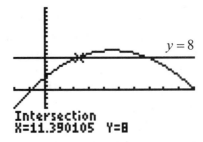

Intersection
X=11.390105 Y=8

[–10, 50] by [–10, 20]

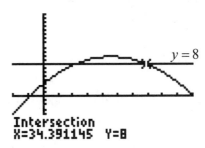

Intersection
X=34.391145 Y=8

[–10, 50] by [–10, 20]

Between 1972 and 1994 inclusive the homicide rate is above 8 per 100,000 people.

52. Applying the intersection of graphs method with the given function and $y = 3$

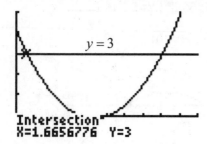

[0, 30] by [−1, 5]

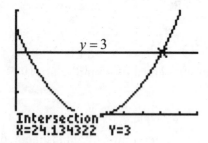

[0, 30] by [−1, 5]

Between 2000 and 2022 inclusive the percentage change in hotel room supply is below 3%. In the context of the question, between 2000 and 2010 inclusive, the percentage change in hotel room supply is below 3%.

53. **a.** Let $x = 1998 - 1980 = 18$

$y = 161,488.4931(18)^{-0.4506}$

$y = 161,488.4931(0.2718780893)$

$y = 43,905.18295$

$y \approx 43,905$

The model predicts 43,905 injuries in 1998.

b.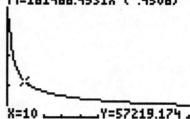

[0, 100] by [−10, 200,000]

In 1990 there are 57,219 predicted injuries.

54. **a.** Yes. The exponent is greater than one.

b.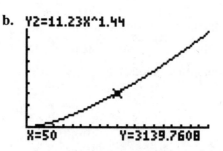

[0, 100] by [−1000, 10,000]

c. The graph is concave up.

d. $S = 11.23(95)^{1.44} \approx 7912.28$

In 1995 the per capita tax is $7912.28.

56. **a.** $\overline{C}(x) = \dfrac{30x + 3150}{x}$

b.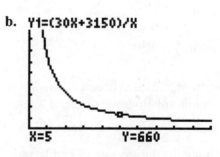

[0, 20] by [−500, 5000]

57. a. Since there is a variable in the denominator $p = \dfrac{30,000}{q} - 20$ is a shifted reciprocal function.

b.

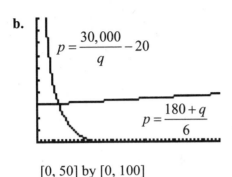

[0, 50] by [0, 100]

58. a. Since there is a variable in the denominator of $p = \dfrac{2555}{q+5}$, the demand function is a shifted reciprocal function.

b.

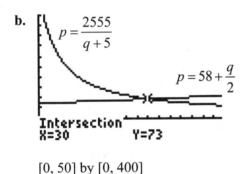

[0, 50] by [0, 400]

59. a.
$$P(90) = 2.320(90)^2 - 389(90) + 21,762$$
$$= 18,792 - 35,010 + 21,762$$
$$= 5544$$

The Indiana population in 1990 is 5544 thousand people or 5,544,000 people.

b.
$$P(x) = 46.0x + 1411.667$$
$$6000 = 46.0x + 1411.667$$
$$46.0x = 4588.333$$
$$x = \frac{4588.333}{46.0} \approx 99.75$$

X	Y₁
96	5827.7
97	5873.7
98	5919.7
99	5965.7
100	6011.7
101	6057.7
102	6103.7

X=100

In 2000, the population exceeds 6 million. Since the model describes population data between 1980 and 1997, the calculation is an extrapolation. The solution is valid only if the model remains valid in 2000.

60. a. For years between 1979 and 1988 inclusive, the function is quadratic.

b.
$$M(79) = -\frac{1}{12}\left(79 - \frac{789}{10}\right)^2 + \frac{15,541}{1200}$$
$$= -\frac{1}{12}(0.1)^2 + \frac{15,541}{1200}$$
$$= 12.95$$

$$M(80) = -\frac{1}{12}\left(80 - \frac{789}{10}\right)^2 + \frac{15,541}{1200}$$
$$= -\frac{1}{12}(1.1)^2 + \frac{15,541}{1200}$$
$$= 12.85$$

$$M(88) = -\frac{1}{12}\left(88 - \frac{789}{10}\right)^2 + \frac{15,541}{1200}$$
$$= -\frac{1}{12}(9.1)^2 + \frac{15,541}{1200}$$
$$= 6.05$$

In 1979, 12.95% of people 12 years of age or older used marijuana at least once in the month prior to being surveyed. In 1980, 12.85% of people 12 years of age or older used marijuana at least once in the month prior to being surveyed. In 1988, 6.05% of people 12 years of age or older used marijuana at least once in the month prior to being surveyed.

c.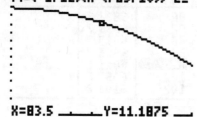

Y1=(-1/12)(X-(789/10))^2_

X=83.5 _____ Y=11.1875 _____

[79, 88] by [0, 15]

d. The domain is $\{79,80,81,82,...,87,88\}$.

e. The maximum occurs in 1979.
Marijuana use decreased until 1988.

61. a. Trade Balance = Exports - Imports
Let $T(x)$ = Trade Balance
$$T(x) = E(x) - I(x)$$
$$= \left(-191.73x^2 + 39,882.69x + 377,849.85\right) - \left(61,487.24x + 435,606.84\right)$$
$$= -191.73x^2 - 21,604.55x - 57,756.99$$

b. Let $x = 10$
$$T(10) = -191.73(10)^2 - 21,604.55(10) - 57,756.99$$
$$= -19,173 - 216,045.5 - 57,756.99$$
$$= -292,975.49$$

The trade deficit in 2000 is $292,975.49 million.

62. a. $R\left(E(t)\right) = 0.165\left(0.017t^2 + 2.164t + 8.061\right) - 0.226$
$$= 0.002805t^2 + 0.35706t + 1.330065 - 0.226$$
$$= 0.002805t^2 + 0.35706t + 1.104065$$

The function calculates the revenue for Southwest Airlines given the number of years past 1990.

b. $R\left(E(t)\right) = 0.002805t^2 + 0.35706t + 1.104065$

$R\left(E(3)\right) = 0.002805(3)^2 + 0.35706(3) + 1.104065$

$R\left(E(3)\right) = 2.2008995 \approx 2.2$

In 1993 Southwest Airlines had revenue of $2.2 billion.

c. $E(7) = 0.017(7)^2 + 2.164(7) + 8.061$
$$= 24.042$$

In 1997 Southwest Airlines has 24,042 employees.

d. $R(E(t)) = 0.002805t^2 + 0.35706t + 1.104065$

$R(E(7)) = 0.002805(7)^2 + 0.35706(7) + 1.104065$

$R(E(7)) = 3.74093 \approx 3.7$

In 1997 Southwest Airlines had revenue of $3.7 billion.

63. a. $f(x) = 0.554x - 2.886$

$y = 0.554x - 2.886$

$x = 0.554y - 2.886$

$0.554y = x + 2.886$

$y = \dfrac{x + 2.886}{0.554}$

$f^{-1}(x) = \dfrac{x + 2.886}{0.554}$

b. The inverse function calculates the mean length of the original prison sentence given the mean time spent in prison.

64. a. $C(x) = -0.093x + 9.929$

$y = -0.093x + 9.929$

$x = -0.093y + 9.929$

$-0.093y = x - 9.929$

$y = \dfrac{x - 9.929}{-0.093}$

$C^{-1}(x) = \dfrac{9.929 - x}{0.093}$

b. Given the number of cows used for milk production, the inverse function will calculate the number of years past 1990.

c. $C^{-1}(C(8)) = 8$

65. a.

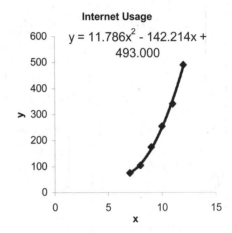

Internet Usage

$y = 11.786x^2 - 142.214x + 493.000$

b. See part a) above.

c. Yes. The function seems to fit the data very well.

d. $y = 11.786x^2 - 142.214x + 493$

Let $y = 550$

$550 = 11.786x^2 - 142.214x + 493$

Using the unrounded model and applying the intersection of graphs method

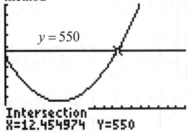

$y = 550$

Intersection
X=12.454974 Y=550

[0, 20] by [-200, 1000]

The model predicts that Internet users will reach 550 billion in 2002. Internet users will exceed 550 billion in 2003.

66. a.

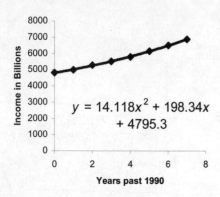

Personal Income

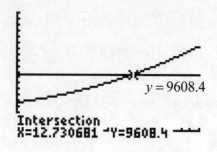

[0, 20] by [−1000, 20,000]
Personal income doubles during 2003.

b. Using the unrounded model and applying the intersections of graphs method

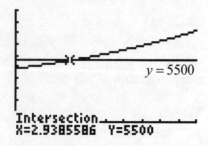

[0, 10] by [−1000, 10,000]

Personal income reaches $5500 billion in 1993.

c. The personal income level in 1990 is $4804.2 billion. Twice that amount is $9608.4 billion.

Using the unrounded model and applying the intersections of graphs method

67.

Year	Age 15-19 Population (thousands)	Male Population (%)	15-19 Male Population (thousands)
1980	21,168	48.6	10,287.648
1985	18,727	48.7	9120.049
1990	17,890	48.7	8712.430
1995	18,152	48.9	8876.328
1997	19,068	49.0	9343.320

Years past 1980	15-19 Male Population (thousands)
0	21,168*0.486 = 10,287.648
5	18,727*0.487 = 9120.049
10	17,890*0.487 = 8712.430
15	18152*.0489 = 8876.328
17	19,068*0.490 = 9343.320

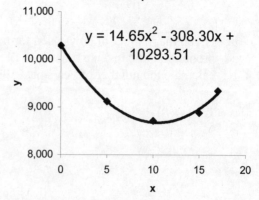

In the quadratic model, x represents the years past 1980 while y represents the 15-19 male population in thousands.

68. a.

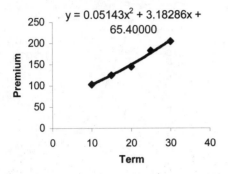

Insurance Premiums

$y = 0.05143x^2 + 3.18286x + 65.40000$

b. Applying the intersection of graphs method using unrounded model and $y = 130$

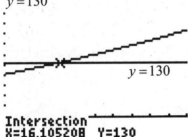

$y = 130$

Intersection
X=16.105208   Y=130

[10, 30] by [−50, 250]

A premium of $130 would purchase a term of 16 years.

69. a.

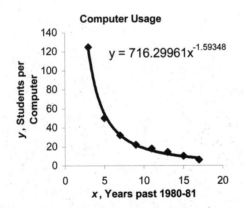

Computer Usage

$y = 716.29961x^{-1.59348}$

The model seems to fit the data well.

b. Yes. The function is decreasing and will eventually reach a value of one, representing one student per computer. Considering the graph below, this

occurs when $x \approx 61.4$ or in school year 2042-2043.

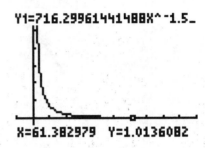

Y1=716.29961441488X^-1.5_

X=61.382979   Y=1.0136082

[−10, 100] by [−25, 150]

c. $C(t) = \dfrac{1}{t}$

The basic function is shifted 0.3 units left and 15 units down and is stretched vertically by a factor of 380.

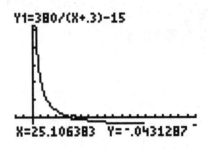

Y1=380/(X+.3)-15

X=25.106383   Y=-.0431287

[−10, 100] by [-10, 100]

d. The power function is the better fit for the situation. It remains positive for all school years.

e. Power function (unrounded model)

$y = 716.29961x^{-1.59348}$

$y = 716.3(18)^{-1.5935} \approx 7.1588$

Rational Function

$C(t) = \dfrac{380}{t + 0.3} - 15$

$C(18) = \dfrac{380}{18 + 0.3} - 15$

$= \dfrac{380}{18.3} - 15 = 5.77$

$C(t)$ is the better fit for $1998 - 99$.

70. a.

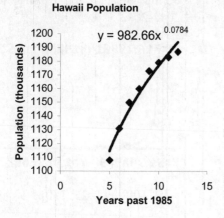

b. Using the table feature of the TI-83 calculator along with the unrounded model

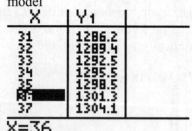

In 2021 the Hawaiian population first exceeds 1.3 million people.

71. a. Between 1980 and 1991 inclusive,

$$y = 0.013x^2 - 0.2225x + 7.2974.$$

c. Combining the solutions to part a) and b) to create a piecewise model:

$$f(x) = \begin{cases} 0.013x^2 - 0.2225x + 7.2974 & 0 \le x \le 11 \\ -0.26x + 9.34 & 12 \le x \le 16 \end{cases}$$

d. i. $f(7) = 0.013(7)^2 - 0.2225(7) + 7.2974$

Using the TI-83 and substituting into the unrounded quadratic model:

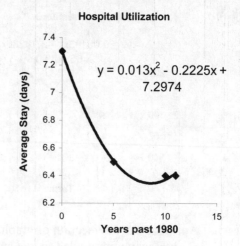

b. Beyond 1991, $y = -0.26x + 9.34$.

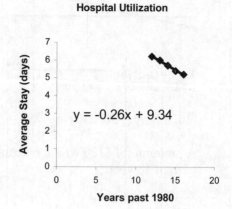

```
7→X
                    7
.0129934924078X^
2+-.222537960954
45X+7.2973969631
236
          6.376312364
```

In 1987 the average hospital stay is predicted to be 6.4 days.

ii. $f(x) = 6.1$

The year is greater than 1991.

$-0.26x + 9.34 = 6.1$

$-0.26x = -3.24$

$x = 12.46154$

The average hospital stay is 6.1 in 1992.

iii. $f(11) = 0.013(11)^2 - 0.2225(11) + 7.2974$

Using the TI-83 and substituting into the unrounded quadratic model:

```
11→X
                    11
.0129934924078X^
2+-.222537960954
45X+7.2973969631
236
          6.421691974
```

In 1997 the average hospital stay is 6.4 days.

72. a.

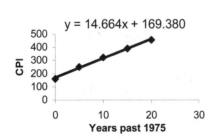

The model fits the data well.

b.

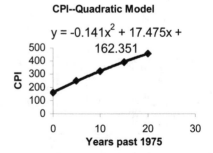

The model fits the data well. The quadratic model seems a little better than the linear model.

c. Using the linear model,

$y = 14.664(25) + 169.380$

Using the TI-83 and substituting into the unrounded linear model:

```
25→X
                    25
14.664X+169.38
                535.98
```

$y = 535.98$

Using the quadratic model,

$y = -0.141(25)^2 + 17.475(25) + 162.351$

Using the TI-83 and substituting into the unrounded quadratic model:

```
25→X
                    25
-.14057142857143
X^2+17.475428571
429X+162.3514285
7143
                511.38
```

$y = 511.38$

The quadratic model does a better job of predicting the CPI value for 2000, estimated to be 511.5.

d.
```
   X  | Y1     | Y2
  26  | 521.69 | 550
  27  | 531.71 | 550
  28  | 541.46 | 550
  29  | 550.92 | 550
  30  | 560.1  | 550
  31  | 569    | 550
  32  | 577.62 | 550
 X=29
```

The CPI reaches 550 in 2004.

73. $P(x) = -0.01x^2 + 62x - 12,000$

$P(x) > 0$

$-0.01x^2 + 62x - 12,000 > 0$

Applying the x-intercept method

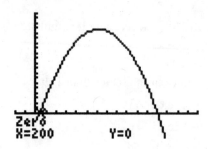

[−1000, 8000] by [−25,000, 100,000]

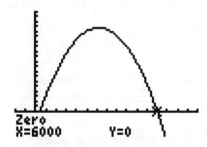

[−1000, 8000] by [−25,000, 100,000]

Manufacturing and producing between 200 and 6000 units will result in a profit.

74. Applying the intersection of graphs method

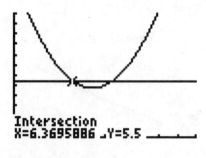

[0, 20] by [0, 12]

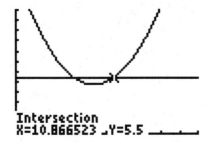

Intersection
X=10.866523 _Y=5.5

[0, 20] by [0, 12]

In the intervals $(-\infty, 6.37)$ and $(10.87, \infty)$, the unemployment rate is above 5.5%. That implies that prior to 1987 and after 1990 exclusive the unemployment rate is greater than 5.5%.

Group Activity/Extended Application I

1.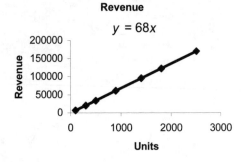

Revenue

$y = 68x$

Cost

$y = 0.01x^2 + 28x + 30000$

2. a.
$$P(x) = R(x) - C(x)$$
$$= 68x - \left(0.01x^2 + 28x + 30{,}000\right)$$
$$= -0.01x^2 + 40x - 30{,}000$$

b.

0	−30,000
100	−26,100
600	−9600
1600	8400
2000	10,000
2500	7500

3. $R(x) = C(x)$

$$68x = 0.01x^2 + 28x + 30{,}000$$

$$0.01x^2 - 40x + 30{,}000 = 0$$

$$x = \frac{-b \pm \sqrt{b^2 - 4ac}}{2a}$$

$$x = \frac{-(-40) \pm \sqrt{(-40)^2 - 4(0.01)(30{,}000)}}{2(0.01)}$$

$$x = \frac{40 \pm \sqrt{1600 - 1200}}{0.02}$$

$$x = \frac{40 \pm \sqrt{400}}{0.02}$$

$$x = \frac{40 \pm 20}{0.02}$$

$$x = 3000, x = 1000$$

Producing and selling either 3000 or 1000 units forces the profit to equal the cost.

4.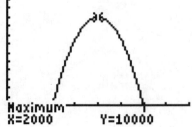

[0, 4000] by [−2000, 12,000]

Producing and selling 2000 units results in a maximum profit of $10,000.

5. a. $\overline{C}(x) = \dfrac{C(x)}{x} = \dfrac{0.01x^2 + 28x + 30{,}000}{x}$

b.

1	30,028.01
100	329.00
300	131.00
1400	63.43
2000	63.00
2500	65.00

c.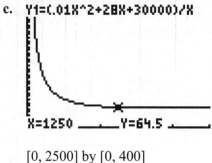

[0, 2500] by [0, 400]

6.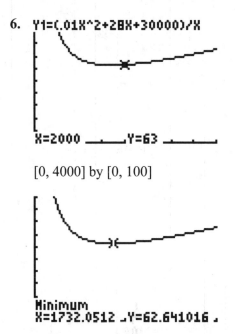

[0, 4000] by [0, 100]

[0, 4000] by [0, 100]

The minimum average cost is $62.64 which
occurs when 1732 units are produced and
sold.

7. The values are different. However, they are
relatively close together. The number of
units that maximizes profit is most
important. While keeping costs low is
important, the key to a successful business is
generating profit.

Chapter 3
Exponential and Logarithmic Functions

Algebra Toolbox

1. $3^{-2} = \dfrac{1}{3^2} = \dfrac{1}{9}$

2. $2^{-3} = \dfrac{1}{2^3} = \dfrac{1}{8}$

3. $-4^{-3} = -\left(\dfrac{1}{4^3}\right) = -\dfrac{1}{64}$

4. $-4^{-2} = -\left(\dfrac{1}{4^2}\right) = -\dfrac{1}{16}$

5. $\dfrac{1}{3^{-3}} = \dfrac{1}{\frac{1}{3^3}} = 1 \div \dfrac{1}{3^3} = 1 \cdot 3^3 = 27$

6. $\dfrac{1}{5^{-2}} = \dfrac{1}{\frac{1}{5^2}} = 1 \div \dfrac{1}{5^2} = 1 \cdot 5^2 = 25$

7. $\left(\dfrac{2}{3}\right)^{-2} = \left(\dfrac{3}{2}\right)^2 = \dfrac{9}{4}$

8. $\left(\dfrac{3}{2}\right)^{-3} = \left(\dfrac{2}{3}\right)^3 = \dfrac{2^3}{3^3} = \dfrac{8}{27}$

9. $10^{-2} \cdot 10^0 = \dfrac{1}{10^2} \cdot 1 = \dfrac{1}{100}$

10. $8^{-2} \cdot 8^0 = \dfrac{1}{8^2} \cdot 1 = \dfrac{1}{64}$

11. $\left(2^{-1}\right)^3 = 2^{-1\cdot 3} = 2^{-3} = \dfrac{1}{2^3} = \dfrac{1}{8}$

12. $\left(4^{-2}\right)^2 = 4^{-2\cdot 2} = 4^{-4} = \dfrac{1}{4^4} = \dfrac{1}{256}$

13. $10^{5^0} = 10^{(5\cdot 0)} = 10^0 = 1$

14. $4^{2^2} = 4^{2\cdot 2} = 4^4 = 256$

15. $x^{-4} \cdot x^{-3} = x^{-4+-3} = x^{-7} = \dfrac{1}{x^7}$

16. $y^{-5} \cdot y^{-3} = y^{-5+-3} = y^{-8} = \dfrac{1}{y^8}$

17. $\left(c^{-6}\right)^3 = c^{-6\cdot 3} = c^{-18} = \dfrac{1}{c^{18}}$

18. $\left(x^{-2}\right)^4 = x^{-2\cdot 4} = x^{-8} = \dfrac{1}{x^8}$

19. $\dfrac{a^{-4}}{a^{-5}} = a^{-4-(-5)} = a^1 = a$

20. $\dfrac{b^{-6}}{b^{-8}} = b^{-6-(-8)} = b^2$

21. $\left(x^{-\frac{1}{2}}\right)\left(x^{\frac{2}{3}}\right) = x^{-\frac{1}{2}+\frac{2}{3}} = x^{-\frac{3}{6}+\frac{4}{6}} = x^{\frac{1}{6}}$

22. $\left(y^{-\frac{1}{3}}\right)\left(y^{\frac{2}{5}}\right) = y^{-\frac{1}{3}+\frac{2}{5}} = y^{-\frac{5}{15}+\frac{6}{15}} = y^{\frac{1}{15}}$

23. $\left(3a^{-3}b^2\right)\left(2a^2b^{-4}\right)$

$= 6a^{-3+2}b^{2+-4}$

$= 6a^{-1}b^{-2}$

$= \dfrac{6}{ab^2}$

24. $\left(4a^{-2}b^3\right)\left(-2a^4b^{-5}\right)$

$= -8a^{-2+4}b^{3+-5}$

$= -8a^2b^{-2}$

$= \dfrac{-8a^2}{b^2}$

25. $\left(\dfrac{2x^{-3}}{x^2}\right)^{-2} = \left(2x^{-3-2}\right)^{-2}$

$= \left(2x^{-5}\right)^{-2}$

$= (2)^{-2}\left(x^{-5}\right)^{-2}$

$= \dfrac{1}{2^2}x^{-5\cdot-2}$

$= \dfrac{1}{4}x^{10}$

$= \dfrac{x^{10}}{4}$

26. $\left(\dfrac{3y^{-4}}{2y^2}\right)^{-3} = \left(\dfrac{2y^2}{3y^{-4}}\right)^3$

$= \dfrac{\left(2y^2\right)^3}{\left(3y^{-4}\right)^3}$

$= \dfrac{8y^6}{27y^{-12}}$

$= \dfrac{8y^{6-(-12)}}{27}$

$= \dfrac{8y^{18}}{27}$

27. $\dfrac{28a^4b^{-3}}{-4a^6b^{-2}} = -7a^{4-6}b^{-3-(-2)}$

$= -7a^{-2}b^{-1}$

$= \dfrac{-7}{a^2b}$

28. $\dfrac{36x^5y^{-2}}{-6x^6y^{-4}} = -6x^{5-6}y^{-2-(-4)}$

$= -6x^{-1}y^2$

$= \dfrac{-6y^2}{x}$

29. 4.6×10^7

30. 8.62×10^{11}

31. 9.4×10^{-5}

32. 2.78×10^{-6}

33. $437,200$

34. $7,910,000$

35. 0.00056294

36. 0.0063478

37. $\left(6.25 \times 10^7\right)\left(5.933 \times 10^{-2}\right)$

$(6.25 \cdot 5.933) \times 10^{7+-2}$

37.08125×10^5

Rewriting in scientific notation

3.708125×10^6

38. $\dfrac{2.961 \times 10^{-2}}{4.583 \times 10^{-4}}$

$\dfrac{2.961}{4.583} \times 10^{-2-(-4)}$

0.6460833515×10^2

Rewriting in scientific notation

6.460833515×10^1

39. Simple Interest

$I = Prt$

$I = (2000)(0.06)(5)$

$I = 600$ or \$600

Future Value

$S = P + I$

$S = 2000 + 600$

$S = 2600$ or \$2600

40. Simple Interest

$I = Prt$

$I = (8500)(0.075)(10)$

$I = 6375$ or \$6375

Future Value

$S = P + I$

$S = 8500 + 6375$

$S = 14,875$ or \$14,875

41. Simple Interest

$I = Prt$

$I = (3500)(0.12)\left(\dfrac{6}{12}\right)$

$I = 210$ or \$210

Future Value

$S = P + I$

$S = 3500 + 210$

$S = 3710$ or \$3710

42. Simple Interest

$I = Prt$

$I = (5600)(0.06)\left(\dfrac{6}{12}\right)$

$I = 168$ or \$168

Future Value

$S = P + I$

$S = 5600 + 168$

$S = 5768$ or \$5768

Section 3.1 Skills Check

1. Functions c) and e) represent exponential functions. They both fit the form $y = a^x$, where a is a constant greater than zero and $a \neq 1$.

3. a.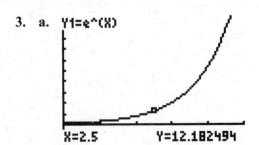

   [0, 5] by [–2, 100]

   b. $f(1) = e^1 = e \approx 2.718$

   $f(-1) = e^{-1} = \dfrac{1}{e} \approx 0.368$

   $f(4) = e^4 \approx 54.598$

   c. $y = 0$

   d. $(0, 1)$ since $f(0) = 1$.

5.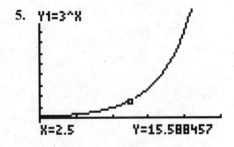

   [0, 5] by [–2, 100]

7.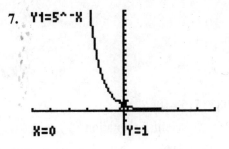

   [–5, 5] by [–5, 20]

9.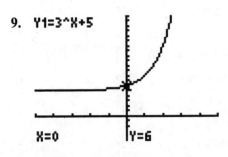

   [–5, 5] by [–5, 20]

11.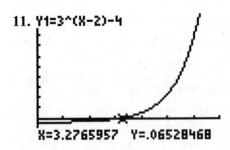

   [0, 7] by [–6, 100]

13. In comparison to 4^x, the graph has the same shape but shifted 2 units up.

   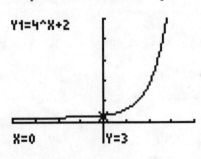

   [–4, 4] by [–10, 50]

15. In comparison to 4^x, the graph has the same shape but reflected with respect to the *x*-axis.

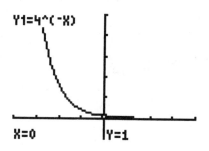

[–4, 4] by [–10, 50]

17. In comparison to 4^x, the graph is stretched vertically by a factor of 3.

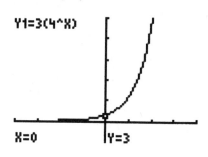

[–4, 4] by [–10, 50]

19. Both graphs are stretched vertically by a factor of 3 in comparison with 4^x. Therefore, the graph in Exercise 18 has the same shape as the graph in Exercise 17, but it is shifted 2 units right and 3 units down.

20. All the functions are increasing except for the functions in Exercises 15 and 16, which are decreasing.

21. a.

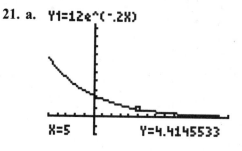

[–5, 15] by [–10, 60]

b. $f(10) = 12e^{-0.2(10)} = 12e^{-2} = \dfrac{12}{e^2} \approx 1.624$

$f(-10) = 12e^{-0.2(-10)} = 12e^2 \approx 88.669$

c. Since the function is decreasing, it represents decay.

Section 3.1 Exercises

23. a. Let $x = 0$ and solve for y.

$$y = 12{,}000\left(2^{-0.08 \cdot 0}\right)$$
$$= 12{,}000\left(2^{0}\right)$$
$$= 12{,}000(1)$$
$$= 12{,}000$$

At the end of the ad campaign, sales were $12,000 per week.

b. Let $x = 6$ and solve for y.

$$y = 12{,}000\left(2^{-0.08 \cdot 6}\right)$$
$$= 12{,}000\left(2^{-0.48}\right)$$
$$= 12{,}000(0.716977624)$$
$$\approx 8603.73$$

Six weeks after the ad campaign, sales were $8603.73 per week.

c. No. Sales approach a level of zero but never actually reach that level. Consider the graph of the model below.

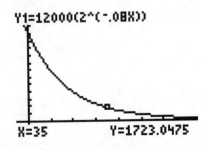

$[-5, 75]$ by $[-2000, 15{,}000]$

25. a.

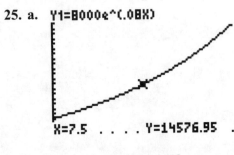

$[0, 15]$ by $[50, 30{,}000]$

b.

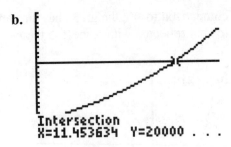

$[0, 15]$ by $[50, 30{,}000]$

The future value will be $20,000 in approximately 11.45 years.

27. a.

$$A(10) = 500e^{-0.02828(10)}$$
$$= 500e^{-0.2828}$$
$$= 500(0.7536705069)$$
$$\approx 376.84$$

Approximately 376.84 grams remain after 10 years.

b.

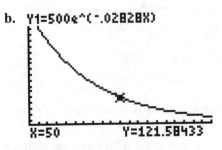

$[0, 100]$ by $[-50, 500]$

c.

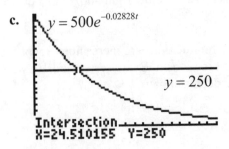

$[0, 100]$ by $[-50, 500]$

The half-life is approximately 24.5 years.

29. a.

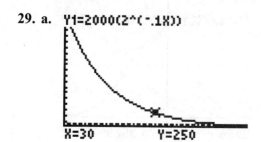

Y1=2000(2^(-.1X))

X=30   Y=250

$[0, 60]$ by $[-200, 2000]$

b.

X	Y1	Y2
6	1319.5	0
7	1231.1	0
8	1148.7	0
9	1071.8	0
10	1000	0
11	933.03	0
12	870.55	0

X=10

Y1=2000(2^(-.1X))

X=10   Y=1000

$[0, 60]$ by $[-200, 2000]$

Ten weeks after the campaign ended, the weekly sales were $1000.

c. Weekly sales drop by half, from $2000 to $1000, ten weeks after the end of the ad campaign. It is important for this company to advertise.

31. a. $P = 40,000(0.95^{20})$

$= 40,000(0.3584859224)$

$= 14,339.4369$

$\approx 14,339.44$

The purchasing power will be $14,339.44.

b. A person with $40,000 in retirement income at age 50 would be receiving the

equivalent of approximately $14,339 twenty years later at age 70. Note that answers to part *b* could vary.

33. a. $y = 100,000e^{0.05(4)}$

$= 100,000e^{0.2}$

$= 100,000(1.221402758)$

$= 122,140.2758$

$\approx 122,140.28$

The value of property after 4 years will be $122,140.28.

b.

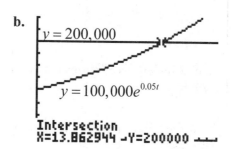

$y = 200,000$

$y = 100,000e^{0.05t}$

Intersection
X=13.862944 Y=200000

$[0, 20]$ by $[-5000, 250,000]$

The value of the property doubles in 13.86 years or approximately 14 years.

35. a. Increasing. The exponent is positive for all values of $t \geq 0$.

b. $P(5) = 53,000e^{0.015(5)}$

$= 53,000e^{0.075}$

$= 53,000(1.077884151)$

$\approx 57,128$

The population is 57,128 in 2005.

c. $P(10) = 53,000e^{0.015(10)}$

$= 53,000e^{0.15}$

$= 53,000(1.161834243)$

$\approx 61,577$

The populatin is 61,577 in 2010.

d. $\dfrac{y_2 - y_1}{x_2 - x_1} = \dfrac{61{,}577 - 53{,}000}{10 - 0}$

$\phantom{\dfrac{y_2 - y_1}{x_2 - x_1}} = \dfrac{8577}{10}$

$\phantom{\dfrac{y_2 - y_1}{x_2 - x_1}} = 857.7$

The average rate of growth in population between 2000 and 2010 is 857.7 people per year.

37. a. $y = 100e^{-0.00012378(1000)}$

$ = 100e^{-0.12378}$

$ = 100(0.8835742058)$

$ \approx 88.36$

Appriximately 88.36 atoms remain after 1000 years.

b.

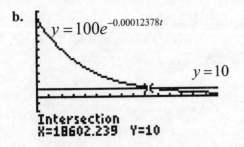

[0, 30,000] by [−50, 110]

After approximately 18,602 years, 10 grams of Carbon-14 remains.

39. a.

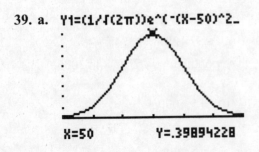

[47, 53] by [0, 0.5]

b.

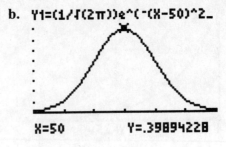

[47, 53] by [0, 0.5]

The average score is 50.

Section 3.2 Skills Check

1. $y = \log_3 x \Leftrightarrow 3^y = x$

3. $y = \ln(2x) = \log_e(2x) \Leftrightarrow e^y = 2x$

5. $x = 4^y \Leftrightarrow \log_4 x = y$

7. $32 = 2^5 \Leftrightarrow \log_2 32 = 5$

9. 0.845

11. 4.454

13. 4.806

15. $y = \log_2 32 \Leftrightarrow 2^y = 32$
Therefore, $y = 5$.

17. $y = \log_3 27 \Leftrightarrow 3^y = 27$
Therefore, $y = 3$.

19. $y = \log_5 625 \Leftrightarrow 5^y = 625$
$5^y = 5^4$
Therefore, $y = 4$.

21. $y = \log_9 27 \Leftrightarrow 9^y = 27$
$3^{2y} = 3^3$
$2y = 3$
$y = \dfrac{3}{2}$

23. $y = \ln(e^3) = \log_e(e^3) \Leftrightarrow e^y = e^3$
Therefore, $y = 3$.

25. $y = \log_3\left(\dfrac{1}{27}\right) \Leftrightarrow 3^y = \dfrac{1}{27}$
$3^y = \dfrac{1}{3^3}$
$3^y = 3^{-3}$
Therefore, $y = -3$.

27. $y = \ln(e) = \log_e(e) \Leftrightarrow e^y = e$
$e^y = e^1$
Therefore, $y = 1$.

29. $y = \log_3 x = \dfrac{\ln x}{\ln 3}$

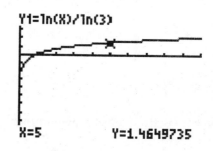

[0, 10] by [−10, 5]

31.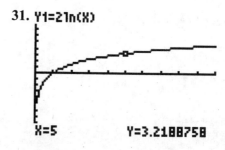

[0, 10] by [−10, 10]

33. Y1=log(X+1)+2

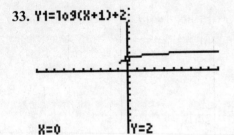

X=0   Y=2

[0, 10] by [–10, 10]

35. a. $y = 4^x$

$x = 4^y \Leftrightarrow \log_4 x = y$

Therefore, the inverse function is

$y = \log_4 x.$

b.

$y = 4^x$

$y = \log_4 x$

[–5, 5] by [–5, 5]

The graphs are symmetric with respect to the $y = x$ line.

37. $\log_a a = x \Leftrightarrow a^x = a$

If $a > 0$ and $a \neq 1$, then $x = 1.$

39. $\log_2 x = 3 \Leftrightarrow 2^3 = x$

$x = 8$

41. $5 + 2\ln x = 8$

$2\ln x = 3$

$\ln x = \dfrac{3}{2}$

$\log_e x = \dfrac{3}{2} \Leftrightarrow e^{\frac{3}{2}} = x$

$x = e^{\frac{3}{2}} \approx 4.4817$

Section 3.2 Exercises

43. a. In 1925, $x = 1925 - 1900 = 25$.

$f(25) = 12.734\ln(25) + 17.875$

$f(25) = 58.8642$

In 1925, the expected life span is 59 years.

In 1996, $x = 1996 - 1900 = 96$.

$f(96) = 12.734\ln(96) + 17.875$

$f(96) = 75.9974$

In 1996, the expected life span is 76 years.

b. Based on the model, life span increased tremendously between 1925 and 1996. The increase could be due to multiple factors, including improved healthcare and nutrition.

45. a. In 1990, $x = 1990 - 1980 = 10$.

$f(10) = 282.1666 + 2771.0125\ln(10)$

$f(10) = 6662.66$

In 1990, the official single poverty level is approximately $6662.

In 1999, $x = 1999 - 1980 = 19$.

$f(19) = 282.1666 + 2771.0125\ln(19)$

$f(19) = 8441.24$

In 1999, the official single poverty level is approximately $8441.

b. Based on the solutions to part a), the function seems to be increasing.

c. Over time, inflation causes the minimum poverty level to rise.

47. $\dfrac{\ln 2}{0.10} \approx 6.9$ years

49. $n = \dfrac{\log 2}{0.0086}$

$n = 35.0035 \approx 35$

Since it takes approximately 35 quarters for an investment to double under this scenario, then in terms of years the time to double is

approximately $\dfrac{35}{4} = 8.75$ years.

51.

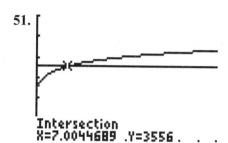

```
Intersection
X=7.0044689 .Y=3556. . .
```

[0, 40] by [3200, 3800]

When the cost is $3556, approximately 7 units are produced.

53. a.

$$f(x) = 12.734\ln(x) + 17.875$$
$$y = 12.734\ln(x) + 17.875$$
$$y - 17.875 = 12.734\ln(x)$$
$$\frac{y - 17.875}{12.734} = \frac{12.734\ln(x)}{12.734}$$
$$\ln(x) = \frac{y - 17.875}{12.734}$$

b. $\ln(x) = \dfrac{y - 17.875}{12.734}$

$\log_e x = \dfrac{y - 17.875}{12.734} \Leftrightarrow e^{\frac{y - 17.875}{12.734}} = x$

c. Let $y = 75$ and solve for x.

$x = e^{\frac{y - 17.875}{12.734}}$

$x = e^{\frac{(75) - 17.875}{12.734}}$

$x = e^{4.486021674}$

$x = 88.76759607 \approx 89$

In approximately 1989, the expected life span will be 75 years.

d.

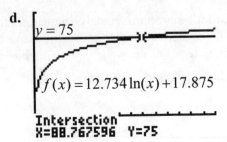

$y = 75$

$f(x) = 12.734 \ln(x) + 17.875$

Intersection
X=88.767596   Y=75

[0, 150] by [−20, 100]

As expected, the answers to parts c) and d) are the same.

55. $R = \log\left(\dfrac{I}{I_0}\right)$

$R = \log\left(\dfrac{25,000 I_0}{I_0}\right)$

$R = \log(25,000) = 4.3979 \approx 4.4$

The earthquake measures 4.4 on the Richter scale.

57. a. $R = \log\left(\dfrac{I}{I_0}\right) = \log_{10}\left(\dfrac{I}{I_0}\right)$

$R = \log_{10}\left(\dfrac{I}{I_0}\right) \Leftrightarrow 10^R = \dfrac{I}{I_0}$

b. $10^R = \dfrac{I}{I_0}$

$I = 10^R I_0$

$I = 10^{7.1} I_0$

$I = 12,589,254 I_0$

59. The difference in the Richter scale measurements is $8.25 - 7.1 = 1.15$. Therefore, the intensity of the 1906 earthquake was $10^{1.15} \approx 14.13$ times stronger than the intensity of the 1989 earthquake.

61. $L = 10 \log\left(\dfrac{I}{I_0}\right)$

$L = 10 \log\left(\dfrac{20,000 I_0}{I_0}\right)$

$L = 10 \log(20,000) \approx 43$

The decibel level is approximately 43.

63. $L = 10 \log\left(\dfrac{I}{I_0}\right)$

$40 = 10 \log_{10}\left(\dfrac{I}{I_0}\right)$

$\log_{10}\left(\dfrac{I}{I_0}\right) = 4 \Leftrightarrow 10^4 = \dfrac{I}{I_0}$

$\dfrac{I}{I_0} = 10^4$

$I = 10^4 I_0 = 10,000 I_0$

65. $L_1 = 10 \log\left(\dfrac{115 I_0}{I_0}\right)$

$= 10 \log(115)$

≈ 20.6

$L_2 = 10 \log\left(\dfrac{9,500,000 I_0}{I_0}\right)$

$= 10 \log(9,500,000)$

≈ 69.8

The decibel level on a busy street is approximately 49 more than the decibel level of a whisper.

67. $pH = -\log\left[H^+\right]$

$= -\log(0.0000631)$

$= 4.2$

69. $pH = -\log\left[H^+\right]$

$-pH = \log_{10}\left[H^+\right] \Leftrightarrow 10^{-pH} = H^+$

$H^+ = 10^{-pH}$

If $pH = 1, H^+ = 10^{-1} = \dfrac{1}{10} = 0.1$

If $pH = 14, H^+ = 10^{-14} = \dfrac{1}{10^{14}}$

$\dfrac{1}{10^{14}} \le H^+ \le \dfrac{1}{10}$

Section 3.3 Skills Check

1.
$$1600 = 10^x$$
$$\log(1600) = \log(10^x)$$
$$x = \log(1600) \approx 3.204$$

3.
$$2500 = e^x$$
$$\ln(2500) = \ln(e^x)$$
$$x = \ln(2500) \approx 7.824$$

5.
$$8900 = e^{5x}$$
$$\ln(8900) = \ln(e^{5x})$$
$$5x = \ln(8900)$$
$$x = \frac{\ln(8900)}{5} \approx 1.819$$

7.
$$4000 = 200e^{8x}$$
$$20 = e^{8x}$$
$$\ln(20) = \ln(e^{8x})$$
$$8x = \ln(20)$$
$$x = \frac{\ln(20)}{8} \approx 0.374$$

9.
$$8000 = 500(10^x)$$
$$16 = 10^x$$
$$\log(16) = \log(10^x)$$
$$x = \log(16) \approx 1.204$$

11.
$$\log 10^{14} = \log_{10} 10^{14}$$
$$= 14 \log_{10} 10$$
$$= 14(1) = 14$$

13. $10^{\log_{10} 12} = 12$

15.
$$\log_a(100) = \log_a(20 \cdot 5)$$
$$= \log_a(20) + \log_a(5)$$
$$= 1.4406 + 0.7740$$
$$= 2.2146$$

17.
$$\log_a 5^3 = 3\log_a 5$$
$$= 3(0.7740)$$
$$= 2.322$$

19. $\ln\left(\dfrac{3x-2}{x+1}\right) = \ln(3x-2) - \ln(x+1)$

21. $\log_3 \dfrac{\sqrt[4]{4x+1}}{4x^2}$
$$= \log_3\left(\sqrt[4]{4x+1}\right) - \log_3\left(4x^2\right)$$
$$= \log_3\left[(4x+1)^{\frac{1}{4}}\right] - \left[\log_3(4) + \log_3(x^2)\right]$$
$$= \frac{1}{4}\log_3(4x+1) - \left[\log_3(4) + 2\log_3(x)\right]$$
$$= \frac{1}{4}\log_3(4x+1) - \log_3(4) - 2\log_3(x)$$

23. $3\log_2 x + \log_2 y$
$$= \log_2 x^3 + \log_2 y$$
$$= \log_2\left(x^3 y\right)$$

25. $4\ln(2a) - \ln(b)$
$$= \ln(2a)^4 - \ln(b)$$
$$= \ln\left(\frac{16a^4}{b}\right)$$

27. $\log_6(18) = \dfrac{\ln(18)}{\ln(6)} = 1.6131$

29. $\log_8\left(\sqrt{2}\right) = \log_8(2)^{\frac{1}{2}}$

$\qquad = \frac{1}{2}\log_8(2)$

$\qquad = \left(\frac{1}{2}\right)\left(\frac{\ln(2)}{\ln(8)}\right)$

$\qquad = 0.1667$

31. $\qquad 8^x = 1024$

$\ln\left(8^x\right) = \ln(1024)$

$x\ln(8) = \ln(1024)$

$\qquad x = \frac{\ln(1024)}{\ln(8)}$

$\qquad x = 3.\overline{3} = \frac{10}{3}$

33. $\quad 2\left(5^{3x}\right) = 31,250$

$\qquad 5^{3x} = 15,625$

$\ln\left(5^{3x}\right) = \ln(15,625)$

$3x\ln(5) = \ln(15,625)$

$\qquad x = \frac{\ln(15,625)}{3\ln(5)}$

$\qquad x = 2$

35. $\qquad 5^{x-2} = 11.18$

$\ln\left(5^{x-2}\right) = \ln(11.18)$

$(x-2)\ln(5) = \ln(11.18)$

$\qquad x - 2 = \frac{\ln(11.18)}{\ln(5)}$

$\qquad x = \frac{\ln(11.18)}{\ln(5)} + 2$

$\qquad x \approx 3.5$

37. $\qquad 18,000 = 30\left(2^{12x}\right)$

$\qquad 600 = 2^{12x}$

$\log(600) = \log\left(2^{12x}\right)$

$12x\log(2) = \log(600)$

$\qquad x = \frac{\log(600)}{12\log(2)}$

$\qquad x \approx 0.769$

39. $\qquad 5 + \ln(8x) = 23 - 2\ln(x)$

$\ln(8x) + 2\ln(x) = 23 - 5$

$\ln(8x) + \ln(x)^2 = 18$

$\ln\left(8x \cdot x^2\right) = 18$

$\qquad e^{\ln\left(8x^3\right)} = e^{18}$

$\qquad 8x^3 = e^{18}$

$\qquad x = \sqrt[3]{\frac{e^{18}}{8}}$

$\qquad x = \frac{e^6}{2}$

$\qquad x \approx 201.7$

41. $2\log(x) - 2 = \log(x-25)$

$\log\left(x^2\right) - \log(x-25) = 2$

$\log\left(\frac{x^2}{x-25}\right) = 2$

$10^{\log\left(\frac{x^2}{x-25}\right)} = 10^2$

$\frac{x^2}{x-25} = 100$

$x^2 = 100(x-25)$

$x^2 = 100x - 2500$

$x^2 - 100x + 2500 = 0$

$(x-50)(x-50) = 0$

$x = 50$

43. $3^x < 243$

$\ln\left(3^x\right) < \ln\left(243\right)$

$x\ln\left(3\right) < \ln\left(243\right)$

$x < \dfrac{\ln\left(243\right)}{\ln\left(3\right)}$

$x < 5$

45. $5\left(2^x\right) \ge 2560$

$\ln\left(2^x\right) \ge \ln\left(\dfrac{2560}{5}\right)$

$x\ln\left(2\right) \ge \ln\left(512\right)$

$x \ge \dfrac{\ln\left(512\right)}{\ln\left(2\right)}$

$x \ge 9$

Section 3.3 Exercises

47. $10,880 = 340\left(2^q\right)$

$2^q = \dfrac{10,880}{340}$

$2^q = 32$

$\ln\left(2^q\right) = \ln\left(32\right)$

$q\ln(2) = \ln(32)$

$q = \dfrac{\ln(32)}{\ln(2)}$

$q = 5$

When the price is \$10,880, the quantity supplied is 5.

49. a. $s = 25,000e^{-0.072x}$

$\dfrac{s}{25,000} = e^{-0.072x}$

$\Leftrightarrow \log_e\left(\dfrac{s}{25,000}\right) = -0.072x$

$\ln\left(\dfrac{s}{25,000}\right) = -0.072x$

b. $\ln\left(\dfrac{16,230}{25,000}\right) = -0.072x$

$x = \dfrac{\ln\left(\dfrac{16,230}{25,000}\right)}{-0.072}$

$x = 6$

Six weeks after the completion of the campaign, the sales fell to \$16,230.

51. a. $S = 3200e^{-0.08(0)} = 3200e^0 = 3200$

At the end of the ad campaign, daily sales were \$3200.

b.
$$S = 3200e^{-0.08x}$$
$$1600 = 3200e^{-0.08x}$$
$$\frac{1}{2} = e^{-0.08x}$$
$$\ln\left(\frac{1}{2}\right) = \ln\left(e^{-0.08x}\right)$$
$$-0.08x = \ln\left(\frac{1}{2}\right)$$
$$x = \frac{\ln\left(\frac{1}{2}\right)}{-0.08} = 8.664$$

Approximately 9 days after the completion of the ad campaign, sales dropped below half their level the day the campaign ended.

53. a.
$$B(10) = 1.337e^{0.718(10)}$$
$$= 1.337e^{7.18}$$
$$= 1755.358341$$
$$\approx 1755.36$$

Based on the model, in 1995, Snapple's revenue was approximately $1755.36 million.

b. Applying the intersection of graphs method

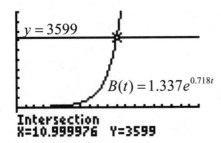

$y = 3599$

$B(t) = 1.337e^{0.718t}$

Intersection
X=10.999976 Y=3599

[0, 20] by [−1500, 5000]

Revenues would reach $3599 in 1996.

55.
$$20,000 = 40,000e^{-0.05t}$$
$$e^{-0.05t} = 0.5$$
$$\ln\left(e^{-0.05t}\right) = \ln\left(0.5\right)$$
$$-0.05t = \ln\left(0.5\right)$$
$$t = \frac{\ln\left(0.5\right)}{-0.05}$$
$$t = 13.86294361$$

It will take approximately 13.86 years for the $40,000 pension to decrease to $20,000 in purchasing power.

57.
$$200,000 = 100,000e^{0.03t}$$
$$2 = e^{0.03t}$$
$$\ln\left(2\right) = \ln\left(e^{0.03t}\right)$$
$$\ln\left(2\right) = 0.03t$$
$$t = \frac{\ln\left(2\right)}{0.03}$$
$$t = 23.1049$$

It will take approximately 23.1 years for the value of the property to double.

59. $S = Pe^{rt}$
Note that the initial investment is P and that double the initial investment is $2P$.
$$2P = Pe^{rt}$$
$$2 = e^{rt}$$
$$\ln\left(2\right) = \ln\left(e^{rt}\right)$$
$$rt = \ln\left(2\right)$$
$$t = \frac{\ln\left(2\right)}{r}$$

The time to double is $\ln\left(2\right)$ divided by the interest rate.

61. a. $n = \log_{1.02} 2$

$n = \dfrac{\log 2}{\log 1.02}$

$n = 35.0027 \approx 35$

b. Since it takes approximately 35 quarters for an investment to double under this scenario, then in terms of years the time to double is $\dfrac{35}{4} = 8.75$ years.

63. $t = \log_{1.05} 2$

$t = \dfrac{\log 2}{\log 1.05}$

$t = 14.20669908$

$t \approx 14.2$

The future value will be $40,000 in approximately 14.2 years.

65. $40{,}000 = 10{,}000\left(1 + \dfrac{0.08}{12}\right)^{12t}$

$4 = \left(1 + \dfrac{0.08}{12}\right)^{12t}$

$\ln(4) = \ln\left[\left(1 + \dfrac{0.08}{12}\right)^{12t}\right]$

$12t \ln\left(1.00\overline{6}\right) = \ln(4)$

$t = \dfrac{\ln(4)}{12\ln\left(1.00\overline{6}\right)}$

$t \approx 17.3864$

It will take approximately 17.4 years for the initial investment of $10,000 to grow to $40,000.

67. a. $A(0) = 500e^{-0.02828(0)} = 500e^0 = 500$
The initial quantity is 500 grams.

b. $250 = 500e^{-0.02828t}$

$0.5 = e^{-0.02828t}$

$\ln(0.5) = \ln\left(e^{-0.02828t}\right)$

$-0.02828t = \ln(0.5)$

$t = \dfrac{\ln(0.5)}{-0.02828} = 24.51$

The half-life, the time it takes the initial quantity to become half, is approximately 24.5 years.

69. Applying the intersection of graphs method:

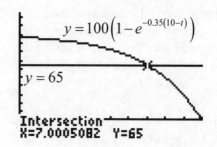

[0, 10] by [−20, 125]

After approximately 7 hours, the percent of the maximum dosage present is 65%.

71. $50 = 100e^{-0.00002876t}$

$0.5 = e^{-0.00002876t}$

$\ln(0.5) = \ln\left(e^{-0.00002876t}\right)$

$-0.00002876t = \ln(0.5)$

$t = \dfrac{\ln(0.5)}{-0.00002876}$

$t \approx 24{,}101$

The half-life is approximately 24,101 years.

73. Applying the intersection of graphs method:

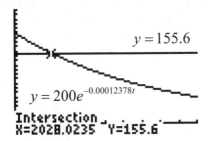

$[0, 6000]$ by $[0, 400]$

After approximately 2028 years 155.6 grams of carbon-14 remains.

75. a.
$$S = 600e^{-0.05x}$$

$$\frac{S}{600} = e^{-0.05x}$$

$$\ln\left(\frac{S}{600}\right) = \ln\left(e^{-0.05x}\right)$$

$$-0.05x = \ln\left(\frac{S}{600}\right)$$

Let $S = 269.60$

$$-0.05x = \ln\left(\frac{269.60}{600}\right)$$

$$x = \frac{\ln\left(\dfrac{269.60}{600}\right)}{-0.05}$$

$$x \approx 16$$

Sixteen weeks after the end of the campaign, sales dropped below $269.60 thousand.

b.

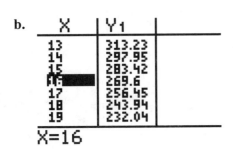

77. Applying the intersection of graphs method:

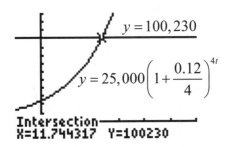

$[-5, 30]$ by $[-20{,}000, 130{,}000]$

After 12 years, the future value of the investment will be greater than $100,230.

79.
$$S = P(1.10)^n$$

$$P(1.10)^n = S$$

$$1.10^n = \frac{S}{P}$$

$$\log\left(1.10^n\right) = \log\left(\frac{S}{P}\right)$$

$$n\log(1.10) = \log\left(\frac{S}{P}\right)$$

$$n = \frac{\log\left(\dfrac{S}{P}\right)}{\log(1.10)}$$

Let $S = 2P$, since the investment doubles.

$$n = \frac{\log\left(\dfrac{2P}{P}\right)}{\log(1.10)}$$

$$n = \frac{\log(2)}{\log(1.10)}$$

$$n = \log_{1.10} 2$$

Section 3.4 Skills Check

1.

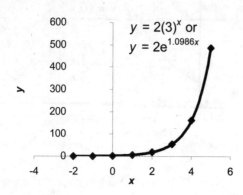

$y = 2(3)^x$ or
$y = 2e^{1.0986x}$

3.

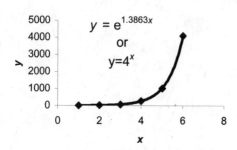

$y = e^{1.3863x}$

or

$y=4^x$

5. a.

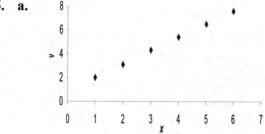

b.

x	y	First Differences	Percent Change
1	2		
2	3.1	1.1	55.00%
3	4.3	1.2	38.71%
4	5.4	1.1	25.58%
5	6.5	1.1	20.37%
6	7.6	1.1	16.92%

Considering the scatter plot from part a) and the chart above, a linear model fits the data best. The first differences are very close to being constant.

7.

x	y	First Differences	Percent Change
1	2		
2	6	4	200.00%
3	14	8	133.33%
4	34	20	142.86%
5	81	47	138.24%

Since the percent change is approximately constant and the first differences vary, an exponential function will fit the data best.

9. Using technology yields, $y = 0.876e^{0.914x}$ or $y = 0.876(2.494)^x$.

11. a.

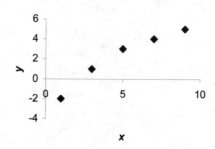

b. Based on the scatter plot, it appears that a logarithmic model fits the data best.

Section 3.4 Exercises

13. a. $y = a(1+r)^x$

$y = 30,000(1+0.04)^x$

$y = 30,000(1.04)^x$

b. $y = 30,000(1.04)^t$

$y = 30,000(1.04)^{15} \approx 54,028.31$

In 2010, the retail price of the automobile is predicted to be $54,028.31.

15. a. $y = a(1+r)^x$

$y = 20,000(1-0.02)^x$

$y = 20,000(0.98)^x$

b. $y = 20,000(0.98)^t$

$= 20,000(0.98)^5$

$= 18,078.42$

In five weeks the sales are predicted to decline to $18,078.42.

17. a.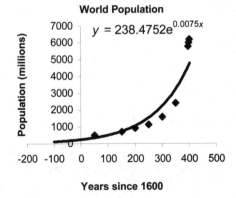

The model is

$y = 238.4752e^{0.0075x}$ or

$y = 238.4752(1.0075)^x$.

b. See part a).

19. a.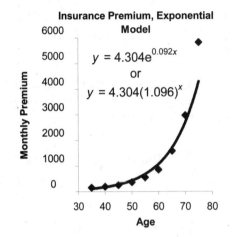

b.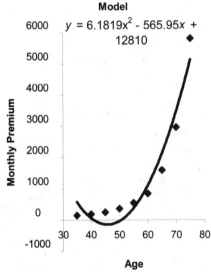

c. Considering parts a) and b), the exponential model is the best fit.

21. a.

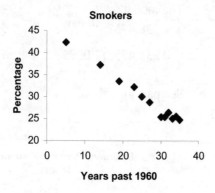

b. Yes, an exponential function could be used. A linear function would also fit the data well.

c.

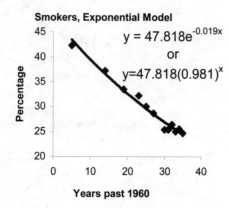

d.

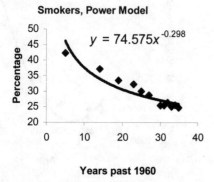

e. The exponential model appears to be best.

23. a.

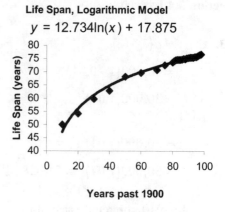

b.

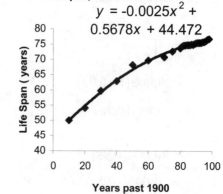

c. Based on the graphs in parts a) and b), it appears that the quadratic function is the better fit.

d. In 2010, $x = 110$.
Using the logarithmic function,
$y = 12.734 \ln(110) + 17.875$.

Substituting into the unrounded model yields $y \approx 76.73$.

```
110→X
              110
17.874767463979+
12.7344139134211
n(X)
      77.73263003
```

Using the quadratic function,

$$y = -0.0025(110)^2$$
$$+0.5678(110) + 44.472.$$

Substituting into the unrounded model yields $y \approx 76.77$.

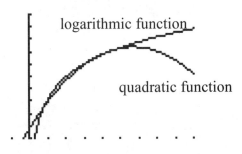

The logarithmic function produces a better prediction. For years beyond 2010, the logarithmic function continues to produce better predictions. The logarithmic function increases while the quadratic function begins to decrease.

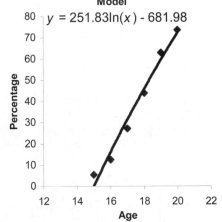

[–20, 200] by [40, 90]

25. a.

Sexually Active Girls, Logarithmic Model

$y = 251.83\ln(x) - 681.98$

b. $y = 251.83\ln(x) - 681.98$

$y = 251.83\ln(17) - 681.98$

Substituting into the unrounded model yields $y \approx 31.5$.

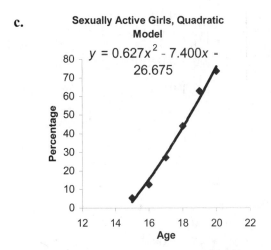

The percentage of girls 17 or younger who have been sexually active is 31.5%.

c.

Sexually Active Girls, Quadratic Model

$y = 0.627x^2 - 7.400x - 26.675$

d. Based on the graphs in parts a) and c), the quadratic function seems to be the better fit.

Section 3.5 Skills Check

1. $15,000e^{0.06(20)}$
$= 15,000e^{1.2}$
$= 15,000(3.320116923)$
$= 49,801.75$

3. $3000(1.06)^{30}$
$= 3000(5.743491173)$
$= 17,230.47$

5. $12,000\left(1+\dfrac{0.10}{4}\right)^{(4)(8)}$
$= 12,000(1+.025)^{32}$
$= 12,000(1.025)^{32}$
$= 12,000(2.203756938)$
$= 26,445.08$

7. $P\left(1+\dfrac{r}{k}\right)^{kn}$
$= 3000\left(1+\dfrac{0.08}{2}\right)^{(2)(18)}$
$= 3000(1.04)^{36}$
$= 12,311.80$

9. $300\left[\dfrac{1.02^{240}-1}{0.02}\right]$
$= 300\left[\dfrac{115.8887352-1}{0.02}\right]$
$= 300\left[\dfrac{114.8887352}{0.02}\right]$
$= 300[5744.436758]$
$= 1,723,331.03$

11. $g(2.5)=1123.60$
$g(3)=1191.00$
$g(3.5)=1191.00$

13. $S = P\left(1+\dfrac{r}{k}\right)^{kn}$
$P\left(1+\dfrac{r}{k}\right)^{kn} = s$
$P = \dfrac{S}{\left(1+\dfrac{r}{k}\right)^{kn}}$
$P = S\left(1+\dfrac{r}{k}\right)^{-kn}$

Section 3.5 Exercises

15. a. $S = P\left(1 + \dfrac{r}{k}\right)^{kt}$

$P = 8800,\ r = 0.08,\ k = 1,\ t = 8$

$S = 8800\left(1 + \dfrac{0.08}{1}\right)^{(1)(8)}$

$S = 8800(1.08)^8$

$S = 16,288.19$

The future value is $16,288.19.

b. $S = P\left(1 + \dfrac{r}{k}\right)^{kt}$

$P = 8800,\ r = 0.08,\ k = 1,\ t = 30$

$S = 8800\left(1 + \dfrac{0.08}{1}\right)^{(1)(30)}$

$S = 8800(1.08)^{30}$

$S = 88,551.38$

The future value is $88,551.38.

17. a. Y1=3300(1.10)^X

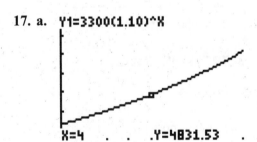

X=4 . . .Y=4831.53 .

[0, 8] by [2500, 9000]

b.

S = 6600

$S = 3300(1.10)^x$

Intersection
X=7.2725409 .Y=6600 . .

[0, 8] by [2500, 9000]

The initial investment doubles in approximately 7.3 years. After 8 years compounded annually, the initial investment will be more than doubled.

19. $S = P\left(1 + \dfrac{r}{k}\right)^{kt}$

$P = 10,000,\ r = 0.12,\ k = 4,\ t = 10$

$S = 10,000\left(1 + \dfrac{0.12}{4}\right)^{(4)(10)}$

$S = 10,000(1.03)^{40}$

$S = 32,620.38$

The future value is $32,620.38.

21. a. $S = P\left(1 + \dfrac{r}{k}\right)^{kt}$

$P = 10,000,\ r = 0.12,\ k = 365,\ t = 10$

$S = 10,000\left(1 + \dfrac{0.12}{365}\right)^{(365)(10)}$

$S = 10,000(1.0003287671233)^{3650}$

$S = 33,194.62$

The future value is $33,194.62.

b. Since the compounding occurs more often in Exercise 21 than in Exercise 19, the future value in Exercise 21 is greater.

23. $S = P\left(1 + \dfrac{r}{k}\right)^{kt}$

$P = 10,000,\ r = 0.12,\ k = 12,\ t = 15$

$S = 10,000\left(1 + \dfrac{0.12}{12}\right)^{(12)(15)}$

$S = 10,000(1.01)^{180}$

$S = 59,958.02$

The future value is $59,958.02. The interest earned is the future value minus the present value. In this case, $59,958.02 - 10,000 = \$49,958.02$.

25. **a.** $S = Pe^{rt}$

   $P = 10,000, r = 0.06, t = 12$

   $S = 10,000e^{(0.06)(12)}$

   $S = 10,000e^{0.72}$

   $S = 20,544.33$

   The future value is $20,544.33.

   b. $S = Pe^{rt}$

   $P = 10,000, r = 0.06, t = 18$

   $S = 10,000e^{(0.06)(18)}$

   $S = 10,000e^{1.08}$

   $S = 29,446.80$

   The future value is $29,446.80.

27. **a.** $S = P\left(1 + \dfrac{r}{k}\right)^{kt}$

   $P = 10,000, r = 0.06, k = 1, t = 18$

   $S = 10,000\left(1 + \dfrac{0.06}{1}\right)^{(1)(18)}$

   $S = 10,000(1.06)^{18}$

   $S \approx 28,543.39$

   The future value is $28,543.39.

   b.   Continuous compounding yields a higher future value, $29,446.80 - 28,543.39 = 903.41$ additional dollars.

29. **a.** $S = P\left(1 + \dfrac{r}{k}\right)^{kt}$

   Doubling the investment implies

   $S = 2P.$

   $2P = P\left(1 + \dfrac{r}{k}\right)^{kt}$

   $\dfrac{2P}{P} = \dfrac{P\left(1 + \dfrac{r}{k}\right)^{kt}}{P}$

   $2 = \left(1 + \dfrac{r}{k}\right)^{kt}$

   $k = 1, r = 0.10$

   $2 = \left(1 + \dfrac{0.10}{1}\right)^{(1)t}$

   $2 = (1.10)^{t}$

   Applying the intersection of graphs method:

   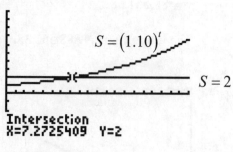

   Intersection
   X=7.2725409 Y=2

   [0, 20] by [−5, 10]

   The time to double is approximately 7.3 years. In terms of discrete units, the time to double is 8 years.

   b. $S = Pe^{rt}$

   Doubling the investment implies

   $S = 2P.$

   $2P = Pe^{rt}$

   $\dfrac{2P}{P} = \dfrac{Pe^{rt}}{P}$

   $2 = e^{rt}$

   $r = 0.10$

   $2 = e^{0.10t}$

Applying the intersection of graphs method:

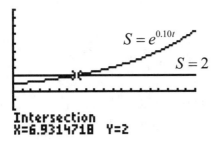

[0, 20] by [–5, 10]

The time to double is approximately 6.9 years. In terms of discrete units, the time to double is 7 years.

31. a. $S = P\left(1 + \dfrac{r}{k}\right)^{kt}$

$P = 2000,\ r = 0.05,\ k = 1,\ t = 8$

$S = 2000\left(1 + \dfrac{0.05}{1}\right)^{(1)(8)}$

$S = 2000(1.05)^8$

$S = 2954.91$

The future value is $2954.91.

b. $S = P\left(1 + \dfrac{r}{k}\right)^{kt}$

$P = 2000,\ r = 0.05,\ k = 1,\ t = 18$

$S = 2000\left(1 + \dfrac{0.05}{1}\right)^{(1)(18)}$

$S = 2000(1.05)^{18}$

$S = 4813.24$

The future value is $4813.24.

33. $S = P\left(1 + \dfrac{r}{k}\right)^{kt}$

$P = 3000,\ r = 0.06,\ k = 12,\ t = 112$

$S = 3000\left(1 + \dfrac{0.06}{12}\right)^{(12)(12)}$

$S = 3000(1.005)^{144}$

$S = 6152.25$

The future value is $6152.25.

35. a.

Years	Future Value
0	1000
7	2000
14	4000
21	8000
28	16,000

b. $S = 1000\left(1 + \dfrac{0.10}{4}\right)^{4t}$

$S = 1000(1.025)^{4t}$

$S = 1000\left((1.025)^4\right)^{t}$

$S = 1000(1.104)^{t}$

c. After five years, the investment is worth
$S = 1000(1.104)^5 = \$1640.01$.
After 10.5 years, the investment is worth
$S = 1000(1.104)^{10.5} = \$2826.02$.

Section 3.6 Skills Check

1.
$$S = P(1+i)^n$$

$$\frac{S}{(1+i)^n} = \frac{P(1+i)^n}{(1+i)^n}$$

$$P = \frac{S}{(1+i)^n}$$

3.
$$A \cdot i = R\left[1-(1+i)^{-n}\right]$$

$$\frac{A \cdot i}{i} = \frac{R\left[1-(1+i)^{-n}\right]}{i}$$

$$A = R\left[\frac{1-(1+i)^{-n}}{i}\right]$$

5.
$$A = R\left[\frac{1-(1+i)^{-n}}{i}\right]$$

$$i \cdot A = i\left(R\left[\frac{1-(1+i)^{-n}}{i}\right]\right)$$

$$iA = R\left[1-(1+i)^{-n}\right]$$

$$\frac{iA}{\left[1-(1+i)^{-n}\right]} = \frac{R\left[1-(1+i)^{-n}\right]}{\left[1-(1+i)^{-n}\right]}$$

$$R = \frac{iA}{\left[1-(1+i)^{-n}\right]} = A\left[\frac{i}{1-(1+i)^{-n}}\right]$$

Section 3.6 Exercises

7.
$$S = P\left(1+\frac{r}{k}\right)^{kt}$$

$$10,000 = P\left(1+\frac{0.06}{1}\right)^{(1)(10)}$$

$$10,000 = P(1.06)^{10}$$

$$P = \frac{10,000}{(1.06)^{10}}$$

$$P = 5583.94$$

An initial amount of \$5583.94 will grow to \$10,000 in 10 years if invested at 6% compounded annually.

9.
$$S = P\left(1+\frac{r}{k}\right)^{kt}$$

$$30,000 = P\left(1+\frac{0.10}{12}\right)^{(12)(18)}$$

$$30,000 = P(1.00\overline{83})^{216}$$

$$P = \frac{30,000}{(1.00\overline{83})^{216}}$$

$$P = 4996.09$$

An initial amount of \$4996.09 will grow to \$30,000 in 18 years if invested at 10% compounded monthly.

11. $A = R\left[\dfrac{1-(1+i)^{-n}}{i}\right]$

$A = 1000\left[\dfrac{1-(1+0.07)^{-10}}{0.07}\right]$

$A = 1000\left[\dfrac{1-(1.07)^{-10}}{0.07}\right]$

$A = 1000\left[\dfrac{1-(0.5083492921)}{0.07}\right]$

$A = 1000[7.023581541]$

$A = 7023.58$

Investing $7023.58 initially will produce an income of $1000 per year for 10 years if the interest rate is 7% compounded annually.

13. $A = R\left[\dfrac{1-(1+i)^{-n}}{i}\right]$

$A = 50,000\left[\dfrac{1-(1+0.08)^{-19}}{0.08}\right]$

$A = 50,000\left[\dfrac{1-(1.08)^{-19}}{0.08}\right]$

$A = 50,000\left[\dfrac{1-(0.231712064)}{0.08}\right]$

$A = 50,000[9.6035992]$

$A = 480,179.96$

The formula above calculates the present value of the annuity given the payment made at the end of each period. Twenty total payments were made, but only nineteen occurred at the end of a compounding period. The first payment of $50,000 was made up front. Therefore, the total value of the lottery winnings is
$50,000 + 480,179.96 = \$530,179.96.$

15. $A = R\left[\dfrac{1-(1+i)^{-n}}{i}\right]$

$A = 3000\left[\dfrac{1-\left(1+\dfrac{0.09}{12}\right)^{-(30)(12)}}{\dfrac{0.09}{12}}\right]$

$A = 3000\left[\dfrac{1-(1.0075)^{-360}}{0.0075}\right]$

$A = 3000\left[\dfrac{1-0.0678860074}{0.0075}\right]$

$A = 3000[124.2818657]$

$A = 372,845.60$

The disabled man should seek a lump sum payment of $372,845.60.

17. a. $A = R\left[\dfrac{1-(1+i)^{-n}}{i}\right]$

$A = 122,000\left[\dfrac{1-(1+0.10)^{-9}}{0.10}\right]$

$A = 122,000\left[\dfrac{1-(1.10)^{-9}}{0.10}\right]$

$A = 122,000\left[\dfrac{1-0.4240976184}{0.10}\right]$

$A = 122,000[5.759023816]$

$A = 702,600.91$

The formula above calculates the present value of the annuity given the payment made at the end of each period. Ten total payments were made, but only nine occurred at the end of a compounding period. The first payment of $100,000 was made up front. Therefore, the total value of the sale is
$100,000 + 702,600.91 = \$802,600.91.$

b. $R = A\left[\dfrac{i}{1-(1+i)^{-n}}\right]$

$R = 700,000\left[\dfrac{0.10}{1-(1+0.10)^{-9}}\right]$

$R = 700,000\left[\dfrac{0.10}{1-(1.10)^{-9}}\right]$

$R = 700,000\left[\dfrac{0.10}{1-0.4240976184}\right]$

$R = 700,000[0.1736405391]$

$R = 121,548.38$

The annuity payment is \$121,548.38.

c. The \$100,000 plus the annuity yields a higher present value and therefore would be the better choice. Over the nine year annuity period, the \$100,000 cash plus \$122,000 annuity yields \$452 more per year than investing \$700,000 in cash.

19. a. $A = R\left[\dfrac{1-(1+i)^{-n}}{i}\right]$

$A = 1600\left[\dfrac{1-\left(1+\dfrac{0.09}{12}\right)^{-(30)(12)}}{\dfrac{0.09}{12}}\right]$

$A = 1600\left[\dfrac{1-(1.0075)^{-360}}{0.0075}\right]$

$A = 1600\left[\dfrac{1-0.0678860074}{0.0075}\right]$

$A = 1600[124.2818657]$

$A = 198,850.99$

The couple can afford to pay \$198,850.99 for a house.

b. ($1600 per month) $\times$ (12 months)

$\times$ (30 years) = \$576,000

c. $576,000 - 198,850.99 = \$377,149.01$

21. a. $\dfrac{8}{4} = 2\%$

b. (4 years) $\times$ (4 payments per year)

$= 16$ payments

c. $R = A\left[\dfrac{i}{1-(1+i)^{-n}}\right]$

$R = 10,000\left[\dfrac{0.02}{1-(1+0.02)^{-16}}\right]$

$R = 10,000\left[\dfrac{0.02}{1-(1.02)^{-16}}\right]$

$R = 10,000\left[\dfrac{0.02}{1-0.7284458137}\right]$

$R = 10,000[0.0736501259]$

$R = 736.50$

The quarterly payment is \$736.50.

23. a. $i = \dfrac{0.06}{12} = 0.005, \; n = 360$

$R = A\left[\dfrac{i}{1-(1+i)^{-n}}\right]$

$R = 250,000\left[\dfrac{0.005}{1-(1+0.005)^{-360}}\right]$

$R = 250,000\left[\dfrac{0.005}{1-0.166041928}\right]$

$R = 250,000\left[\dfrac{0.005}{0.833958072}\right]$

$R = 250,000[0.0059955053]$

$R = 1498.88$

The monthly mortgage payment is $1498.88.

b. $(30 \text{ years}) \times (12 \text{ payments per year})$
$\times (\$1498.88) = \$539,596.80$
Including the down payment, the total cost of the house is $639,596.80.

c. $639,596.80 - 350,000 = \$289,596.80$

Section 3.7 Skills Check

1. $$\frac{79.514}{1+0.835e^{-0.0298(80)}}$$

   $$=\frac{79.514}{1+0.835e^{-2.384}}$$

   $$=\frac{79.514}{1+0.835(0.0921811146)}$$

   $$=\frac{79.514}{1.076971231}$$

   $$=73.83112727$$

   $$\approx 73.83$$

3. $1000(0.06)^{0.2^{t}}$

   Let $t = 4$.

   $$1000(0.06)^{0.2^{4}}=1000(0.06)^{0.0016}$$
   $$=1000(0.9955086592)$$
   $$\approx 995.51$$

   Let $t = 6$.

   $$1000(0.06)^{0.2^{6}}=1000(0.06)^{0.000064}$$
   $$=1000(0.9998199579)$$
   $$\approx 999.82$$

5. a. Y1=100/(1+3e^(-X))

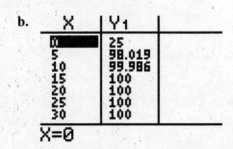

X=7.5   Y=99.83435

[0, 15] by [0, 120]

   b.

X	Y₁
0	25
5	98.019
10	99.986
15	100
20	100
25	100
30	100

   X=0

$f(0) = 25$
$f(10) = 99.986$

   c. The graph is increasing.

   d. Based on the graph, the y-values of the function approach 100. Therefore the limiting value of the function is 100. $y = c = 100$ is a horizontal asymptote of the function.

7. a. Y1=100(.05^(.3^X))

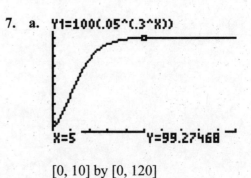

X=5   Y=99.27468

[0, 10] by [0, 120]

   b. Let $x = 0$, and solve for y.
   $$y=100(0.05)^{0.3^{0}}=100(0.05)^{1}=5$$
   The initial value is 5.

   c. The maximum value is c. In this case, $c = 100$.

Section 3.7 Exercises

9. a. Y1=5000/(1+1000e^(-.8X))_

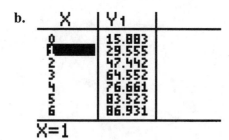

X=7.5   Y=1437.297

[0, 15] by [−1000, 5500]

b. At $x = 0$, the number of infected students is the value of the y-intercept of the function. The y-intercept is
$$\frac{c}{1+a} = \frac{5000}{1+1000} = 4.995 \approx 5.$$

c. The upper limit is $c = 5000$.

11. a. Y1=89.786/(1+4.6531e^(-.8_

X=2.5   Y=56.444738

[0, 5] by [0, 100]

b.

X	Y1
0	15.883
1	29.555
2	47.442
3	64.552
4	76.661
5	83.523
6	86.931

X=1

The model indicates that 29.56% of 16-year old boys have been sexually active

c. Consider the table in part b) above. The model indicates that 86.93% of 21-year old boys have been sexually active

d. The upper limit is $c = 89.786$.

13. a. Let $t = 1$ and solve for N.
$$N = \frac{10,000}{1+100e^{-0.8(1)}}$$
$$= \frac{10,000}{1+100e^{-0.8}}$$
$$= \frac{10,000}{45.393289641}$$
$$\approx 218$$

Approximately 218 people have heard the rumor after the first day.

b. Let $t = 4$ and solve for N.
$$N = \frac{10,000}{1+100e^{-0.8(4)}}$$
$$= \frac{10,000}{1+100e^{-3.2}}$$
$$= \frac{10,000}{5.076220398}$$
$$\approx 1970$$

c.

X	Y1
3	992.87
4	1970
5	3531.6
6	5485.5
7	7300.4
8	8575.2
9	9305.3

X=7

After seven days, 7300 people have heard the rumor.

15. a.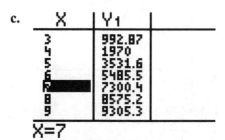

Logistic
y=c/(1+ae^(-bx))
a=4.653058106
b=.8256482181
c=89.78571366

$$y = \frac{89.786}{1+4.653e^{-0.826x}}$$

b. Yes.

c. LinReg
  y=ax+b
  a=14.13714286
  b=17.62380952

$$y = 14.137x + 17.624$$

d. Y1=89.785713657376/(1+4..

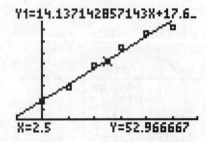

X=2.5   Y=56.447273

[–1, 6] by [–10, 100]

Y1=14.137142857143X+17.6_

X=2.5   Y=52.966667

[–1, 6] by [–10, 100]

The logistic model is a much better fit.

17. a. Let $t = 0$ and solve for N.

$$N = 10{,}000(0.4)^{0.2^0}$$
$$= 10{,}000(0.4)^1$$
$$= 10{,}000(0.4)$$
$$= 4000$$

The initial population size is 4000 students.

b. Let $t = 4$ and solve for N.

$$N = 10{,}000(0.4)^{0.2^4}$$
$$= 10{,}000(0.4)^{0.0016}$$
$$= 10{,}000(0.998535009)$$
$$= 9985.35009$$
$$\approx 9985$$

After four years, the population is approximately 9985 students.

c. Y1=10000(.4)^(.2^X)

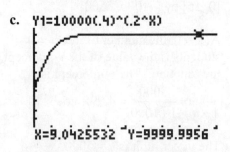

X=9.0425532 ⁻Y=9999.9956 ⁻

[0, 10] by [0, 12,000]

The upper limit appears to be 10,000.

19. a. Let $t = 1$ and solve for N.

$$N = 40{,}000(0.2)^{0.4^1}$$
$$= 40{,}000(0.2)^{0.4}$$
$$= 40{,}000(0.5253055609)$$
$$= 21{,}012.22244$$
$$\approx 21{,}012$$

After one month, the approximately 21,012 units will be sold.

b. Y1=40000(.2)^(.4^X)

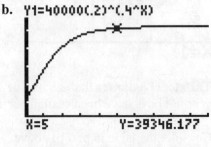

X=5   Y=39346.177

[0, 10] by [0, 50,000]

c. The upper limit appears to be 40,000.

21. a. Let $t = 0$ and solve for N.

$$N = 1000(0.01)^{0.5^0}$$
$$= 1000(0.01)^1$$
$$= 1000(0.01)$$
$$= 10$$

Initially the company had 10 employees.

b. Let $t = 1$ and solve for N.

$$N = 1000(0.01)^{0.5^1}$$
$$= 1000(0.01)^{0.5}$$
$$= 1000(0.1)$$
$$= 100$$

After one year, the company had 100 employees.

c. The upper limit is 1000 employees.

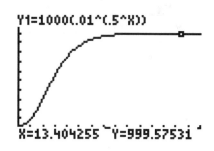

$Y1=1000(.01^(.5^X))$

$X=13.404255 \quad Y=999.57531$

[0, 15] by [0, 1200]

d.

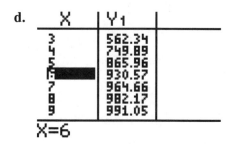

X	Y1
3	562.34
4	749.89
5	865.96
6	930.57
7	964.66
8	982.17
9	991.05

X=6

In the sixth year 930 people were employed by the company.

23.

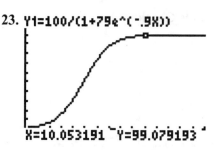

$Y1=100/(1+79 \varepsilon^{(-.9X)})$

$X=10.053191 \quad Y=99.079193$

[0, 15] by [0, 120]

After 10 days, 99 people are infected.

25.

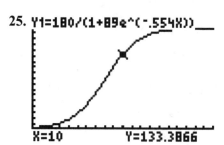

$Y1=180/(1+89 \varepsilon^{(-.554X)})$

$X=10 \quad Y=133.3866$

[0, 20] by [−20, 200]

X	Y1
7	63.345
8	87.452
9	111.93
10	133.39
11	149.9
12	161.38
13	168.81

X=11

In approximately 11 years, the deer population reaches a level of 150.

Chapter 3 Skills Check

1. a.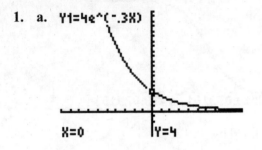

    [–10, 10] by [–5, 20]

    b. $f(-10) = 4e^{-0.3(-10)}$

    $= 4e^3$

    ≈ 80.342

    $f(10) = 4e^{-0.3(10)}$

    $= 4e^{-3}$

    ≈ 0.19915

2. The function in Exercise 1, $f(x) = 4e^{-0.3x}$, is decreasing.

3.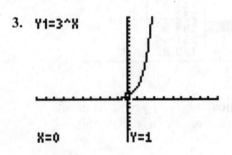

    [–10, 10] by [–10, 20]

4.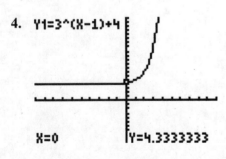

    [–10, 10] by [–10, 20]

5. The graph in Exercise 4 is shift one unit right and three units up in comparison with the graph in Exercise 3.

6. The function is increasing. See the solution to Exercise 4.

7. a. $y = 1000(2)^{-0.1x}$

    $= 1000(2)^{-0.1(10)}$

    $= 1000(2)^{-1}$

    $= 500$

    b.

X	Y₁	
15	353.55	
16	329.88	
17	307.79	
18	287.17	
19	267.94	
20	250	
21	233.26	

    X=20

    When $y = 250$, $x = 20$.

8. $x = 6^y \Leftrightarrow \log_6 x = y$

9. $y = 7^{3x} \Leftrightarrow \log_7 y = 3x$

10. $y = \log_4 x \Leftrightarrow x = 4^y$

11. $y = \log(x) = \log_{10} x$

    $y = \log_{10} x \Leftrightarrow x = 10^y$

12. $y = \ln x = \log_e x$

    $y = \log_e x \Leftrightarrow x = e^y$

13. $y = 4^x$

$x = 4^y$

$x = 4^y \Leftrightarrow \log_4 x = y$

Therefore, the inverse function is
$y = \log_4 x.$

14. $\log 22 = \log_{10} 22 = 1.3424$

15. $\ln 56 = \log_e 56 = 4.0254$

16. $\log 10 = \log_{10} 10 = 1$

17. $\log_2 16$

$y = \log_2 16 \Leftrightarrow 2^y = 16$

$y = 4$

18. $\ln\left(e^4\right) = \log_e\left(e^4\right)$

$y = \log_e\left(e^4\right) \Leftrightarrow e^y = e^4$

$y = 4$

19. $\log(0.001) = \log_{10}\left(\dfrac{1}{1000}\right)$

$y = \log_{10}\left(\dfrac{1}{1000}\right) \Leftrightarrow 10^y = \dfrac{1}{1000}$

$10^y = \dfrac{1}{1000} = \dfrac{1}{10^3} = 10^{-3}$

$y = -3$

20. $\log_3(54) = \dfrac{\ln(54)}{\ln(3)} = 3.6039$

21. $\log_8(56) = \dfrac{\ln(56)}{\ln(8)} = 1.9358$

22.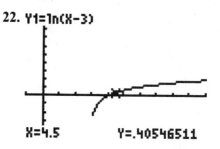

$[-1, 10]$ by $[-5, 10]$

23.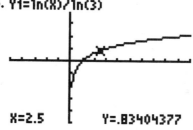

$[-5, 10]$ by $[-5, 5]$

24. $340 = e^x$

$\ln(340) = \ln\left(e^x\right)$

$x = \ln(340)$

$x \approx 5.8289$

25. $1500 = 300e^{8x}$

$\dfrac{1500}{300} = \dfrac{300e^{8x}}{300}$

$5 = e^{8x}$

$\ln(5) = \ln\left(e^{8x}\right)$

$8x = \ln(5)$

$x = \dfrac{\ln(5)}{8}$

$x \approx 0.2012$

26.
$$9200 = 23\left(2^{3x}\right)$$
$$\frac{9200}{23} = \frac{23\left(2^{3x}\right)}{23}$$
$$2^{3x} = 400$$
$$\ln\left(2^{3x}\right) = \ln\left(400\right)$$
$$3x\ln\left(2\right) = \ln\left(400\right)$$
$$x = \frac{\ln\left(400\right)}{3\ln\left(2\right)}$$
$$x \approx 2.8813$$

27.
$$4\left(3^x\right) = 36$$
$$3^x = 9$$
$$\log\left(3^x\right) = \log\left(9\right)$$
$$x\log\left(3\right) = \log\left(9\right)$$
$$x = \frac{\log\left(9\right)}{\log\left(3\right)}$$
$$x = 2$$

28.
$$\ln\left[\frac{\left(2x-5\right)^3}{x-3}\right]$$
$$= \ln\left(2x-5\right)^3 - \ln\left(x-3\right)$$
$$= 3\ln\left(2x-5\right) - \ln\left(x-3\right)$$

29.
$$6\log_4 x - 2\log_4 y$$
$$= \log_4 x^6 - \log_4 y^2$$
$$= \log_4\left(\frac{x^6}{y^2}\right)$$

30.

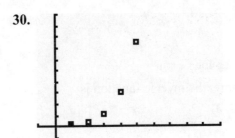

$[-1, 10]$ by $[-10, 100]$

The data is best modeled by an exponential function.

```
ExpReg
 y=a*b^x
 a=.809857516
 b=2.469570215
```

$$y = 0.810\left(2.470\right)^x$$

31.
$$P\left(1+\frac{r}{k}\right)^{kn}$$
$$= 1000\left(1+\frac{0.08}{12}\right)^{(12)(20)}$$
$$= 1000\left(1.00\overline{6}\right)^{240}$$
$$\approx 4926.80$$

32.
$$1000\left[\frac{1-1.03^{-240+120}}{0.03}\right]$$
$$= 1000\left[\frac{1-1.03^{-120}}{0.03}\right]$$
$$= 1000\left[\frac{0.9711906782}{0.03}\right]$$
$$= 1000\left[32.37302261\right]$$
$$= 32,373.02$$

33. a.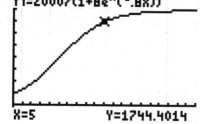

[0, 10] by [0, 2300]

b.
$$f(0) = \frac{2000}{1 + 8e^{-0.8(0)}}$$
$$= \frac{2000}{1 + 8e^{0}}$$
$$= \frac{2000}{9}$$
$$\approx 222.22$$

$$f(8) = \frac{2000}{1 + 8e^{-0.8(8)}}$$
$$= \frac{2000}{1 + 8e^{-6.4}}$$
$$= \frac{2000}{1.013292458}$$
$$\approx 1973.76$$

c. The limiting value of the function is 2000.

34. a.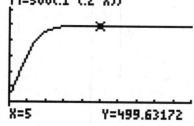

[0, 10] by [0, 700]

b.
$$y = 500(0.1)^{0.2^{0}}$$
$$= 500(0.1)^{1}$$
$$= 500(0.1)$$
$$= 50$$

c. The limiting value is 500.

Chapter 3 Review Exercises

35. Let $x = 30$.

$$y = 84.7518(1.0746)^{30}$$

$$= 733.75997$$

$$\approx 733.760$$

In 1990, the total number of prisoners was approximately 733,760.

36. Let $x = 4$.

$$y = 2000(2)^{-0.1(4)}$$

$$= 2000(2)^{-0.4}$$

$$= 2000(0.7578582833)$$

$$\approx 1515.72$$

Four weeks after the end of the advertising campaign, the daily sales in dollars will be $1515.72.

37.

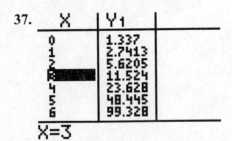

Annual revenue exceeded $10 million during 1988.

38. a. $R = \log\left(\dfrac{I}{I_0}\right)$

$$R = \log\left(\dfrac{1000 I_0}{I_0}\right)$$

$$R = \log(1000) = 3$$

The earthquake measures 3 on the Richter scale.

b. $10^R = \dfrac{I}{I_0}$

$$I = 10^R I_0$$

$$I = 10^{6.5} I_0$$

$$I = 3{,}162{,}277.66 I_0$$

39. The difference in the Richter scale measurements is $7.9 - 4.8 = 3.1$. Therefore the intensity of the Indian earthquake was $10^{3.1} \approx 1259$ times stronger than the intensity of the U.S. earthquake.

40. $t = \log_{1.12} 3 = \dfrac{\ln 3}{\ln 1.12} \approx 9.69$

The investment will triple in approximately 10 years.

41. a. $S = 1000(2)^{\left(\frac{x}{7}\right)}$

$$\dfrac{S}{1000} = (2)^{\left(\frac{x}{7}\right)} \Leftrightarrow \log_2\left(\dfrac{S}{1000}\right) = \dfrac{x}{7}$$

$$x = 7\log_2\left(\dfrac{S}{1000}\right)$$

b. $x = 7\log_2\left(\dfrac{19{,}504}{1000}\right)$

$$= 7\log_2(19.504)$$

$$= 7\left(\dfrac{\ln 19.504}{\ln 2}\right)$$

$$\approx 29.99989$$

In about 30 years the future value will be $19,504.

42. $1000 = 2000(2)^{-0.1x}$

$0.5 = 2^{-0.1x}$

$\ln(0.5) = \ln\left(2^{-0.1x}\right)$

$\ln(0.5) = -0.1x \ln 2$

$x = \dfrac{\ln 0.5}{-0.1 \ln 2}$

$x = 10$

In 10 days, sales will decay by half.

43.

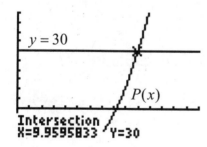

X = 29

In 1989, the rate is 275 per 100,000.

44. a. $P(x) = R(x) - C(x)$

$P(x) = 10(1.26)^x - (2x + 50)$

$= 10(1.26)^x - 2x - 50$

b. Applying the intersection of graphs method:

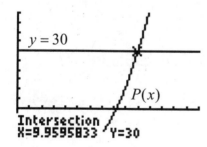

[0, 15] by [–15, 50]

Selling at least 10 mobile homes produces a profit of at least $30,000.

45. Applying the intersection of graphs method:

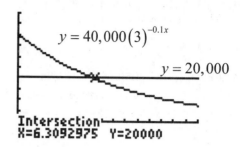

[0, 15] by [–7500, 50,000]

After seven weeks, sales will be less than half.

46. a. $y = 100e^{-0.00012378(5000)}$

$= 100e^{-0.6189}$

$= 100(0.5385365021)$

≈ 53.85

After 5000 years, approximately 53.85 grams of carbon-14 remains.

b. $36\% y_0 = y_0 e^{-0.00012378t}$

$0.36 = e^{-0.00012378t}$

$\ln(0.36) = \ln\left(e^{-0.00012378t}\right)$

$\ln(0.36) = -0.00012378t$

$-0.00012378t = \ln(0.36)$

$t = \dfrac{\ln(0.36)}{-0.00012378}$

$t \approx 8253.77$

The wood was cut approximately 8254 years ago.

47.

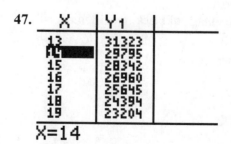

X = 14

After approximately 14 years, the purchasing power will be less than half of the original $60,000 income.

48. $S = 2000e^{0.08(10)} = 2000e^{0.8} \approx 4451.08$
The future value is approximately $4451.08 after 10 years.

49. $13,784.92 = 3300(1.10)^x$

$(1.10)^x = 4.177248485$

$\ln\left[(1.10)^x\right] = \ln[4.177248485]$

$x\ln(1.10) = \ln(4.177248485)$

$x = \dfrac{\ln(4.177248485)}{\ln(1.10)}$

$x \approx 15$

The investment reaches the indicated value in 15 years.

50. a.

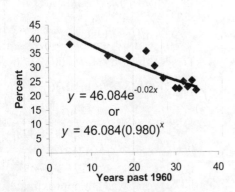

Smokers

$y = 46.084e^{-0.02x}$
or
$y = 46.084(0.980)^x$

b.   In 1972, the percentage of black female smokers is 36.3%. In 1996, the percentage of black female smokers is 22.5%.

```
36→X
                    36
46.084225150784*
.98022276354031^
X
         22.45145586

12→X
                    12
46.084225150784*
.98022276354031^
X
         36.26182819
```

51. a.

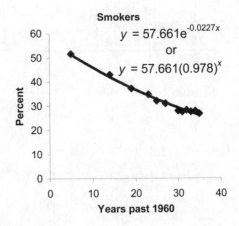

Smokers

$y = 57.661e^{-0.0227x}$
or
$y = 57.661(0.978)^x$

b.   In 2003 the percentage of U.S. smokers is 21.8%.

```
43→X
                    43
57.66092840212*
.97760181058669^
X
         21.76945277
```

52. a. `ExpReg`
`y=a*b^x`
`a=11.25228244`
`b=1.045136856`

$$y = 11.252(1.045)^x$$

b. `Logistic`
`y=c/(1+ae^(-bx)`
`a=12.33134353`
`b=.0844440721`
`c=96.36411134`

$$y = \frac{96.364}{1 + 12.331e^{-0.084x}}$$

c. exponential model

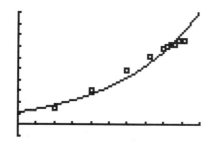

[0, 50] by [−10, 100]

logistic model

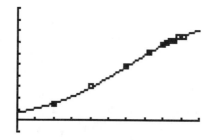

[0, 50] by [−10, 100]

It appears the logistic model fits the data points better.

53. a. `ExpReg`
`y=a*b^x`
`a=2102.964558`
`b=1.026726773`

$$y = 2102.96(1.0267)^x$$

b. `Logistic`
`y=c/(1+ae^(-bx)`
`a=4.722514478`
`b=.0623453003`
`c=8672.062996`

$$y = \frac{8672.06}{1 + 4.723e^{-0.0623x}}$$

c. exponential model

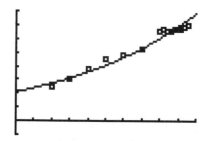

[0, 50] by [−1000, 8000]

logistic model

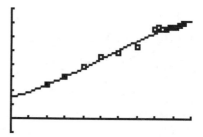

[0, 50] by [−1000, 8000]

The logistic model fits the data best.

54.
```
LnReg
  y=a+blnx
  a=27.94521723
  b=8.840292063
```

$$y = 27.945 + 8.84 \ln x$$

55. a.

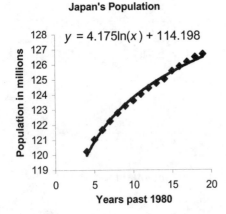

Japan's Population

$y = 4.175\ln(x) + 114.198$

b.
$$y = 114.198 + 4.175\ln(2004 - 1980)$$
$$= 114.198 + 4.175\ln(24)$$
$$\approx 127.4663747$$

In 2004 Japan's population is approximately 127,466,375. Using the unrounded model yields 127,466,485.

```
24→X
               24
114.19830610294+
4.17493825671681
n(X)
          127.4664846
```

56. a.

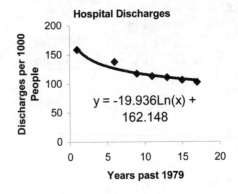

Hospital Discharges

$y = -19.936\text{Ln}(x) + 162.148$

b. See part a) above.

c.
```
26→X
               26
162.14845792466+
-19.936339398596
ln(X)
          97.19393955
```

In 2005 the number of discharges per 1000 people is approximately 97.

57. $S = Pe^{rt}$
$$S = 12,500e^{(0.05)(10)}$$
$$= 12,500e^{0.5} \approx 20,609.02$$
The future value is $20,609.02.

58. $S = P\left(1 + \dfrac{r}{k}\right)^{kt}$
$$S = 20,000\left(1 + \dfrac{0.06}{1}\right)^{(1)(7)}$$
$$S = 20,000(1.06)^7 \approx 30,072.61$$
The future value is $30,072.61.

59. $S = R\left[\dfrac{(1+i)^n - 1}{i}\right]$

$S = 1000\left[\dfrac{\left(1+\dfrac{0.12}{4}\right)^{(4)(6)} - 1}{\dfrac{0.12}{4}}\right]$

$S = 1000\left[\dfrac{(1.03)^{24} - 1}{0.03}\right]$

$S = 1000\left[\dfrac{(1.03)^{24} - 1}{0.03}\right]$

$S = 1000(34.42647022) \approx 34,426.47$

The future value is $34,426.47.

60. $S = R\left[\dfrac{(1+i)^n - 1}{i}\right]$

$S = 1500\left[\dfrac{\left(1+\dfrac{0.08}{12}\right)^{(12)(10)} - 1}{\dfrac{0.08}{12}}\right]$

$S = 1500\left[\dfrac{(1.00\overline{6})^{120} - 1}{0.00\overline{6}}\right]$

$S = 1500(182.9460352)$

$S \approx 274,419.05$

The future value is $274,419.05.

61. $A = R\left[\dfrac{1-(1+i)^{-n}}{i}\right]$

$A = 2000\left[\dfrac{1-\left(1+\dfrac{0.08}{12}\right)^{-(12)(15)}}{\dfrac{0.08}{12}}\right]$

$A = 2000\left[\dfrac{1-(1.00\overline{6})^{-180}}{0.00\overline{6}}\right]$

$A = 2000[104.6405922] \approx 209,281.18$

The formula above calculates the present value of the annuity given the payment made at the end of each period. The present value is $209,218.18.

62. $A = R\left[\dfrac{1-(1+i)^{-n}}{i}\right]$

$A = 500\left[\dfrac{1-\left(1+\dfrac{0.10}{2}\right)^{-(2)(12)}}{\dfrac{0.10}{2}}\right]$

$A = 500\left[\dfrac{1-(1.05)^{-24}}{0.05}\right]$

$A = 500[13.79864179]$

$A \approx 6899.32$

The formula above calculates the present value of the annuity given the payment made at the end of each period. The present value is $6899.32.

63. $R = A\left[\dfrac{i}{1-(1+i)^{-n}}\right]$

$R = 2000\left[\dfrac{\dfrac{0.12}{12}}{1-\left(1+\dfrac{0.12}{12}\right)^{-36}}\right]$

$R = 2000\left[\dfrac{0.01}{1-(1.01)^{-36}}\right]$

$R = 2000[0.0332143098]$

$R \approx 66.43$

The monthly payment is \$66.43.

64. $R = A\left[\dfrac{i}{1-(1+i)^{-n}}\right]$

$R = 120{,}000\left[\dfrac{\dfrac{0.06}{12}}{1-\left(1+\dfrac{0.06}{12}\right)^{-(12)(25)}}\right]$

$R = 120{,}000\left[\dfrac{0.005}{1-(1.005)^{-300}}\right]$

$R = 120{,}000[0.006443014]$

$R \approx 773.16$

The monthly payment is \$773.16.

65. a. In 1990, $x = 1990 - 1950 = 40$.

$y = \dfrac{96.3641}{1+12.3313e^{-0.0844(40)}}$

$= \dfrac{96.3641}{1+12.3313e^{-3.376}}$

$= \dfrac{96.3641}{1+0.4215321394}$

≈ 67.79

Based on the model the percentage of out-of-wedlock births in 1990 is 67.79%.

In 1996, $x = 1996 - 1950 = 46$.

$y = \dfrac{96.3641}{1+12.3313e^{-0.0844(46)}}$

$= \dfrac{96.3641}{1+12.3313e^{-3.8824}}$

$= \dfrac{96.3641}{1+0.2540410897}$

≈ 76.84

Based on the model the percentage of out-of-wedlock births in 1996 is 76.84%.

b. The upper limit on the percentage of out-of-wedlock births is 96.3641%.

66. a. Let $x = 14$.

$y = \dfrac{1400}{1+200e^{-0.5(14)}}$

$= \dfrac{1400}{1+200e^{-7}}$

$= \dfrac{1400}{1+0.1823763931}$

≈ 1184.06

After 14 days approximately 1184 students are infected.

b.

X	Y1
13	1076.4
14	1184.1
15	1260.6
16	1312
17	1345.3
18	1366.3
19	1379.4

X=16

After 16 days 1312 students are infected.

67. a.

$$N = 4000(0.06)^{0.4^{(2-1)}}$$

$$= 4000(0.06)^{0.4^1}$$

$$= 4000(0.06)^{0.4}$$

$$\approx 1298.13$$

After two years, the enrollment will be approximately 1298 students.

b.

$$N = 4000(0.06)^{0.4^{(10-1)}}$$

$$= 4000(0.06)^{0.4^9}$$

$$= 4000(0.06)^{0.000262144}$$

$$\approx 3997.05$$

After ten years the enrollment will be approximately 3997 students.

c. The upper limit on the number of students based on the model is 4000.

68. a.

$$N = 18,000(0.03)^{0.4^{10}}$$

$$= 18,000(0.03)^{0.0001048576}$$

$$\approx 17,993.38$$

After ten months the number of units sold in a month will be approximately 17,993.

b. The upper limit on the number of units sold per month is 18,000.

69. a.

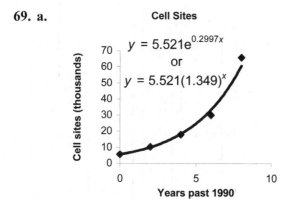

Cell Sites

$y = 5.521e^{0.2997x}$

or

$y = 5.521(1.349)^x$

b.

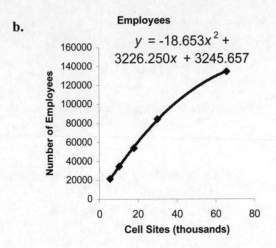

Employees

$$y = -18.653x^2 + 3226.250x + 3245.657$$

c. $C(t) = 5.521(1.349)^t$

$E(C) = -18.653C^2 + 3226.250C + 3245.657$

$E(C(t)) = -18.653\left(5.521(1.349)^t\right)^2 + 3226.250\left(5.521(1.349)^t\right) + 3245.657$

$E(C(t))$ calculates the number of employees given the number of years past 1990.

d. $E(C(7)) = -18.653\left(5.521(1.349)^7\right)^2 + 3226.250\left(5.521(1.349)^7\right) + 3245.657$

$\quad\quad\quad = -18.653(44.88501709)^2 + 3226.250(44.88501709) + 3245.657$

$\quad\quad\quad = -18.653(2014.664759) + 3226.250(44.88501709) + 3245.657$

$\quad\quad\quad = 110,476.4016$

e. In 1997 the cellular telephone industry employed approximately 110,476 people.

Group Activity/Extended Applications

1. The first person on the list receives $36. Each of the original six people on the list sends their letter to six people. Therefore, 36 people receive letters with the original six names, and each of the 36 forwards a dollar to the first person on the original list.

2. The 36 people receiving the first letter place their name on the bottom of the list, shift up the second person to first place. The 36 people send out six letters each, for a total of $36 \cdot 6 = 216$ letters. Therefore the second person on the original list receives $216.

3.

Cycle Number	Money Sent to the Person on Top of the List
1	$6^2 = 36$
2	$6^3 = 216$
3	$6^4 = 1296$
4	$6^5 = 7776$
5	$6^6 = 46,656$

4. Position 5 generates the most money!

5. ```
QuadReg
 y=ax²+bx+c
 a=5914.285714
 b=-25405.71429
 c=22356
```

```
PwrReg
 y=a*x^b
 a=20.33965715
 b=4.338874682
```

```
ExpReg
 y=a*b^x
 a=6
 b=6
```

The exponential model, $y = 6(6)^x = 6^{x+1}$, fits the data exactly.

6. $y = 6^{6+1} = 6^7 = 279,936$

The sixth person on the original list receives $279,936.

7. The total number of responses on the sixth cycle would be
$6 + 36 + 216 + 1296 + 7776 + 46,656 + 279,936 = 335,922$

8. $y = 6^{10+1} = 6^{11} = 362,797,056$

On the tenth cycle 362,797,056 people receive the chain letter and are suppose to respond with $1.00 to the first name on the list.

9. The answer to problem 8 is larger than the U.S. population. There is no unsolicited person in the U.S. to whom to send the letter.

10. Chain letters are illegal since people entering lower on the chain have a very small chance of earning money from the scheme.
```

Chapter 4
Higher-Degree Polynomial and Rational Functions

Algebra Toolbox

1. a. The polynomial is 4th degree.

   b. The leading coefficient is 3.

2. a. The polynomial is 3rd degree.

   b. The leading coefficient is 5.

3. a. The polynomial is 5th degree.

   b. The leading coefficient is -14.

4. a. The polynomial is 6th degree.

   b. The leading coefficient is -8.

5. $4x^3 - 8x^2 - 140x$

   $= 4x\left(x^2 - 2x - 35\right)$

   $= 4x(x-7)(x+5)$

6. $4x^2 + 7x^3 - 2x^4$

   $= -2x^4 + 7x^3 + 4x^2$

   $= -1x^2\left(2x^2 - 7x - 4\right)$

   $= -x^2(2x+1)(x-4)$

7. $x^4 - 13x^2 + 36$

   $= \left(x^2 - 9\right)\left(x^2 - 4\right)$

   $= (x+3)(x-3)(x+2)(x-2)$

8. $x^4 - 21x^2 + 80$

   $= \left(x^2 - 16\right)\left(x^2 - 5\right)$

   $= (x+4)(x-4)\left(x^2 - 5\right)$

9. $2x^4 - 8x^2 + 8$

   $= 2\left(x^4 - 4x^2 + 4\right)$

   $= 2\left(x^2 - 2\right)\left(x^2 - 2\right)$

   $= 2\left(x^2 - 2\right)^2$

10. $3x^5 - 24x^3 + 48x$

   $= 3x\left(x^4 - 8x^2 + 16\right)$

   $= 3x\left(x^2 - 4\right)\left(x^2 - 4\right)$

   $= 3x(x+2)(x-2)(x+2)(x-2)$

   $= 3x(x+2)^2(x-2)^2$

11. $\dfrac{x-3y}{3x-9y} = \dfrac{x-3y}{3(x-3y)} = \dfrac{1}{3}$

12. $\dfrac{x^2-9}{4x+12} = \dfrac{(x+3)(x-3)}{4(x+3)} = \dfrac{x-3}{4}$

13. $\dfrac{2y^3 - 2y}{y^2 - y}$

   $= \dfrac{2y\left(y^2 - 1\right)}{y(y-1)}$

   $= \dfrac{2y(y+1)(y-1)}{y(y-1)}$

   $= 2(y+1)$

14. $\dfrac{4x^3 - 3x}{x^2 - x}$

$= \dfrac{x\left(4x^2 - 3\right)}{x\left(x - 1\right)}$

$= \dfrac{4x^2 - 3}{x - 1}$

15. $\dfrac{x^2 - 6x + 8}{x^2 - 16}$

$= \dfrac{\left(x - 4\right)\left(x - 2\right)}{\left(x + 4\right)\left(x - 4\right)}$

$= \dfrac{x - 2}{x + 4}$

16. $\dfrac{3x^2 - 7x - 6}{x^2 - 4x + 3}$

$= \dfrac{\left(3x + 2\right)\left(x - 3\right)}{\left(x - 3\right)\left(x - 1\right)}$

$= \dfrac{3x + 2}{x - 1}$

17. $\dfrac{x - 3}{x^3} \cdot \dfrac{x\left(x - 4\right)}{\left(x - 4\right)\left(x - 3\right)}$

$= \dfrac{x}{x^3}$

$= \dfrac{1}{x^2}$

18. $\left(x + 2\right)\left(x - 2\right)\left(\dfrac{2x - 3}{x + 2}\right)$

$= \left(x - 2\right)\left(2x - 3\right)$

$= 3x^2 - 7x + 6$

19. $\dfrac{4x + 4}{x - 4} \div \dfrac{8x^2 + 8x}{x^2 - 6x + 8}$

$= \dfrac{4x + 4}{x - 4} \cdot \dfrac{x^2 - 6x + 8}{8x^2 + 8x}$

$= \dfrac{4\left(x + 1\right)}{x - 4} \cdot \dfrac{\left(x - 2\right)\left(x - 4\right)}{8x\left(x + 1\right)}$

$= \dfrac{x - 2}{2x}$

20. $\dfrac{6x^2}{4x^2 y - 12xy} \div \dfrac{3x^2 + 12x}{x^2 + x - 12}$

$= \dfrac{6x^2}{4x^2 y - 12xy} \cdot \dfrac{x^2 + x - 12}{3x^2 + 12x}$

$= \dfrac{6x^2}{4xy\left(x - 3\right)} \cdot \dfrac{\left(x + 4\right)\left(x - 3\right)}{3x\left(x + 4\right)}$

$= \dfrac{1}{2y}$

21. $3 + \dfrac{1}{x^2} - \dfrac{2}{x^3}$ LCM: x^3

$= \dfrac{3x^3}{x^3} + \dfrac{x}{x^3} - \dfrac{2}{x^3}$

$= \dfrac{3x^3 + x - 2}{x^3}$

22. $\dfrac{5}{x} - \dfrac{x - 2}{x^2} + \dfrac{4}{x^3}$ LCM: x^3

$= \dfrac{5x^2}{x^3} - \dfrac{x\left(x - 2\right)}{x^3} + \dfrac{4}{x^3}$

$= \dfrac{5x^2 - \left(x^2 - 2x\right) + 4}{x^3}$

$= \dfrac{5x^2 - x^2 + 2x + 4}{x^3}$

$= \dfrac{4x^2 + 2x + 4}{x^3}$

23. $\dfrac{a}{a^2-2a} - \dfrac{a-2}{a^2} = \dfrac{a}{a(a-2)} - \dfrac{a-2}{a^2}$ $\qquad$ LCM: $a^2(a-2)$

$$= \frac{a(a)}{a(a)(a-2)} - \frac{(a-2)(a-2)}{a^2(a-2)}$$

$$= \frac{a^2}{a^2(a-2)} - \frac{a^2-4a+4}{a^2(a-2)}$$

$$= \frac{a^2 - (a^2-4a+4)}{a^2(a-2)}$$

$$= \frac{a^2 - a^2 + 4a - 4}{a^2(a-2)}$$

$$= \frac{4a-4}{a^2(a-2)}$$

$$= \frac{4(a-1)}{a^2(a-2)}$$

$$= \frac{4a-4}{a^3-2a^2}$$

24. $\dfrac{5x}{x^4-16} + \dfrac{8x}{x+2} = \dfrac{5x}{(x^2+4)(x^2-4)} + \dfrac{8x}{x+2}$

$$= \frac{5x}{(x^2+4)(x+2)(x-2)} + \frac{8x}{x+2} \qquad \text{LCM: } (x^2+4)(x+2)(x-2)$$

$$= \frac{5x}{(x^2+4)(x+2)(x-2)} + \frac{8x(x^2+4)(x-2)}{(x^2+4)(x+2)(x-2)}$$

$$= \frac{5x + 8x(x^3-2x^2+4x-8)}{(x^2+4)(x+2)(x-2)}$$

$$= \frac{5x + 8x^4 - 16x^3 + 32x^2 - 64x}{(x^2+4)(x+2)(x-2)}$$

$$= \frac{8x^4 - 16x^3 + 32x^2 - 59x}{(x^2+4)(x+2)(x-2)}$$

$$= \frac{8x^4 - 16x^3 + 32x^2 - 59x}{x^4-16}$$

25. $\dfrac{x-1}{x+1} - \dfrac{2}{x(x+1)}$

$\{\text{LCM: } x(x+1)\}$

$= \dfrac{x(x-1)}{x(x+1)} - \dfrac{2}{x(x+1)}$

$= \dfrac{x^2 - x - 2}{x(x+1)}$

$= \dfrac{(x-2)(x+1)}{x(x+1)}$

$= \dfrac{x-2}{x}$

26. $\dfrac{2x+1}{2(2x-1)} + \dfrac{5}{2x} - \dfrac{x+1}{x(2x-1)}$

$\{\text{LCM: } 2x(2x-1)\}$

$= \dfrac{x(2x+1)}{2x(2x-1)} + \dfrac{5(2x-1)}{2x(2x-1)} - \dfrac{2(x+1)}{2x(2x-1)}$

$= \dfrac{2x^2 + x + (10x - 5) - (2x + 2)}{2x(2x-1)}$

$= \dfrac{2x^2 + x + 10x - 5 - 2x - 2}{2x(2x-1)}$

$= \dfrac{2x^2 + 9x - 7}{2x(2x-1)}$

27.

$$
\begin{array}{r}
x^4 - x^3 + 2x^2 - 2x + 2 \\
x+1 \overline{\smash{)}\, x^5 + 0x^4 + x^3 + 0x^2 - 0x - 1} \\
\underline{-x^5 - x^4} \\
-x^4 + x^3 \\
\underline{+x^4 + x^3} \\
2x^3 + 0x^2 \\
\underline{-2x^3 - 2x^2} \\
-2x^2 - 0x \\
\underline{2x^2 + 2x} \\
2x - 1 \\
\underline{-2x - 2} \\
-3
\end{array}
$$

$x^4 - x^3 + 2x^2 - 2x + 2 \ R - 3$

or

$x^4 - x^3 + 2x^2 - 2x + 2 - \dfrac{3}{x+1}$

28.

$$
\begin{array}{r}
a^3 + a^2 \\
a+2 \overline{\smash{)}\, a^4 + 3a^3 + 2a^2} \\
\underline{-a^4 - 2a^3} \\
a^3 + 2a^2 \\
\underline{-a^3 - 2a^2} \\
0
\end{array}
$$

$a^3 + a^2$

29.

$$
\begin{array}{r}
3x^3 - x^2 + 6x - 2 \\
x^2 - 2 \overline{\smash{\big)}\ 3x^5 - x^4 + 0x^3 + 0x^2 + 5x - 1} \\
\underline{-3x^5 \qquad\quad + 6x^3} \\
-x^4 + 6x^3 + 0x^2 \\
\underline{x^4 \qquad\quad - 2x^2} \\
6x^3 - 2x^2 + 5x \\
\underline{-6x^3 \qquad\quad + 12x} \\
-2x^2 + 17x - 1 \\
\underline{2x^2 \qquad\quad - 4} \\
17x - 5
\end{array}
$$

$\left(3x^3 - x^2 + 6x - 2\ \right) R\ \left(17x - 5\right)$

or

$3x^3 - x^2 + 6x - 2 + \dfrac{17x - 5}{x^2 - 2}$

30.

$$
\begin{array}{r}
x^2 + 1 \\
3x^2 - 1 \overline{\smash{\big)}\ 3x^4 + 0x^3 + 2x^2 + 0x + 1} \\
\underline{-3x^4 \qquad\quad + x^2} \\
3x^2 + 0x + 1 \\
\underline{-3x^2 \qquad\quad + 1} \\
2
\end{array}
$$

$x^2 + 1\ R\ 2$

or

$x^2 + 1 + \dfrac{2}{3x^2 - 1}$

Section 4.1 Skills Check

1. a. Y1=3X^3+5X^2-X-10

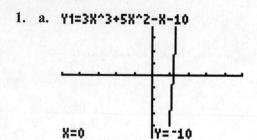

X=0   Y=-10

[–5, 5] by [–5, 5]

b. Y1=3X^3+5X^2-X-10

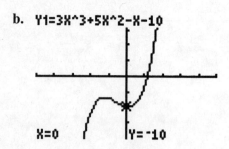

X=0   Y=-10

[–5, 5] by [–20, 20]

View b) is best.

3. a. Y1=3X^4-12X^2

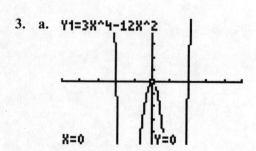

X=0   Y=0

[–5, 5] by [–5, 5]

b.

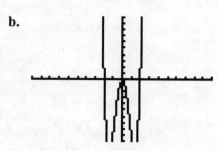

[–10, 10] by [–10, 10]

c.

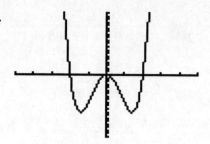

[–5, 5] by [–20, 20]

View c) is best.

5. a. The x-intercepts appear to be –3, –1, and 2.

    b. The leading coefficient is positive since the graph rises to the right.

    c. The polynomial is at least 3rd degree since the curve has two turns and three x-intercepts.

7. a. The x-intercepts appear to be –1, 3, and 5.

    b. The leading coefficient is negative since the graph falls to the right.

    c. The polynomial is at least 3rd degree since the curve has two turns and three x-intercepts.

9. a. The polynomial is 3rd degree, and the leading coefficient is 2.

    b. The graph rises right and falls left because the leading coefficient is positive and the function is cubic.

c. Y1=2X^3-X

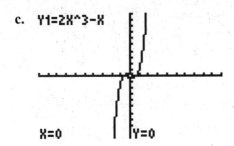

X=0   Y=0

[−10, 10] by [−10, 10]

11. a. The polynomial is 3ʳᵈ degree, and the leading coefficient is −2.

b. The graph falls right and rises left because the leading coefficient is negative and the function is cubic.

c. Y1=-2(X-1)(X^2-4)

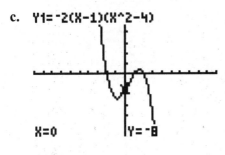

X=0   Y=-8

[−10, 10] by [−30, 30]

13. a. Y1=X^3-3X^2-X+3

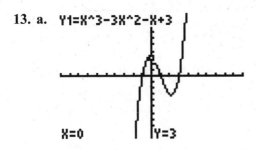

X=0   Y=3

[−10, 10] by [−10, 10]

b. Yes, the graph is complete. As suggested by the degree of the cubic function, three x-intercepts, one y-intercept, and two turns are displayed on the graph.

15. a. Y1=25X-X^3

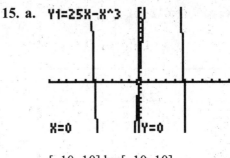

X=0   Y=0

[−10, 10] by [−10, 10]

b. Y1=25X-X^3

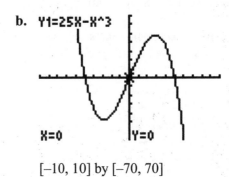

X=0   Y=0

[−10, 10] by [−70, 70]

17. a. Y1=X^4-4X^3+4X^2

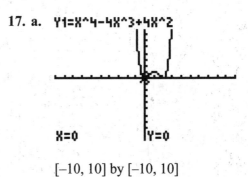

X=0   Y=0

[−10, 10] by [−10, 10]

b. Y1=X^4-4X^3+4X^2

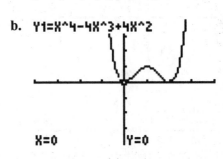

X=0   Y=0

[−4, 4] by [−4, 4]

c. The graph in part b) yields the best view of the turning points.

19. a.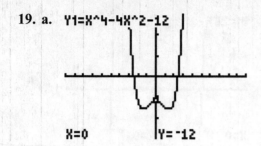

[–10, 10] by [–30, 30]

b. The graph has three turning points.

c. Since the polynomial is degree 4, it has at most three turning points.

21. Y1=-.5(X-1)^3+3

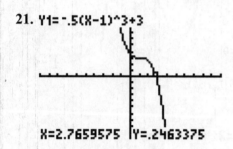

X=2.7659575   Y=.2463375

[–10, 10] by [–10, 10]

Answers will vary.

23. Y1=X^4-3X^2-4

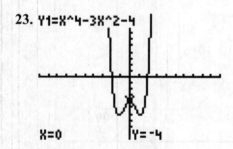

X=0   Y=-4

[–10, 10] by [–10, 10]

Answers will vary.

25. a.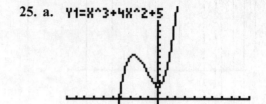

X=0   Y=5

[–10, 10] by [–10, 30]

b.

Maximum
X=-2.666667   Y=14.481481

[–10, 10] by [–10, 30]

The local maximum is approximately $(-2.67, 14.48)$.

c.

Minimum
X=1.3792E-6   Y=5

[–10, 10] by [–10, 30]

The local minimum is $(0, 5)$.

27. $y = x^4 - 4x^3 + 4x^2$

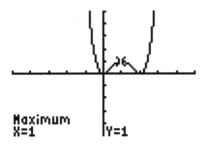

Maximum
X=1   Y=1

$[-5, 5]$ by $[-5, 5]$

The local maximum is $(1,1)$.

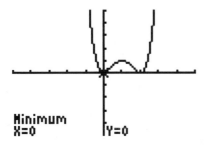

Minimum
X=0   Y=0

$[-5, 5]$ by $[-5, 5]$

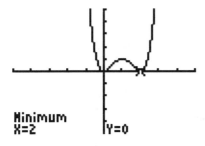

Minimum
X=2   Y=0

$[-5, 5]$ by $[-5, 5]$

The local minimums are $(0,0)$ and $(2,0)$.

Section 4.1 Exercises

29. a. Y1= -.1X^3+11X^2-100X

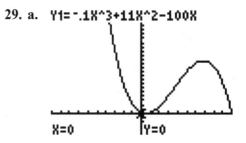

X=0   Y=0

$[-100, 100]$ by $[-5000, 25{,}000]$

There are two turning points.

b. Based on the physical context of the problem, both x and y need to be greater than or equal to zero.

c. Y1= -.1X^3+11X^2-100X

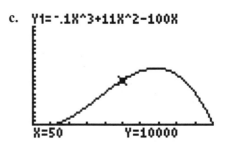

X=50   Y=10000

$[0, 100]$ by $[0, 25{,}000]$

d. Y1= -.1X^3+11X^2-100X

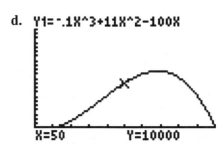

X=50   Y=10000

$[0, 100]$ by $[0, 25{,}000]$

Fifty units produce revenue of \$10,000.

31. a.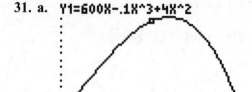

[0, 100] by [0, 30,000]

b.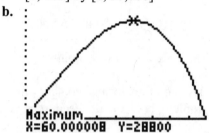

[0, 100] by [0, 30,000]

Selling 60 units produces a maximum daily revenue of $28,800.

c.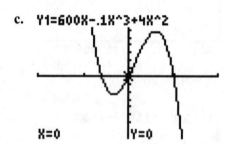

[−200, 200] by [−40,000, 40,000]

Answers will vary.

d. The graph in part a) best represents the physical situation. Both the number of units produced and the revenue must be greater than or equal to zero.

e. The graph is increasing on the interval $(0, 60)$.

33. a.

r, rate	S, future value
0	$2,000.00
5	$2,315.25
10	$2,662.00
15	$3,041.75
20	$3,456.00

b.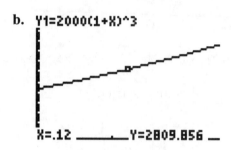

[0, 0.24] by [0, 5000]

c. At the 20% rate the investment yields $3456. At the 10% rate the investment yields $2662. Therefore the 20% rate yields an extra $794.

d. The 10% rate is more realistic.

35. a.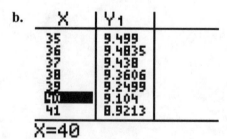

[0, 50] by [0, 12]

b.

X	Y₁
35	9.499
36	9.4835
37	9.438
38	9.3606
39	9.2499
40	9.104
41	8.9213

X=40

In 1990, when $x = 40$, the homicide rate was approximately 9 per 100,000 people.

c.

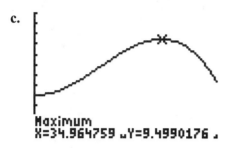

[0, 50] by [0, 12]

An *x*-value of 35 corresponds with the year 1985. The number of homicides is at a maximum in 1985.

d.   Consider the following table

X	Y₁
15	6.079
20	7.264
25	8.369
30	9.184
35	9.499
40	9.104
45	7.789

X=20

In 1970 the number of homicides per 100,000 people is approximately 7.264. In 1990 the number of homicides per 100,000 people is approximately 9.104. Calculating the average rate of change:

$$\frac{y_2 - y_1}{x_2 - x_1}$$

$$= \frac{9.104 - 7.264}{40 - 20}$$

$$= \frac{1.84}{20}$$

$$= 0.092$$

The average rate of change for 1970-1990 is approximately 0.09 homicides per 100,000 people per year.

37. a.   Y1=⁻.6182X^3+54.956X^2-1_

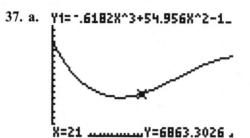

X=21 ....Y=6863.3026 .

[0, 42] by [0, 20,000]

b.

X	Y₁
26	8287
27	8627
28	8976.7
29	9332.5
30	9690.68
31	10047
32	10399

Y₁=9690.676

In 1980, when *x* = 30, the young adult population was approximately 9,690,676.

c.

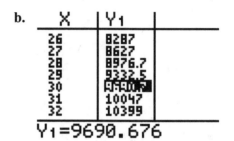

Minimum
X=15.732033 ⌐Y=6238.2623 .

[0, 42] by [0, 20,000]

When *x* is 15.73, the year is 1966. The population reaches a minimum in 1966.

39. a.   Y1=⁻1.832X^3+22.111X^2-5_

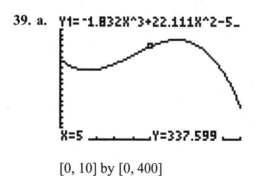

X=5 ....Y=337.599 .__

[0, 10] by [0, 400]

b.

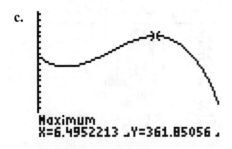

X	Y1
6	358.74
7	358.15
8	
9	247.82
10	116.09
11	‾81.34
12	‾355.5

Y1=324.843

In 1998, when $x = 8$, the national debt is approximately \$324.843 million.

c.

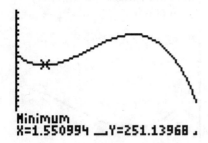

Maximum
X=6.4952213 Y=361.85056

[0, 10] by [0, 400]

Minimum
X=1.550994 Y=251.13968

[0, 10] by [0, 400]

The minimum national debt of approximately \$251 million occurs in 1992, while the maximum national debt of approximately \$362 million occurs in 1997.

41. a. $P(x) = R(x) - C(x)$

$$= \left(120x - 0.015x^2\right) - \left(10,000 + 60x - 0.03x^2 + 0.00001x^3\right)$$

$$= -0.00001x^3 + 0.015x^2 + 60x - 10,000$$

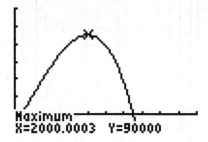

[0, 5000] by [−20,000, 120,000]

2000 units produced and sold produces a maximum profit.

b. The maximum profit is $90,000.

Section 4.2 Skills Check

1.

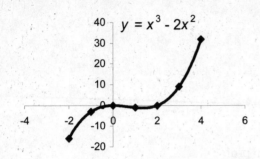

$y = x^3 - 2x^2$

3.

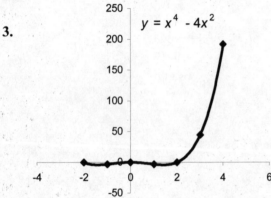

$y = x^4 - 4x^2$

5. a.

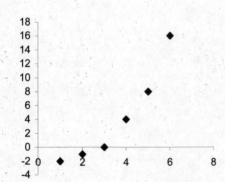

b. It appears based on the scatter plot that a cubic model will fit the data better.

7. a.

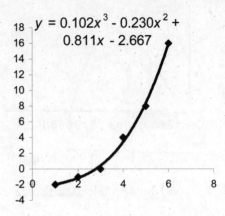

$y = 0.102x^3 - 0.230x^2 + 0.811x - 2.667$

b.

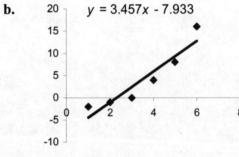

$y = 3.457x - 7.933$

c. Clearly the cubic model is better.

9. a.

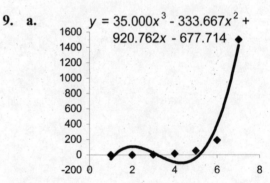

$y = 35.000x^3 - 333.667x^2 + 920.762x - 677.714$

b. $y = 12.515x^4 - 165.242x^3 + 748.000x^2 - 1324.814x + 738.286$

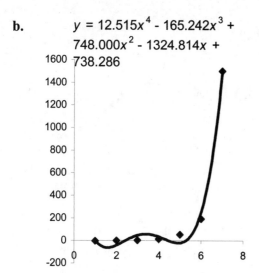

15. $y = 0.565x^3 + 2.425x^2 - 4.251x + 0.556$

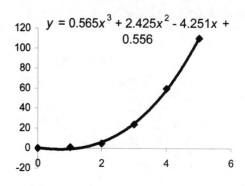

11. $y = x^4 - 4x^2 - 3x + 1$

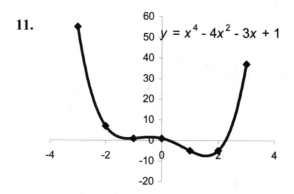

13.

x	$f(x)$	First Difference	Second Difference	Third Difference
0	0			
1	1	1		
2	5	4	3	
3	24	19	15	12
4	60	36	17	2
5	110	50	14	−3

The function f is not exactly cubic.

Section 4.2 Exercises

17. a.

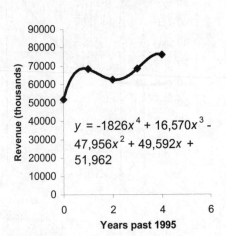

Hotel Revenue

The equation is
$$y = -1826x^4 + 16,570x^3 - 47,956x^2$$
$$+49,592x + 51,962.$$

b. $y = -1826x^4 + 16,570x^3$
$$-47,956x^2 + 49,592x + 51,962$$
$$y = -1826(3)^4 + 16,570(3)^3$$
$$-47,956(3)^2 + 49,592(3) + 51,962$$
$$y = 68,618$$

The hotel revenue is predicted as $68,618,000. The prediction is the same as the actual data point.

Using the unrounded model:

```
-1825.9999999952
X^4+16569.999999
962X^3+-47955.99
9999908X^2+49591
.999999934X+5196
2.000000002
              68618
```

19. a.

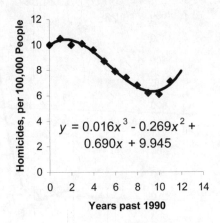

Homicide Rate

$y = 0.016x^3 - 0.269x^2 + 0.690x + 9.945$

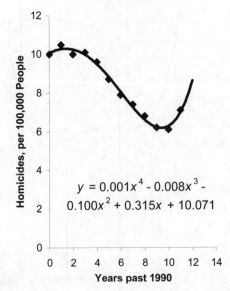

Homicide Rate

$y = 0.001x^4 - 0.008x^3 - 0.100x^2 + 0.315x + 10.071$

While both models appear to fit the data well, the quartic model is better.

b. Consider part a) above. The cubic model is
$$y = 0.0165x^3 - 0.269x^2 + 0.690x$$
$$+9.945.$$

c. Lower. The 9/11 terrorist attack skewed the number of deaths higher.

21. a.

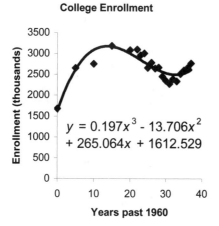

College Enrollment

$y = 0.197x^3 - 13.706x^2 + 265.064x + 1612.529$

b. The predicted college enrollment in 2000 is 2,869,492.

Using the unrounded model:

```
                      40
.19662594898303X
^3+ -13.706024235
179X^2+265.06350
514012X+1612.529
4030614
          2869.491567
```

23. a.

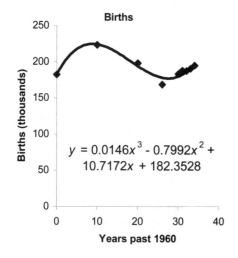

Births

$y = 0.0146x^3 - 0.7992x^2 + 10.7172x + 182.3528$

b.

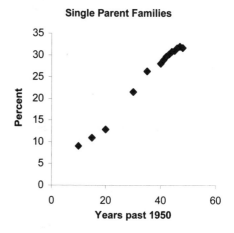

$Y_1 = 196.7008$

In 1994 the number of births is approximately 196,701.

25. a.

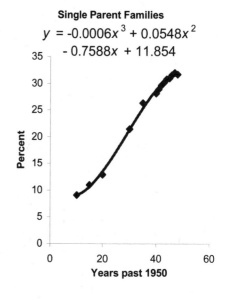

Single Parent Families

b.

Single Parent Families

$y = -0.0006x^3 + 0.0548x^2 - 0.7588x + 11.854$

c.

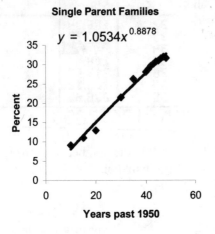

Single Parent Families

$y = 1.0534x^{0.8878}$

d. The cubic model appears to be a better fit.

27. a.

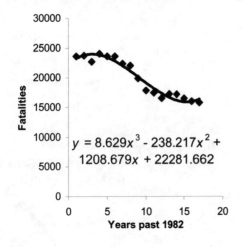

Alcohol Related Accidents

$y = 8.629x^3 - 238.217x^2 + 1208.679x + 22281.662$

b.

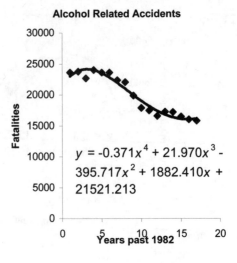

Alcohol Related Accidents

$y = -0.371x^4 + 21.970x^3 - 395.717x^2 + 1882.410x + 21521.213$

c. It appears that, based on parts a) and b), the quartic model is slightly better, although both models fit the data reasonably well.

29. a.

Movie Tickets

$y = 0.0003x^3 + 0.0156x^2 - 0.0804x + 4.2579$

b.

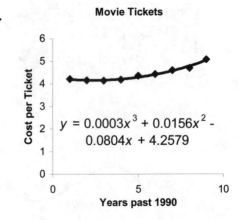

X	Y₁	Y₂
8	4.7805	5.7
9	5.0364	5.7
10	5.3412	5.7
11	5.6969	5.7
12	6.1055	5.7
13	6.5689	5.7
14	7.0891	5.7

X=11

In about 11 years, in the year 2001, the price of a movie will be $5.70.

31. a.

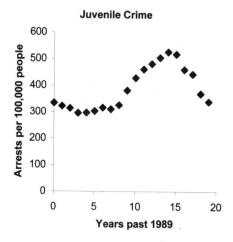

Based on the scatter plot, it appears that cubic model will fit the data well.

b.

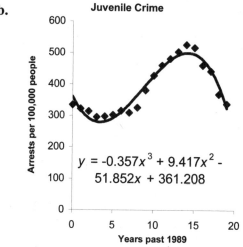

c. See part b) above.

d.

X	Y₁
7	337.14
10	427.13
13	493.69
16	478.96
19	325.05
22	-25.91
25	-631.8

X=10

In 1990, $f(10) = 427.13$.

In 1999, $f(19) = 325.05$.

The ratio is $\dfrac{f(19)}{f(10)} = \dfrac{325.05}{427.13} \approx 76.1\%$

33. a.

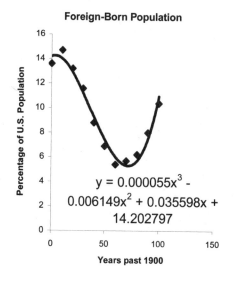

b.

```
                    75
5.4681429681425E
-5X^3+-.00614860
13986X^2+.035598
29059826X+14.202
797202798
         5.355514277
```

In 1975 the percentage of the U.S. population that is foreign-born is approximately 5.36%

Section 4.3 Skills Check

1. $x^3 - 16x = 0$

$x(x^2 - 16) = 0$

$x(x+4)(x-4) = 0$

$x = 0, x = -4, x = 4$

Checking graphically

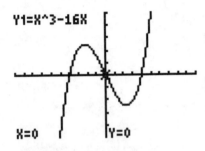

[-10, 10] by [-50, 50]

3. $x^4 - 4x^3 + 4x^2 = 0$

$x^2(x^2 - 4x + 4) = 0$

$x^2(x-2)(x-2) = 0$

$x^2 = 0 \Rightarrow x = 0$

$x = 0, x = 2$

Checking graphically

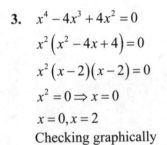

[-5, 5] by [-5, 10]

5. $4x^3 - 4x = 0$

$4x(x^2 - 1) = 0$

$4x(x+1)(x-1) = 0$

$x = 0, x = -1, x = 1$

Checking graphically

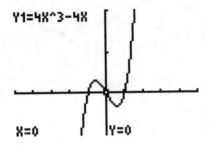

[-5, 5] by [-5, 10]

7. $x^3 - 4x^2 - 9x + 36 = 0$

$(x^3 - 4x^2) + (-9x + 36) = 0$

$x^2(x-4) + (-9)(x-4) = 0$

$(x-4)(x^2 - 9) = 0$

$(x-4)(x+3)(x-3) = 0$

$x = 4, x = -3, x = 3$

9. $3x^3 - 4x^2 - 12x + 16 = 0$

$(3x^3 - 4x^2) + (-12x + 16) = 0$

$x^2(3x-4) + (-4)(3x-4) = 0$

$(3x-4)(x^2 - 4) = 0$

$(3x-4)(x+2)(x-2) = 0$

$x = \dfrac{4}{3}, x = -2, x = 2$

11. $2x^3 - 16 = 0$

$2x^3 = 16$

$x^3 = 8$

$\sqrt[3]{x^3} = \sqrt[3]{8}$

$x = 2$

13. $\dfrac{1}{2}x^4 - 8 = 0$

$$\frac{1}{2}x^4 = 8$$

$$2\left(\frac{1}{2}x^4\right) = 2(8)$$

$$x^4 = 16$$

$$\sqrt[4]{x^4} = \pm\sqrt[4]{16}$$

$$x = \pm 2$$

15. $4x^4 - 8x^2 = 0$

$$4x^2\left(x^2 - 2\right) = 0$$

$$4x^2 = 0,\ x^2 - 2 = 0$$

$$4x^2 = 0 \Rightarrow x = 0$$

$$x^2 - 2 = 0 \Rightarrow x^2 = 2$$

$$\sqrt{x^2} = \pm\sqrt{2}$$

$$x = \pm\sqrt{2}$$

$$x = \pm\sqrt{2},\ x = 0$$

17. $0.5x^3 - 12.5x = 0$

$$0.5x\left(x^2 - 25\right) = 0$$

$$0.5x\left(x + 5\right)\left(x - 5\right) = 0$$

$$x = 0, x = -5, x = 5$$

19. $x^4 - 6x^2 + 9 = 0$

$$\left(x^2 - 3\right)\left(x^2 - 3\right) = 0$$

$$x^2 - 3 = 0,\ x^2 - 3 = 0$$

$$x^2 - 3 = 0 \Rightarrow x^2 = 3$$

$$\sqrt{x^2} = \pm\sqrt{3}$$

$$x = \pm\sqrt{3}$$

21. $f(x) = 0$ implies $x = -3, x = 1, x = 4$. Note that the x-intercepts are the solutions.

23. The x-intercepts (zeros) are the solutions.

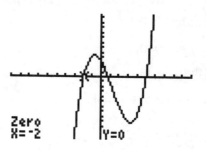

$[-10, 10]$ by $[-125, 125]$

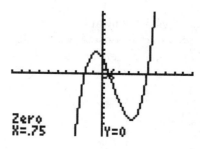

$[-10, 10]$ by $[-125, 125]$

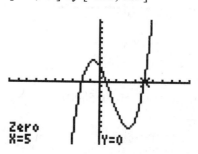

$[-10, 10]$ by $[-125, 125]$

$$x = -2, x = 0.75, x = 5$$

Section 4.3 Exercises

25. a. $R = 400x - x^3$

$400x - x^3 = 0$

$x\left(400 - x^2\right) = 0$

$x(20 - x)(20 + x) = 0$

$x = 0, \ 20 - x = 0, \ 20 + x = 0$

$x = 0, \ -x = -20, \ x = -20$

$x = 0, \ x = 20, \ x = -20$

In the physical context of the problem, selling zero units or selling 20 units will yield revenue of zero dollars.

b.   Yes.

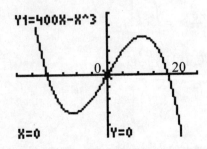

Y1=400X-X^3

X=0   Y=0

[−30, 30] by [−5000, 5000]

27. a. $R = \left(100{,}000 - 0.1x^2\right)x$

$\left(100{,}000 - 0.1x^2\right)x = 0$

$x = 0, \ 100{,}000 - 0.1x^2 = 0$

$-0.1x^2 = -100{,}000$

$x^2 = 1{,}000{,}000$

$x = \pm\sqrt{1{,}000{,}000}$

$x = 0, \ x = 1000, \ x = -1000$

In the physical context of the problem, selling zero units or selling 1000 units will yield revenue of zero dollars.

b.   Yes.

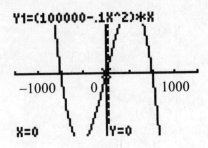

Y1=(100000−.1X^2)*X

−1000   0   1000

X=0   Y=0

[−2000, 2000] by
[−35,000,000, 35,000,000]

29. a.

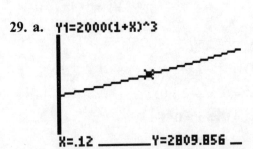

Y1=2000(1+X)^3

X=.12 ———— Y=2809.856 —

[0, 0.24] by [0, 5000]

b. $2662 = 2000\left(1 + r\right)^3$

$\left(1 + r\right)^3 = \dfrac{2662}{2000}$

$\sqrt[3]{\left(1 + r\right)^3} = \sqrt[3]{\dfrac{2662}{2000}}$

$1 + r = \sqrt[3]{1.331}$

$r = \sqrt[3]{1.331} - 1$

$r \approx 0.10 = 10\%$

c. $3456 = 2000\left(1 + r\right)^3$

$\left(1 + r\right)^3 = \dfrac{3456}{2000}$

$\sqrt[3]{\left(1 + r\right)^3} = \sqrt[3]{\dfrac{3456}{2000}}$

$1 + r = \sqrt[3]{1.728}$

$r = \sqrt[3]{1.728} - 1$

$r = 0.20 = 20\%$

31. a.   The height is x inches, since the distance cut is x units and that distance when folded forms the height of the box.

b. The length and width of box will be what is left after the corners are cut. Since each corner measures x inches square then the length and the width are $18-2x$.

c. $V = lwh$

$V = (18-2x)(18-2x)x$

$V = (324 - 36x - 36x + 4x^2)x$

$V = 324x - 72x^2 + 4x^3$

d. $V = 0$

$0 = 324x - 72x^2 + 4x^3$

From part c) above:

$0 = (18-2x)(18-2x)x$

$18 - 2x = 0, \ x = 0$

$18 - 2x = 0 \Rightarrow 2x = 18 \Rightarrow x = 9$

$x = 0, \ x = 9$

e. A box will not exist for either of the values calculated in part d) above. For both values of x, no tab will exist to fold up to form the box.

33. $400 = -x^3 + 2x^2 + 400x - 400$

$-1(x^3 - 2x^2 - 400x + 400) = 400$

$x^3 - 2x^2 - 400x + 400 = -400$

$x^3 - 2x^2 - 400x + 800 = 0$

$(x^3 - 2x^2) + (-400x + 800) = 0$

$x^2(x-2) + (-400)(x-2) = 0$

$(x-2)(x^2 - 400) = 0$

$(x-2)(x+20)(x-20) = 0$

$x = 2, x = -20, x = 20$

The negative answer does not make sense in the physical context of the question. Producing and selling 2 units or 20 units leads to a profit of $40,000.

35. a.

X	Y1
0	810
.1	240
.2	30
.3	0
.4	-30
.5	-240
.6	-810

X=0

t	0	0.1	0.2	0.3
s (cm/sec)	810	240	30	0

b. $0 = 30(3-10t)^3$

$(3-10t)^3 = \dfrac{0}{30}$

$\sqrt[3]{(3-10t)^3} = \sqrt[3]{0}$

$3 - 10t = 0$

$t = \dfrac{-3}{-10}$

$t = 0.3$

The solution in the table is the same as the solution found by the root method.

37. a.

X	Y1
0	838.95
1	861.51
2	856.95
3	835.16
4	806
5	779.37
6	765.12

Y1=779.365

In 1995 the estimated number of arrests is 779,365.

b.

X	Y1
3	835.16
4	806
5	779.37
6	765.12
7	773.14
8	813.31
9	895.5

Y1=813.31

In 1998 the estimated number of arrests is 813,310.

39. a. Applying the intersection of graphs method:

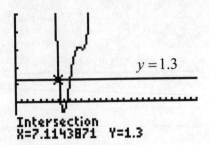

```
Intersection
X=7.1143871  Y=1.3
```

[0, 30] by [−2, 5]

After 7.11 years, in 1998, the percent change is 1.3%.

b. The model will be 1.3% again. Note that there is a second intersection point on the graph in part a).

Section 4.4 Skills Check

1.

$$3\overline{)\begin{array}{ccccc} 1 & -4 & 0 & 3 & 10 \\ & 3 & -3 & -9 & -18 \\ \hline 1 & -1 & -3 & -6 & -8 \end{array}}$$

$$x^3 - x^2 - 3x - 6 - \frac{8}{x-3}$$

3.

$$1\overline{)\begin{array}{ccccc} 2 & -3 & 0 & 1 & -7 \\ & 2 & -1 & -1 & 0 \\ \hline 2 & -1 & -1 & 0 & -7 \end{array}}$$

$$2x^3 - x^2 - x - \frac{7}{x-1}$$

5.

$$3\overline{)\begin{array}{ccccc} 2 & -4 & 0 & 3 & 18 \\ & 6 & 6 & 18 & 63 \\ \hline 2 & 2 & 6 & 21 & 81 \end{array}}$$

Since the remainder is not zero, 3 is not a solution of the equation.

7.

$$-3\overline{)\begin{array}{ccccc} -1 & 0 & -9 & 3 & 0 \\ & 3 & -9 & 54 & 171 \\ \hline -1 & 3 & -18 & 57 & 171 \end{array}}$$

Since the remainder is not zero, $x + 3$ is not a factor.

9.

$$-1\overline{)\begin{array}{cccc} -1 & 1 & 1 & -1 \\ & 1 & -2 & 1 \\ \hline -1 & 2 & -1 & 0 \end{array}}$$

One factor is $x - 1$. The new polynomial is $-x^2 + 2x - 1$. Solve $-x^2 + 2x - 1 = 0$.

$$x^2 - 2x + 1 = 0$$

$$(x-1)(x-1) = 0$$

$$x = 1, x = 1$$

The solutions are $x = -1$ and $x = 1$ (repeated two times).

11.

$$-5\overline{)\begin{array}{ccccc} 1 & 2 & -21 & -22 & 40 \\ & -5 & 15 & 30 & -40 \\ \hline 1 & -3 & -6 & 8 & 0 \end{array}}$$

One factor is $x + 5$. The new polynomial is $x^3 - 3x^2 - 6x + 8$. Applying the rational solutions test yields:

$$\frac{p}{q} = \pm\left(\frac{1,2,4,8}{1}\right) = \pm(1,2,4,8)$$

$$1\overline{)\begin{array}{cccc} 1 & -3 & -6 & 8 \\ & 1 & -2 & -8 \\ \hline 1 & -2 & -8 & 0 \end{array}}$$

One factor is $x - 1$. The new polynomial is $x^2 - 2x - 8$. Solve $x^2 - 2x - 8 = 0$.

$$(x-4)(x+2) = 0$$

$$x = 4, x = -2$$

The solutions are $x = 4$, $x = -2$, $x = 1$, and $x = -5$.

13. Applying the x-intercept method:

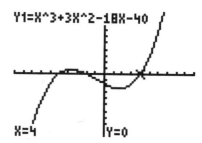

[–10, 10] by [–250, 250]

One solution appears to be $x = 4$.

$$4\overline{)\begin{array}{cccc} 1 & 3 & -18 & -40 \\ & 4 & 28 & 40 \\ \hline 1 & 7 & 10 & 0 \end{array}}$$

One factor is $x - 4$. The new polynomial is $x^2 + 7x + 10$. Solve $x^2 + 7x + 10 = 0$.

$$(x+2)(x+5) = 0$$

$$x = -2, x = -5$$

The solutions are $x = -5, x = -2, x = 4$.

15. Applying the *x*-intercept method:

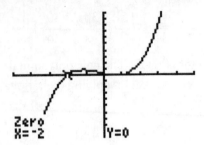

[–5, 5] by [–100, 100]

One solution appears to be $x = -2$.

$$-2\overline{)\ \begin{array}{cccc} 3 & 2 & -7 & 2 \end{array}}$$
$$\begin{array}{cccc} & -6 & 8 & -2 \end{array}$$
$$\overline{\begin{array}{cccc} 3 & -4 & 1 & 0 \end{array}}$$

One factor is $x + 2$. The new polynomial
is $3x^2 - 4x + 1$. Solve $3x^2 - 4x + 1 = 0$.

$$(3x - 1)(x - 1) = 0$$

$$x = \frac{1}{3}, x = 1$$

The solutions are $x = -2, x = 1, x = \frac{1}{3}$.

17. $\dfrac{p}{q} = \pm\left(\dfrac{1,2,3,4,6,12}{1}\right) = \pm(1,2,3,4,6,12)$

19. $\dfrac{p}{q} = \pm\left(\dfrac{1,2,4}{1,3,9}\right)$

$= \pm\left(1,2,4,\dfrac{1}{3},\dfrac{2}{3},\dfrac{4}{3},\dfrac{1}{9},\dfrac{2}{9},\dfrac{4}{9}\right)$

21. Applying the *x*-intercept method:

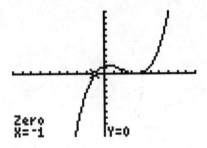

[–10, 10] by [–100, 100]

One solution appears to be $x = -1$.

$$-1\overline{)\ \begin{array}{cccc} 1 & -6 & 5 & 12 \end{array}}$$
$$\begin{array}{cccc} & -1 & 7 & -12 \end{array}$$
$$\overline{\begin{array}{cccc} 1 & -7 & 12 & 0 \end{array}}$$

One factor is $x + 1$. The new polynomial
is $x^2 - 7x + 12$. Solve $x^2 - 7x + 12 = 0$.

$$(x - 3)(x - 4) = 0$$

$$x = 3, x = 4$$

The solutions are $x = -1, x = 3, x = 4$.

23. Applying the *x*-intercept method:

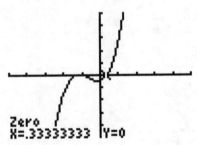

[–5, 5] by [–50, 50]

One solution appears to be $x = \dfrac{1}{3}$.

$$\frac{1}{3} \overline{)\begin{array}{rrrr} 9 & 18 & 5 & -4 \\ & 3 & 7 & 4 \\ \hline 9 & 21 & 12 & 0 \end{array}}$$

One factor is $x - \dfrac{1}{3}$. The new polynomial is

$9x^2 + 21x + 12$. Solve $9x^2 + 21x + 12 = 0$.

$3(3x^2 + 7x + 4) = 0$

$3(3x + 4)(x + 1) = 0$

$3x + 4 = 0, x + 1 = 0$

$x = -\dfrac{4}{3}, x = -1$

The solutions are $x = -\dfrac{4}{3}, x = -1, x = \dfrac{1}{3}$.

25. $x^3 = 10x - 7x^2$

$x^3 + 7x^2 - 10x = 0$

$x(x^2 + 7x - 10) = 0$

$x = 0, x^2 + 7x + 10 = 0$

Applying the quadratic formula:

$x = \dfrac{-7 \pm \sqrt{7^2 - 4(1)(-10)}}{2(1)}$

$x = \dfrac{-7 \pm \sqrt{89}}{2}$

The solutions are $x = 0, x = \dfrac{-7 \pm \sqrt{89}}{2}$.

27. Applying the x-intercept method:

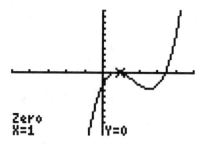

[–5, 5] by [–10, 10]

It appears that $w = 1$ is a zero.

$$1 \overline{)\begin{array}{rrrr} 1 & -5 & 6 & -2 \\ & 1 & -4 & 2 \\ \hline 1 & -4 & 2 & 0 \end{array}}$$

The remaining quadratic factor is

$w^2 - 4w + 2$.

Applying the quadratic formula:

$w = \dfrac{-(-4) \pm \sqrt{(-4)^2 - 4(1)(2)}}{2(1)}$

$w = \dfrac{4 \pm \sqrt{8}}{2}$

$w = \dfrac{4 \pm 2\sqrt{2}}{2}$

$w = 2 \pm \sqrt{2}$

The solutions are $w = 1, \ w = 2 \pm \sqrt{2}$.

29. Applying the x-intercept method:

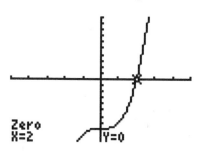

[–5, 5] by [–10, 10]

It appears that $z = 2$ is a zero.

$$2 \overline{)\begin{array}{rrrr} 1 & 0 & 0 & -8 \\ & 2 & 4 & 8 \\ \hline 1 & 2 & 4 & 0 \end{array}}$$

The remaining quadratic factor
is $z^2 + 2z + 4$.

Applying the quadratic formula:

$$z = \frac{-2 \pm \sqrt{(2)^2 - 4(1)(4)}}{2(1)}$$

$$z = \frac{-2 \pm \sqrt{-12}}{2}$$

$$z = \frac{-2 \pm 2i\sqrt{3}}{2}$$

$$z = -1 \pm i\sqrt{3}$$

The solutions are $z = 2, z = -1 \pm i\sqrt{3}$.

Section 4.4 Exercises

31. a.

$$
\begin{array}{r|rrrr}
50) & -0.2 & 66 & -1600 & -60{,}000 \\
    &   & -10 & 2800 & 60{,}000 \\
\hline
    & -0.2 & 56 & 1200 & 0
\end{array}
$$

The quadratic factor of $P(x)$ is $-0.2x^2 + 56x + 1200$.

b. $-0.2x^2 + 56x + 1200 = 0$

$-0.2(x^2 - 280x - 6000) = 0$

$-0.2(x + 20)(x - 300) = 0$

$x = -20, \ x = 300$

In the context of the problem, only the positive solution is reasonable. Producing and selling 300 units results in break-even.

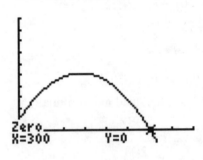

$[-20, 400]$ by $[-1000, 10{,}000]$

33. a.

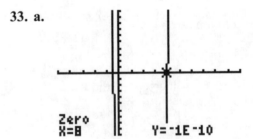

$[-10, 20]$ by $[-100, 100]$

b.   Based on the graph in part a), one x-intercept appears to be $x = 8$.

c.
$$8\overline{)\begin{array}{rrrr} -0.1 & 50.7 & -349.2 & -400 \\ & -0.8 & 399.2 & 400 \\ \hline -0.1 & 49.9 & 50 & 0 \end{array}}$$

The quadratic factor of $P(x)$
is $-0.1x^2 + 49.9x + 50$.

d. Based on parts b) and c), one zero
is $x = 8$. To find more zeros,
solve $-0.1x^2 + 49.9x + 50 = 0$.
$$-0.1\left(x^2 - 499 - 500\right) = 0$$
$$-0.1(x - 500)(x + 1) = 0$$
$$x = 500, \ x = -1$$
The zeros are $x = 500$, $x = -1$, $x = 8$.

e. Based on the context of the problem
producing and selling 8 units or 500
units results in break-even.

35. $R(x) = 9000$
$$1810x - 81x^2 - x^3 = 9000$$
$$x^3 + 81x^2 - 1810x + 9000 = 0$$
Since $x = 9$ is a solution,

$$9\overline{)\begin{array}{rrrr} 1 & 81 & -1810 & 9000 \\ & 9 & 810 & -9000 \\ \hline 1 & 90 & -1000 & 0 \end{array}}$$

The quadratic factor of $R(x)$
is $x^2 + 90x - 1000$. To determine more
solutions, solve $x^2 + 90x - 1000 = 0$.
$$(x + 100)(x - 10) = 0$$
$$x = -100, x = 10$$
Revenue of \$9000 is also achieved by
selling 10 units.

36. $R(x) = 1000$
$$250x - 5x^2 - x^3 = 1000$$
$$x^3 + 5x^2 - 250x + 1000 = 0$$
Since $x = 5$ is a solution,

$$5\overline{)\begin{array}{rrrr} 1 & 5 & -250 & 1000 \\ & 5 & 50 & -1000 \\ \hline 1 & 10 & -200 & 0 \end{array}}$$

The quadratic factor of $R(x)$
is $x^2 + 10x - 200$. To determine more
solutions, solve $x^2 + 10x - 200 = 0$.
$$(x + 20)(x - 10) = 0$$
$$x = -20, x = 10$$
Revenue of \$1000 is also achieved by
selling 10 units.

37. a. $y = 244$

$0.4566x^3 - 14.3085x^2 + 117.2978x + 107.8456 = 244$

$0.4566x^3 - 14.3085x^2 + 117.2978x - 136.1544 = 0$

b.

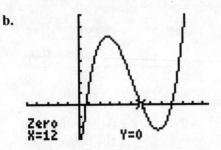

[–10, 25] by [–75,200]

It appears that $x = 12$ is a zero.

c.

$$12 \overline{)\ \begin{array}{rrrr} 0.4566 & -14.3085 & 117.2978 & -136.1544 \\ & 5.4792 & -105.9516 & 136.1544 \\ \hline 0.4566 & -8.8293 & 11.3462 & 0 \end{array}}$$

The quadratic factor of $P(x)$ is $0.4566x^2 - 8.8293x + 11.3462$.

d. $0.4566x^2 - 8.8293x + 11.3462 = 0$

$$x = \frac{-b \pm \sqrt{b^2 - 4ac}}{2a}$$

$$x = \frac{-(-8.8293) \pm \sqrt{(-8.8293)^2 - 4(0.4566)(11.3462)}}{2(0.4566)}$$

$$x = \frac{8.8293 \pm \sqrt{57.23383881}}{0.9132}$$

$$x = 17.953, \ x = 1.384$$

e. Based on the solutions in previous parts, the number of fatalities is 244 in 1982, 1992, and 1998.

39. Applying the intersection of graphs method:

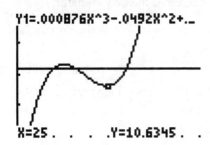

[0, 50] by [9, 13]

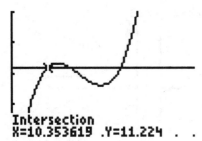

[0, 50] by [9, 13]

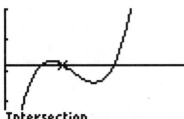

[0, 50] by [9, 13]

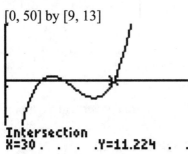

[0, 50] by [9, 13]

Based on the graphs, 11,224 births occur in 1971, 1976, and 1990.

Section 4.5 Skills Check

1. a. To find the vertical asymptote let
$q(x) = 0$.
$x - 5 = 0$
$x = 5$ is the vertical asymptote.

b. The degree of the numerator is less than the degree of the denominator. Therefore, $y = 0$ is the horizontal asymptote.

3. a. To find the vertical asymptote let
$q(x) = 0$.
$5 - 2x = 0$
$-2x = -5$
$x = \dfrac{-5}{-2} = \dfrac{5}{2}$ is the vertical asymptote.

b. The degree of the numerator is equal to the degree of the denominator. Therefore, $y = \dfrac{1}{-2} = -\dfrac{1}{2}$ is the horizontal asymptote.

5. a. To find the vertical asymptote let
$q(x) = 0$.
$x^2 - 1 = 0$
$(x+1)(x-1) = 0$
$x = -1, x = 1$ are the vertical asymptotes.

b. The degree of the numerator is greater than the degree of the denominator. Therefore, there is not a horizontal asymptote.

7. The function in part c) does not have a vertical asymptote. Its denominator cannot be zero. Parts a), b), and d) all have denominators that can equal zero for specific x-values.

9. a. The degree of the numerator is equal to the degree of the denominator. Therefore, $y = \dfrac{1}{1} = 1$ is the horizontal asymptote.

b. To find the vertical asymptote let
$q(x) = 0$.
$x - 2 = 0$
$x = 2$ is the vertical asymptote.

c.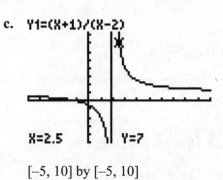

[−5, 10] by [−5, 10]

11. a. The degree of the numerator is less than the degree of the denominator. Therefore, $y = 0$ is the horizontal asymptote.

b. To find the vertical asymptote let
$q(x) = 0$.
$1 - x^2 = 0$
$x^2 = 1$
$\sqrt{x^2} = \pm\sqrt{1}$
$x = \pm 1$
$x = 1, x = -1$ are the vertical asymptotes.

c.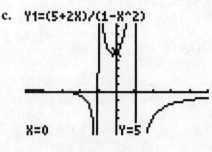

[−5, 5] by [−5, 10]

13. Y1=(X^2-9)/(X-3)

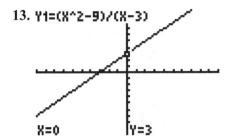

X=0   Y=3

[–10, 10] by [–10, 10]

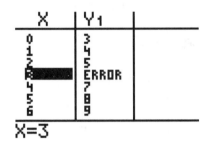

X=3

There is a hole in the graph at $x = 3$.

15. Y1=X^2/(X-1)

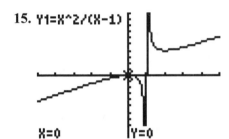

X=0   Y=0

[–5, 5] by [–10, 10]

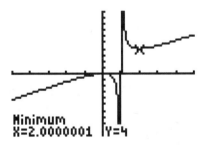

Minimum
X=2.0000001   Y=4

[–5, 5] by [–10, 10]

Based on the graphs, the turning points appear to be (0, 0) and (2, 4).

17.

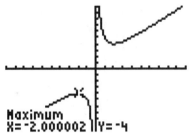

Maximum
X=-2.000002   Y=-4

[–10, 10] by [–10, 10]

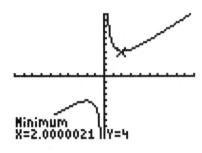

Minimum
X=2.0000021   Y=4

[–10, 10] by [–10, 10]

Based on the graphs, the turning points appear to be (–2, –4) and (2, 4).

19. a. Y1=(1-X^2)/(X-2)

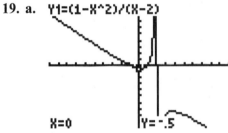

X=0   Y=-.5

[–10, 10] by [–10, 10]

b.

X	Y1
-2	.75
-1	0
0	-.5
1	0
2	ERROR
3	-8
4	-7.5

X=1

Based on the graph and the table, when $x = 1$, $y = 0$ and when $x = 3$, $y = -8$.

c.

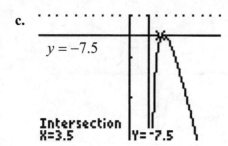

$y = -7.5$

Intersection
X=3.5   Y=-7.5

[–10, 10] by [–10, –7]

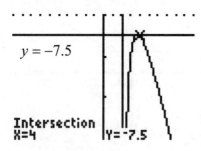

$y = -7.5$

Intersection
X=4   Y=-7.5

[–10, 10] by [–10, –7]

Based on the graphs it appears that if $y = -7.5$, then $x = 3.5$ or $x = 4$.

d. $-7.5 = \dfrac{1-x^2}{x-2}$ $\text{LCM}: x - 2$

$-7.5(x-2) = \left(\dfrac{1-x^2}{x-2}\right)(x-2)$

$-7.5x + 15 = 1 - x^2$

$x^2 - 7.5x + 14 = 0$

$10\left(x^2 - 7.5x + 14\right) = 10(0)$

$10x^2 - 75x + 140 = 0$

$(x-4)(10x-35) = 0$

$x - 4 = 0,\quad 10x - 35 = 0$

$x = 4, x = 3.5$

Both solutions check.

21. a.   Y1=(3-2X)/X

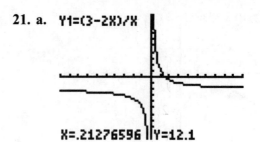

X=.21276596   Y=12.1

[–10, 10] by [–10, 10]

b.

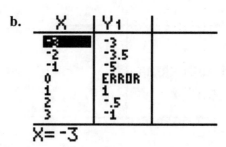

X= -3

Based on the graph and the table, when $x = 3$, $y = -1$ and when $x = -3$, $y = -3$.

c.

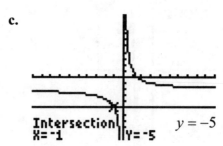

$y = -5$

Intersection
X=-1   Y= -5

[–10, 10] by [–10, 10]

Based on the graph it appears that if $y = -5$, then $x = -1$.

d. $-5 = \dfrac{3-2x}{x}$ $\text{LCM}: x$

$-5x = \left(\dfrac{3-2x}{x}\right)x$

$-5x = 3 - 2x$

$-3x = 3$

$x = -1$

The solution checks.

23. $\dfrac{x^2+1}{x-1}+x=2+\dfrac{2}{x-1}$ LCM: $x-1$

$(x-1)\left(\dfrac{x^2+1}{x-1}+x\right)=(x-1)\left(2+\dfrac{2}{x-1}\right)$

$x^2+1+x(x-1)=2(x-1)+2$

$x^2+1+x^2-x=2x-2+2$

$2x^2-x+1=2x$

$2x^2-3x+1=0$

$(2x-1)(x-1)=0$

$x=\dfrac{1}{2},\;\;x=1$

$x=1$ does not check.

The only solution is $x=\dfrac{1}{2}$.

24. $\dfrac{x}{x-2}-x=1+\dfrac{2}{x-2}$ LCM: $x-2$

$(x-2)\left(\dfrac{x}{x-2}-x\right)=(x-2)\left(1+\dfrac{2}{x-2}\right)$

$x-x(x-2)=1(x-2)+2$

$x-x^2+2x=x-2+2$

$-x^2+3x=x$

$-x^2+2x=0$

$-x(x-2)=0$

$-x=0,\;\;x-2=0$

$x=0,\;\;x=2$

$x=2$ does not check.

The only solution is $x=0$.

Section 4.5 Exercises

25. a. $\overline{C}=\dfrac{400+50(500)+0.01(500)^2}{500}$

$\overline{C}=\dfrac{27,900}{500}=55.8$

The average cost is $55.80 per unit.

b. $\overline{C}=\dfrac{400+50(60)+0.01(60)^2}{60}$

$\overline{C}=\dfrac{3436}{60}=57.2\overline{6}$

The average cost is $57.27 per unit.

c. $\overline{C}=\dfrac{400+50(100)+0.01(100)^2}{100}$

$\overline{C}=\dfrac{5500}{100}=55$

The average cost is $55 per unit.

d. No. Consider the graph of the function.

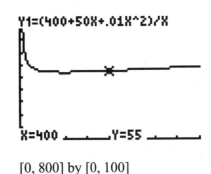

Y1=(400+50X+.01X^2)/X

X=400   Y=55

[0, 800] by [0, 100]

Note that the graph passes through a minimum and then begins to increase.

27. a. $y=\dfrac{400(5)}{5+20}=\dfrac{2000}{25}=80$

$5000 in advertising expenditures results in sales volume of $80,000.

b. When $x=-20$, the denominator is zero and the function is undefined. Since advertising expenditures cannot be negative, x cannot be -20 in the context of the problem.

29. a. Y1=(1000+30X+.1X^2)/X

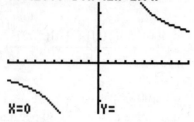

X=0   Y=

[−10, 10] by [−200, 300]

b. Y1=(1000+30X+.1X^2)/X

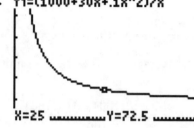

X=25 ⌣⌣⌣⌣⌣Y=72.5 ⌣⌣⌣⌣

[0, 50] by [0, 300]

c. The graph in part b) fits the context of the problem since both the viewing window for both x and $\overline{C}(x)$ are greater than or equal to zero.

d.

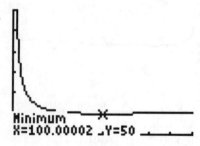

Minimum
X=100.00002 ⌣Y=50 ⌣⌣⌣

[0, 200] by [0, 300]

The minimum average cost of \$50 occurs when 10,000 units are produced.

31. a. Y1=(30+40X)/(5+2X)

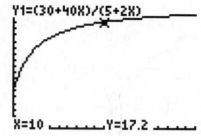

X=10 ⌣⌣⌣⌣Y=17.2 ⌣⌣⌣⌣

[0, 20] by [0, 20]

b. Y1=(30+40X)/(5+2X)

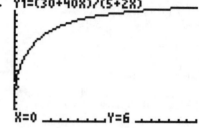

X=0 ⌣⌣⌣⌣Y=6 ⌣⌣⌣⌣

[0, 20] by [0, 20]

$f(0) = 6$. The initial number of employees for the startup company is 6.

c. Y1=(30+40X)/(5+2X)

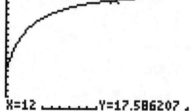

X=12 ⌣⌣⌣⌣Y=17.586207 .

[0, 20] by [0, 20]

$f(12) \approx 17.59$. After 12 months, the number of employees for the startup company is approximately 18.

33. a. To find the vertical asymptote let
$q(x) = 0$.
$100 - p = 0$
$-p = -100$
$p = 100$ is the vertical asymptote.

b. Since $p \neq 100$, 100% of the impurities can not be removed from the waste water.

35. a.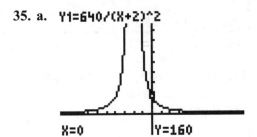

[–10, 10] by [–200, 1000]

X	Y1
-5	71.111
-4	160
-3	640
-2	ERROR
-1	640
0	160
1	71.111

X= -2

Based on the graph and the table, the vertical asymptote occurs at $p = -2$.

b.

Price	Sales Volume
5	13,061
20	1322
50	237
100	62
200	16
500	3

c. The domain of function in the context of the problem is $p \geq 0$. There is no vertical asymptote on the restricted domain.

d. The horizontal asymptote is $V = 0$. As the price grows without bound, the sales volume approaches zero units.

37. a.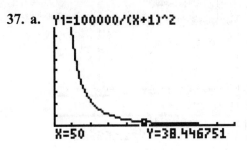

[0, 100] by [0, 1000]

b. The horizontal asymptote is $p = 0$.

c. As the price falls, the quantity demanded increases.

39. a. $S = \dfrac{40}{x} + \dfrac{x}{4} + 10$ LCM: $4x$

$$\left(\frac{40}{x}\right)\left(\frac{4}{4}\right) + \frac{x}{4}\left(\frac{x}{x}\right) + \left(\frac{10}{1}\right)\left(\frac{4x}{4x}\right)$$

$$\frac{160}{4x} + \frac{x^2}{4x} + \frac{40x}{4x}$$

$$\frac{x^2 + 40x + 160}{4x}$$

$$S = \frac{x^2 + 40x + 160}{4x}$$

b. $21 = \dfrac{x^2 + 40x + 160}{4x}$

$$21(4x) = \left(\frac{x^2 + 40x + 160}{4x}\right)(4x)$$

$$84x = x^2 + 40x + 160$$

$$x^2 - 44x + 160 = 0$$

$$(x - 40)(x - 4) = 0$$

$$x = 40, \quad x = 4$$

After 4 hours or 40 hours of training, the monthly sales will be $21,000.

41. a. Y1=(5+3X)/(2X+1)

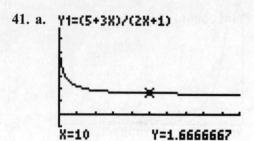

X=10   Y=1.6666667

[0, 20] by [0, 8]

b. Since the degree of the numerator equals the degree of the denominator, the horizontal asymptote is $H = \dfrac{3}{2}$. As the amount of training increases, the time it takes to assemble one unit approaches 1.5 hours.

c.

X	Y1
14.5	1.6167
15	1.6129
15.5	1.6094
16	1.6061
16.5	1.6029
17	1.6
17.5	1.5972

X=17

It takes 17 days of training to reduce the time it takes to assemble one unit to 1.6 hours.

b.

X	Y1
100	37.052
110	31.372
120	25.988
130	20.998
140	16.459
150	12.394
160	8.7996

X=130

The predicted value is 21.0%, while the actual value in the table is 21.2%. The values are relatively close.

43. a.

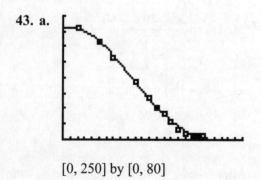

[0, 250] by [0, 80]

Section 4.6 Skills Check

1. Applying the *x*-intercept method:

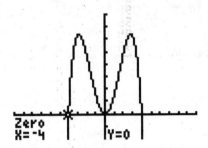

[−10, 10] by [−20, 80]

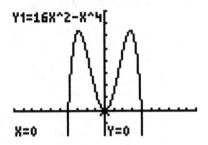

[−10, 10] by [−20, 80]

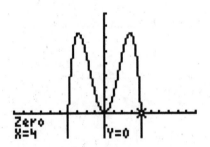

[−10, 10] by [−20, 80]

The function is greater than or equal to zero on the interval $\left[-4,4\right]$ or when $-4 \le x \le 4$.

3. Applying the *x*-intercept method:

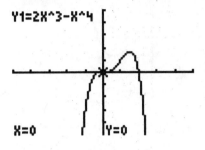

[−5, 5] by [−5, 5]

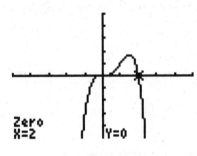

[−5, 5] by [−5, 5]

The function is less than zero on the interval $\left(-\infty,0\right)\cup\left(2,\infty\right)$ or when $x < 0$ or $x > 2$.

5. Applying the *x*-intercept method:

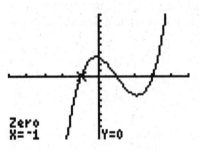

[−5, 5] by [−10, 10]

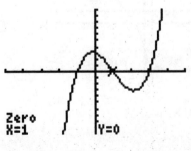

[−5, 5] by [−10, 10]

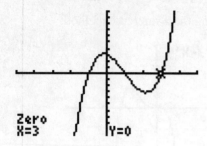

[–5, 5] by [–10, 10]

The function is greater than or equal to zero on the interval $[-1,1]\cup[3,\infty)$ or when $-1\le x\le 1$ or $x\ge 3$.

7. Applying the intersection of graphs method:

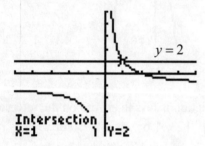

[–5, 5] by [–10, 10]

Note that the graphs intersect when $x = 1$. Also note that a vertical asymptote occurs at $x = 0$. Therefore, $\dfrac{4-2x}{x} > 2$ on the interval $(0,1)$ or when $0 < x < 1$.

9. Applying the intersection of graphs method:

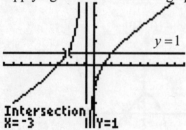

[–10, 10] by [–5, 5]

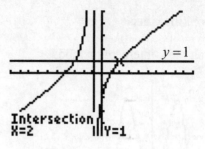

[–10, 10] by [–5, 5]

Note that the graphs intersect when $x = -3$ and $x = 2$. Also note that a vertical asymptote occurs at $x + 1 = 0$ or $x = -1$. Therefore, $\dfrac{x}{2}+\dfrac{x-2}{x+1}\le 1$ on the interval $(-\infty,-3]\cup(-1,2]$ or when $x\le -3$ or $-1 < x\le 2$.

11. Applying the intersection of graphs method:

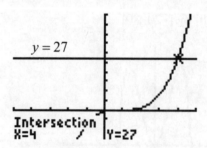

[–5, 5] by [–15, 50]

Note that the graphs intersect when $x = 4$. Therefore $(x-1)^3 > 27$ on the interval $(4,\infty)$ or when $x > 4$.

13. Applying the intersection of graphs method:

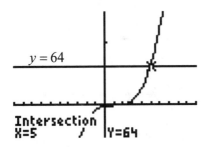

[–10, 10] by [–50, 150]

Note that the graphs intersect when $x = 5$.
Therefore, $(x-1)^3 < 64$ on the interval
$(-\infty, 5)$ or when $x < 5$.

15. Applying the intersection of graphs method:

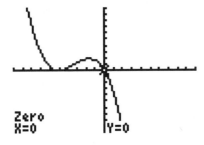

[–10, 10] by [–100, 100]

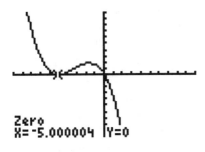

[–10, 10] by [–100, 100]

Note that the graphs intersect when $x = -5$
and $x = 0$. Therefore, $-x^3 - 10x^2 - 25x \le 0$
on the interval $\{-5\} \cup [0, \infty)$ or when
$x = -5$ or $x \ge 0$.

17. a. $f(x) < 0 \Rightarrow x < -3$ or $0 < x < 2$

b. $f(x) \ge 0 \Rightarrow -3 \le x \le 0$ or $x \ge 2$

19. a. $f(x) \ge 2 \Rightarrow \dfrac{1}{2} \le x \le 3$

Section 4.6 Exercises

21. $R = 400x - x^3$

$\qquad = x\left(400 - x^2\right)$

$\qquad = x(20 - x)(20 + x)$

To find the zeros, let $R = 0$ and solve for x.

$R = 0$

$x(20 - x)(20 + x) = 0$

$x = 0, \ 20 - x = 0, \ 20 + x = 0$

$x = 0, \ x = 20, \ x = -20$

Note that since x represents product sales, only positve values of x make sense in the context of the question.

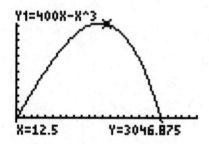

[0, 25] by [-500, 3500]

Based on the graph and the zeros calculated above, the revenue is positive, $R > 0$, in the interval $(0, 20)$ or when $0 < x < 20$. Selling between 0 and 20 units, not inclusive, generates positive revenue.

23. a. $V > 0$

$1296x - 144x^2 + 4x^3 > 0$

Find the zeros:

$1296x - 144x^2 + 4x^3 = 0$

$4x^3 - 144x^2 + 1296x = 0$

$4x\left(x^2 - 36x + 324\right) = 0$

$4x(x - 18)(x - 18) = 0$

$4x = 0, \ x - 18 = 0, \ x - 18 = 0$

$x = 0, \ x = 18, \ x = 18$

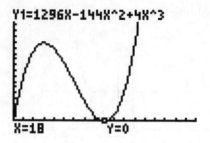

[0, 36] by [-500, 5000]

Based on the graph and the zeros calculated above, the volume is positive in the interval $(0, 18) \cup (18, \infty)$ or when $0 < x < 18$ or $x > 18$.

b. In the context of the question, the largest possible cut is 18 centimeters. Therefore, to generate a positive volume, the size of the cut, x, must be in the interval $(0, 18)$ or $0 < x < 18$.

25. $C(x) \geq 1200$

$3x^3 - 6x^2 - 300x + 1800 \geq 1200$

$3x^3 - 6x^2 - 300x + 600 \geq 0$

Find the zeros:

$3x^3 - 6x^2 - 300x + 600 = 0$

$3\left(x^3 - 2x^2 - 100x + 200\right) = 0$

$3\left[\left(x^3 - 2x^2\right) + \left(-100x + 200\right)\right] = 0$

$3\left[x^2(x - 2) + (-100)(x - 2)\right] = 0$

$3(x - 2)\left(x^2 - 100\right) = 0$

$3(x - 2)(x + 10)(x - 10) = 0$

$x - 2 = 0, \ x + 10 = 0, \ x - 10 = 0$

$x = 2, \ x = -10, \ x = 10$

Sign chart:

Function	---	+++	---	+++
3	+++	+++	+++	+++
$(x-10)$	---	---	---	+++
$(x+10)$	---	+++	+++	+++
$(x-2)$	---	---	+++	+++

$$\quad\quad\quad -10 \quad\quad 2 \quad\quad 10$$

Based on the sign chart, the function is greater than zero on the intervals $(-10,2)$ and $(10,\infty)$. Considering the context of the question, the number of units can not be negative. The endpoints of the interval would be part of the solution because the question uses the phrase "at least." Therefore, total cost is at least $120,000 if $0 \le x \le 2$ or if $x \ge 10$. In interval notation the solution is $[0,2]$ or $[10,\infty)$.

Applying the x-intercept method:

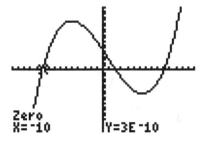

[–15, 15] by [–2000, 2000]

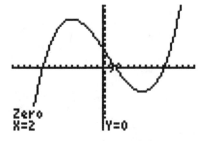

[–15, 15] by [–2000, 2000]

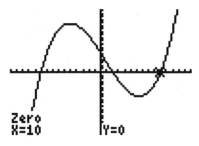

[–15, 15] by [–2000, 2000]

In context of the problem, the number of units must be greater than or equal to zero. Therefore, based on the graphs, the total cost is greater than or equal to $120,000 on the intervals $[0,2]$ or $[10,\infty)$ or when $0 \le x \le 2$ or $x \ge 10$.

27. Applying the intersection of graphs method:

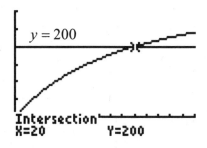

[0, 30] by [–50, 300]

Note that the graphs intersect when $x = 20$. Therefore, $\dfrac{400x}{x+20} \ge 200$ on the interval $[20,\infty)$ or when $x \ge 20$.

Therefore, spending $20,000 or more on advertising creates sales of at least $200,000.

29. Applying the intersection of graphs method

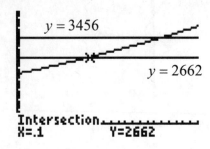

[0, 0.25] by [–500, 4500]

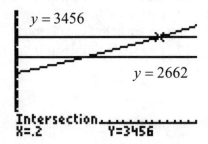

[0, 0.25] by [–500, 4500]

Note that the graphs intersect when $r = 0.10$ or $r = 0.20$. Therefore,

$2662 \leq 2000(1+r)^3 \leq 3456$ on the interval $[0.10, 0.20]$ or when $0.10 \leq r \leq 0.20$.

Therefore, interest rates between 10% and 20% inclusive generate future values between \$2662 and \$3456 inclusive.

31. Applying the intersection of graphs method:

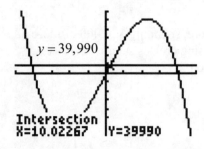

[–250, 250] by [–350,000, 350,000]

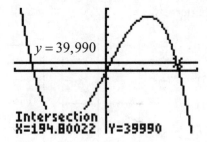

[–250, 250] by [–350,000, 350,000]

Note that the graphs intersect when $x \approx 10.02$ and when $x \approx 194.80$. Therefore, $R = 4000x - 0.1x^3 \geq 39,990$ on the interval $[10.02, 194.80]$ or when $10.02 \leq x \leq 194.80$.

Therefore, producing and selling between 10 units and 195 units inclusive generates revenue of at least \$39,990.

33. Considering the supply function and solving for q:

$6p - q = 180$

$-q = 180 - 6p$

$q = 6p - 180$

Considering the demand function and solving for q:

$(p + 20)q = 30,000$

$q = \dfrac{30,000}{p + 20}$

Supply > Demand

$6p - 180 > \dfrac{30,000}{p + 20}$ LCM: $p + 20$

$(p + 20)(6p - 180) > (p + 20)\left(\dfrac{30,000}{p + 20}\right)$

$(p + 20)(6p - 180) > 30,000$

$(p + 20)(6p - 180) - 30,000 > 0$

Applying the *x*-intercept method:

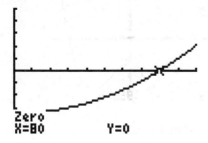

[0, 100] by [−50,000, 50,000]

When the price is at least $80, supply exceeds demand. Note that only positive values of *p* make sense in the context of the question.

35. a. Applying the intersection of graphs method:

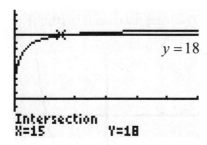

[0, 60] by [−10, 25]

Note that the graphs intersect when
$t = 15$. Therefore, $f(t) = \dfrac{30 + 40t}{5 + 2t} < 18$
on the interval $[0, 15)$ or when
$0 \le t < 15$.

For the first 15 months of the operation, the number of employees is at most 18.

b. The number of employees is less than 18 until month 15. Therefore, the number of employees is less than 18 on the interval $[0, 15)$ or when $0 \le t < 15$.

Thinking in terms of discrete months, for months 0 through 14 the number of employees is less than 18.

Chapter 4 Skills Check

1. The degree of the polynomial is the highest exponent. In this case, the degree of the polynomial is 4.

2. A fourth degree polynomial function is called a quartic function.

3. Y1=-4X^3+4X^2+1

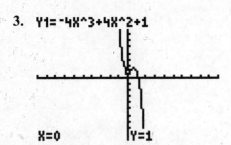

[-10, 10] by [-10, 10]

Yes. The graph is complete on the given viewing window.

4. a. Y1=X^4-4X^2-20

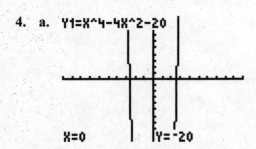

[-10, 10] by [-10, 10]

b. Y1=X^4-4X^2-20

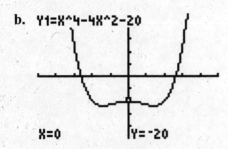

[-5, 5] by [-50, 50]

Viewing windows may vary.

5. a. Y1=X^3-3X^2-4

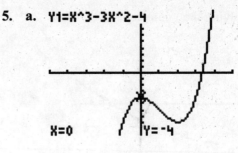

[-5, 5] by [-10, 10]

b.

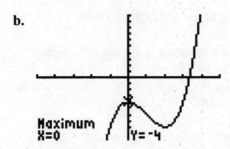

[-5, 5] by [-10, 10]

The local maximum is $(0, -4)$.

c.

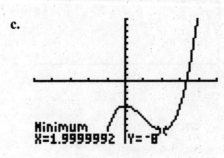

[-5, 5] by [-10, 10]

The local minimum is $(2, -8)$.

6. a. $y = x^3 + 11x^2 - 16x + 40$

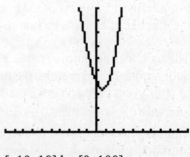

[-10, 10] by [0, 100]

b. No. The graph is not complete.

c.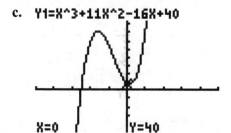

[–25, 25] by [–250, 500]

Viewing windows may vary.

d.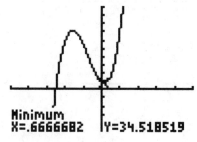

[–25, 25] by [–250, 500]

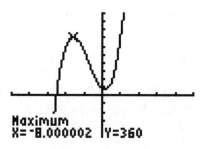

[–25, 25] by [–250, 500]

7. $x^3 - 16x = 0$

$x(x^2 - 16) = 0$

$x(x + 4)(x - 4) = 0$

$x = 0, \quad x + 4 = 0, \quad x - 4 = 0$

$x = 0, \quad x = -4, \quad x = 4$

8. $2x^4 - 8x^2 = 0$

$2x^2(x^2 - 4) = 0$

$2x^2(x + 2)(x - 2) = 0$

$2x^2 = 0, \quad x + 2 = 0, \quad x - 2 = 0$

$x = 0, \quad x = -2, \quad x = 2$

9. $x^4 - x^3 - 20x^2 = 0$

$x^2(x^2 - x - 20) = 0$

$x(x - 5)(x + 4) = 0$

$x = 0, \quad x - 5 = 0, \quad x + 4 = 0$

$x = 0, \quad x = 5, \quad x = -4$

10. $x^3 - 15x^2 + 56x = 0$

$x(x^2 - 15x + 56) = 0$

$x(x - 7)(x - 8) = 0$

$x = 0, \quad x - 7 = 0, \quad x - 8 = 0$

$x = 0, \quad x = 7, \quad x = 8$

11. $4x^3 - 20x^2 - 4x + 20 = 0$

$4(x^3 - 5x^2 - x + 5) = 0$

$4\left[(x^3 - 5x^2) + (-x + 5)\right] = 0$

$4\left[x^2(x - 5) + -1(x - 5)\right] = 0$

$4(x - 5)(x^2 - 1) = 0$

$4(x - 5)(x + 1)(x - 1) = 0$

$x - 5 = 0, \quad x + 1 = 0, \quad x - 1 = 0$

$x = 5, \quad x = -1, \quad x = 1$

12. $12x^3 - 9x^2 - 48x + 36 = 0$

$3\left(4x^3 - 3x^2 - 16x + 12\right) = 0$

$3\left[\left(4x^3 - 3x^2\right) + \left(-16x + 12\right)\right] = 0$

$3\left[x^2\left(4x - 3\right) + \left(-4\right)\left(4x - 3\right)\right] = 0$

$3\left(4x - 3\right)\left(x^2 - 4\right) = 0$

$3\left(4x - 3\right)\left(x + 2\right)\left(x - 2\right) = 0$

$4x - 3 = 0, \quad x + 2 = 0, \quad x - 2 = 0$

$x = \dfrac{3}{4}, \quad x = -2, \quad x = 2$

13. Applying the x-intercept method:

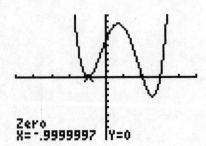

$[-5, 5]$ by $[-10, 10]$

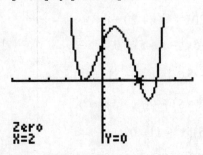

$[-5, 5]$ by $[-10, 10]$

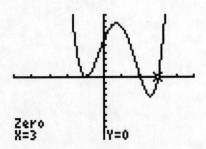

$[-5, 5]$ by $[-10, 10]$

Based on the graphs, the solutions are $x = -1, x = 2,$ and $x = 3$.

14. Applying the x-intercept method:

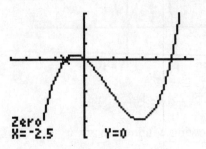

$[-10, 15]$ by $[-2500, 1500]$

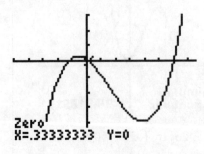

$[-10, 15]$ by $[-2500, 1500]$

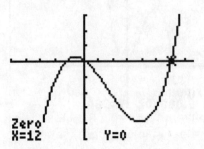

$[-10, 15]$ by $[-2500, 1500]$

Based on the graphs, the solutions are $x = -2.5, x = \dfrac{1}{3},$ and $x = 12$.

15. $\left(x - 4\right)^3 = 8$

$\sqrt[3]{\left(x - 4\right)^3} = \sqrt[3]{8}$

$x - 4 = 2$

$x = 6$

16. $5(x-3)^4 = 80$

$$(x-3)^4 = 16$$

$$\sqrt[4]{(x-3)^4} = \sqrt[4]{16}$$

$$x - 3 = 2$$

$$x = 5$$

17.
$$\begin{array}{r|rrrrr} 2 & 4 & -3 & 0 & 2 & -8 \\ & & 8 & 10 & 20 & 44 \\ \hline & 4 & 5 & 10 & 22 & 36 \end{array}$$

$$4x^3 + 5x^2 + 10x + 22 + \frac{36}{x-2}$$

18.
$$\begin{array}{r|rrrr} 1 & 2 & 5 & -11 & 4 \\ & & 2 & 7 & -4 \\ \hline & 2 & 7 & -4 & 0 \end{array}$$

The remaining polynomial is $2x^2 + 7x - 4$.

Set the polynomial equal to zero and solve.

$$2x^2 + 7x - 4 = 0$$

$$(2x-1)(x+4) = 0$$

$$2x - 1 = 0, x + 4 = 0$$

$$2x - 1 = 0 \Rightarrow 2x = 1 \Rightarrow x = \frac{1}{2}$$

$$x + 4 = 0 \Rightarrow x = -4$$

The solutions are $x = 1, x = -4$,

and $x = \frac{1}{2}$.

19. Applying the x-intercept method:

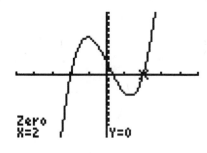

[−5, 5] by [−20, 20]

It appears that $x = 2$ is a zero.

$$\begin{array}{r|rrrr} 2 & 3 & -1 & -12 & 4 \\ & & 6 & 10 & -4 \\ \hline & 3 & 5 & -2 & 0 \end{array}$$

The remaining quadratic factor is $3x^2 + 5x - 2$.

Applying the quadratic formula:

$$x = \frac{-(5) \pm \sqrt{(5)^2 - 4(3)(-2)}}{2(3)}$$

$$x = \frac{-5 \pm \sqrt{49}}{6}$$

$$x = \frac{-5 \pm 7}{6}$$

$$x = \frac{-5 + 7}{6} = \frac{2}{6} = \frac{1}{3}$$

or

$$x = \frac{-5 - 7}{6} = \frac{-12}{6} = -2$$

The solutions are $x = 2, x = -2$,

and $x = \frac{1}{3}$

20. Applying the x-intercept method:

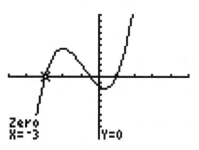

[−5, 5] by [−20, 20]

It appears that $x = -3$ is a zero.

$$\begin{array}{r|rrrr} -3 & 2 & 5 & -4 & -3 \\ & & -6 & 3 & 3 \\ \hline & 2 & -1 & -1 & 0 \end{array}$$

The remaining quadratic factor is $2x^2 - x - 1$.

Applying the quadratic formula:

$$x = \frac{-(-1) \pm \sqrt{(-1)^2 - 4(2)(-1)}}{2(2)}$$

$$x = \frac{1 \pm \sqrt{9}}{4}$$

$$x = \frac{1 \pm 3}{4}$$

$$x = \frac{1+3}{4} = \frac{4}{4} = 1$$

or

$$x = \frac{1-3}{4} = \frac{-2}{4} = -\frac{1}{2}$$

The solutions are $x = -3, x = -\frac{1}{2}$, and $x = 1$.

21. a. To find y-intercepts, let $x = 0$ and solve for y.

$$y = \frac{1-(0)^2}{0+2} = \frac{1}{2}$$

$$\left(0, \frac{1}{2}\right)$$

To find x-intercepts, let the numerator equal zero and solve for x.

$$1 - x^2 = 0$$

$$x^2 = 1$$

$$\sqrt{x^2} = \pm\sqrt{1}$$

$$x = \pm 1$$

$$(-1, 0), (1, 0)$$

b. To find the vertical asymptote let $q(x) = 0$.

$$x + 2 = 0$$

$$x = -2$$

$x = -2$ is the vertical asymptote.

The degree of the numerator is greater than the degree of the denominator. Therefore, there is not a horizontal asymptote.

c.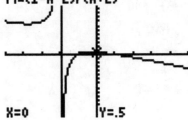

$[-5, 5]$ by $[-15, 15]$

22. a. To find y-intercepts, let $x = 0$ and solve for y.

$$y = \frac{3(0) - 2}{0 - 3} = \frac{-2}{-3} = \frac{2}{3}$$

$$\left(0, \frac{2}{3}\right)$$

To find x-intercepts, let the numerator equal zero and solve for x.

$$3x - 2 = 0$$

$$3x = 2$$

$$x = \frac{2}{3}$$

$$\left(\frac{2}{3}, 0\right)$$

b. To find the vertical asymptote let $q(x) = 0$.

$$x - 3 = 0$$

$$x = 3$$

$x = 3$ is the vertical asymptote.

The degree of the numerator equals the degree of the denominator. Therefore, the ratio of the leading coefficients is the horizontal asymptote.

$$y = \frac{3}{1} = 3$$

c.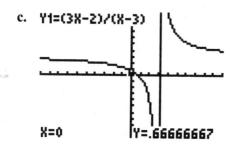

[−10, 10] by [−10, 10]

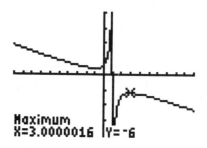

[−10, 10] by [−20, 20

The local maximum is $(3, -6)$, while the local minimum is $(-1, 2)$.

23.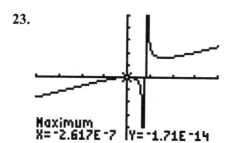

[−20, 20] by [−50, 50]

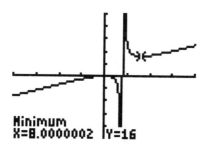

[−20, 20] by [−50, 50]

The local maximum is $(0, 0)$, while the local minimum is $(8, 16)$.

24.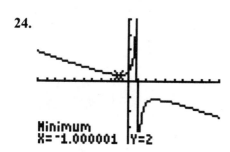

[−10, 10] by [−20, 20]

25. a.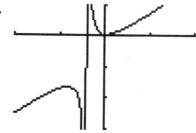

[−10, 10] by [−30, 10]

b.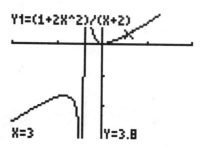

[−10, 10] by [−30, 10]

[−10, 10] by [−30, 10]

When $x = 1$, $y = 1$. When $x = 3$, $y = 3.8$.

c.

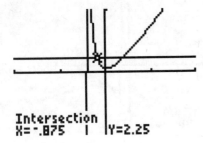

Intersection
X=-.875 | Y=2.25

[–10, 10] by [–10, 10]

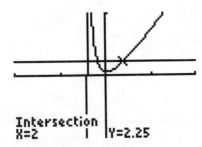

Intersection
X=2 | Y=2.25

[–10, 10] by [–10, 10]

If $y = 2.25$, then $x = -0.875$ or $x = 2$.

d. $\dfrac{9}{4} = \dfrac{1+2x^2}{x+2}$ LCM: $4(x+2)$

$$4(x+2)\left(\dfrac{9}{4}\right) = 4(x+2)\left(\dfrac{1+2x^2}{x+2}\right)$$

$$9(x+2) = 4(1+2x^2)$$

$$9x+18 = 4+8x^2$$

$$8x^2 - 9x - 14 = 0$$

$$(8x+7)(x-2) = 0$$

$$8x+7 = 0 \Rightarrow 8x = -7 \Rightarrow x = -\dfrac{7}{8}$$

$$x-2 = 0 \Rightarrow x = 2$$

The solutions are $x = 2, x = -\dfrac{7}{8}$.

26. $x^4 - 13x^2 + 36 = 0$

$$(x^2 - 9)(x^2 - 4) = 0$$

$$(x+3)(x-3)(x+2)(x-2) = 0$$

$$x+3 = 0, \quad x-3 = 0, \quad x+2 = 0, \quad x-2 = 0$$

$$x = -3, \quad x = 3, \quad x = -2, \quad x = 2$$

27. Applying the x-intercept method:

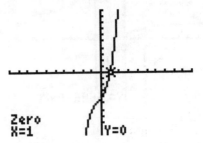

Zero
X=1 | Y=0

[–10, 10] by [–10, 10]

It appears that $x = 1$ is a zero.

$$\begin{array}{r|rrrr}
1) & 1 & 1 & 2 & -4 \\
   &   & 1 & 2 & 4 \\
\hline
   & 1 & 2 & 4 & 0
\end{array}$$

The remaining quadratic factor
is $x^2 + 2x + 4$.
Set the remaining polynomial equal to
zero and solve.

$$x^2 + 2x + 4 = 0$$

$$x = \dfrac{-b \pm \sqrt{b^2 - 4ac}}{2a}$$

$$x = \dfrac{-(2) \pm \sqrt{(2)^2 - 4(1)(4)}}{2(1)}$$

$$x = \dfrac{-2 \pm \sqrt{-12}}{2}$$

$$x = \dfrac{-2 \pm 2i\sqrt{3}}{2}$$

$$x = -1 \pm i\sqrt{3}$$

The solutions are $x = 1, x = -1 \pm i\sqrt{3}$.

28. Applying the *x*-intercept method:

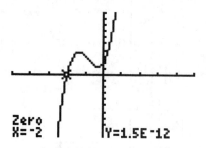

[−5, 5] by [−10, 10]

It appears that $x = -2$ is a zero.

$$
\begin{array}{r|rrrr}
-2 & 4 & 10 & 5 & 2 \\
   &   & -8 & -4 & -2 \\
\hline
   & 4 & 2 & 1 & 0
\end{array}
$$

The remaining quadratic factor is $4x^2 + 2x + 1$.
Set the remaining polynomial equal to zero and solve.

$$4x^2 + 2x + 1 = 0$$

$$x = \frac{-b \pm \sqrt{b^2 - 4ac}}{2a}$$

$$x = \frac{-(2) \pm \sqrt{(2)^2 - 4(4)(1)}}{2(4)}$$

$$x = \frac{-2 \pm \sqrt{-12}}{8}$$

$$x = \frac{-2 \pm 2i\sqrt{3}}{8}$$

$$x = \frac{-1 \pm i\sqrt{3}}{4}$$

$$x = \frac{-1}{4} \pm \frac{\sqrt{3}}{4}i$$

The solutions are $x = -2$, $x = \dfrac{-1}{4} + \dfrac{\sqrt{3}}{4}i$,

and $\dfrac{-1}{4} - \dfrac{\sqrt{3}}{4}i$.

29. $x^3 - 5x^2 \geq 0$

$\quad x^2(x - 5) \geq 0$

$\quad x^2(x - 5) = 0$

$\quad x - 5 = 0, \quad x^2 = 0$

$\quad x = 5, \quad x = 0$

Sign chart:

Function	---	---	+++
x^2	+++	+++	+++
$(x-5)$	---	---	+++

$$ 0 5$$

Based on the sign chart, the function is greater than or equal to zero on the interval $[5, \infty)$ or when $x \geq 5$. In addition, the function is equal to zero when $x = 0$.

30. Applying the *x*-intercept method:

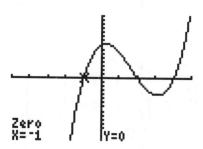

[−5, 5] by [−15, 15]

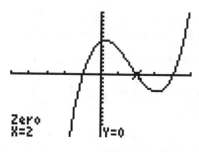

[−5, 5] by [−15, 15]

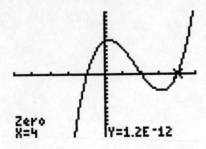

[–5, 5] by [–15, 15]

Based on the graphs, the function is greater than or equal to zero on the intervals $[-1,2]$ or $[4,\infty)$ or when $x \geq 4$ or $-1 \leq x \leq 2$.

31. $2 < \dfrac{4x-6}{x}$

$\dfrac{4x-6}{x} > 2$

Applying the intersection of graphs method:

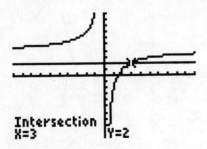

[–10, 10] by [–10, 10]

Note that the graphs intersect when $x = 3$. Also note that a vertical asymptote occurs at $x = 0$. Therefore, $\dfrac{4x-6}{x} > 2$ on the interval $(-\infty, 0) \cup (3, \infty)$ or when $x < 0$ or $x > 3$.

32. $\dfrac{5x-10}{x+1} \geq 20$

Applying the intersection of graphs method:

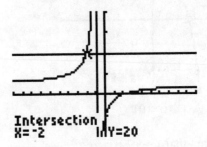

[–10, 10] by [–20, 40]

Note that the graphs intersect when $x = -2$. Also note that a vertical asymptote occurs at $x = -1$. Therefore, $\dfrac{5x-10}{x+1} \geq 20$ on the interval $[-2,-1)$ or when $-2 \leq x < -1$.

Chapter 4 Review Exercises

33. a.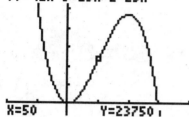

$[-100, 200]$ by $[-2000, 60,000]$

b.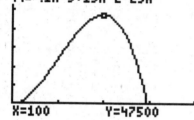

$[0, 200]$ by $[0, 60,000]$

c.

X	Y1
46	20856
47	21578
48	22301
49	23025
50	23750
51	24475
52	25199

Y1=23750

When 50,000 units are produced, the revenue is $23,750.

34.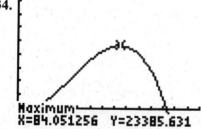

$[0, 150]$ by $[-6000, 40,000]$

When 84,051 units are sold, the maximum revenue of $23,385.63 is generated.

35. a. Using the table feature of a TI-83 graphing calculator:

r (rate)	S (future value)
1%	5307.60
5%	6700.48
10%	8857.81
15%	11,565.30

b.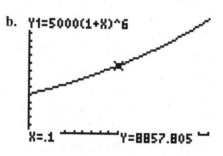

$[0, 0.2]$ by $[-1000, 15,000]$

c.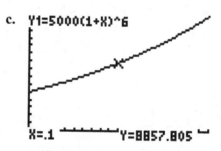

$[0, 0.2]$ by $[-1000, 15,000]$

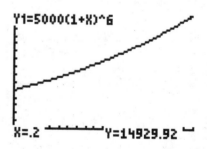

$[0, 0.2]$ by $[-1000, 15,000]$

The difference in the future values is $14,929.92 - 8857.81 = \$6072.11$.

36. a.

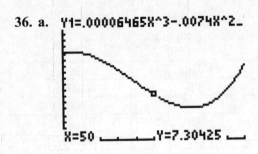

[0, 100] by [0, 20]

b.

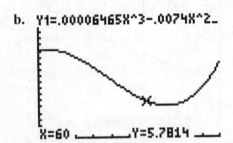

[0, 100] by [0, 20]

In 1960, the model predicts the percentage to be 5.78%. The prediction is relatively close to actual value of 5.4%.

c. Applying intersection of graphs method:

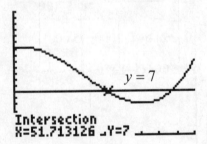

[0, 100] by [0, 20]

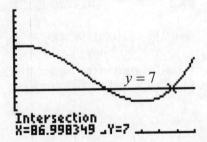

[0, 100] by [0, 20]

Based on the graphs, in approximately 1952 and again in 1987 the percentage is 7%.

37. a.

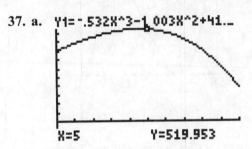

[0, 10] by [−100, 600]

b.

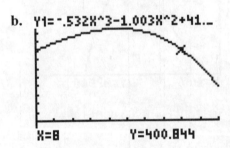

[0, 10] by [−100, 600]

Based on the model, the United Nations debt is approximately $400.844 million in 1998.

c.

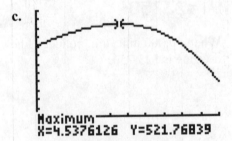

[0, 10] by [−100, 600]

The maximum debt is approximately $521.768 million occurring between 1994 and 1995.

38. a.

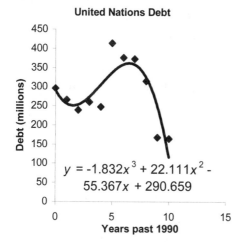

b. See part a) above.

39.

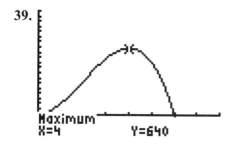

[0, 8] by [0, 1000]

An intensity level of 4 allows the maximum amount of photosynthesis to take place.

40.
$$9261 = 8000(1+r)^3$$
$$(1+r)^3 = \frac{9261}{8000}$$
$$\sqrt[3]{(1+r)^3} = \sqrt[3]{\frac{9261}{8000}}$$
$$1+r = 1.05$$
$$r = 1.05-1$$
$$r = 0.05$$

An interest rate of 5% creates a future value of $9261 after 3 years.

41. a. $V = 0$
$$324x - 72x^2 + 4x^3 = 0$$
$$4x^3 - 72x^2 + 324x = 0$$
$$4x(x^2 - 18x + 81) = 0$$
$$4x(x-9)(x-9) = 0$$
$$4x = 0, \quad x-9=0, \quad x-9=0$$
$$x = 0, \quad x = 9, \quad x = 9$$

b. If the values of x from part a are used to cut squares from corners of a piece of tin, no box can be created. Either no square is cut or the squares encompass all the tin. Therefore the volume of the box is zero.

c. Reasonable values of x would allow for a box to be created. An x-value larger than zero and less than half the length of the edge of the piece of tin would allow for a box to be created. Therefore, reasonable values are
$$0 < x < \frac{18}{2} \quad \text{or} \quad 0 < x < 9.$$

42. a.

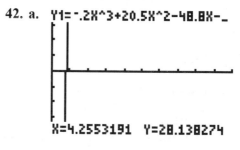

[0, 40] by [−15, 20]

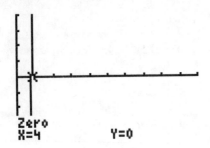

[0, 40] by [-15, 20]

The x-intercept is $(4,0)$.

b.
$$4\overline{\smash{)}\begin{array}{rrrr} -0.2 & 20.5 & -48.8 & -120 \\ & -0.8 & 78.8 & 120 \\ \hline -0.2 & 19.7 & 30 & 0 \end{array}}$$

The remaining quadratic factor is $-0.2x^2 + 19.7x + 30$.

c. Set the remaining polynomial equal to zero and solve.

$-0.2x^2 + 19.7x + 30 = 0$

$x = \dfrac{-b \pm \sqrt{b^2 - 4ac}}{2a}$

$x = \dfrac{-(19.7) \pm \sqrt{(19.7)^2 - 4(-0.2)(30)}}{2(-0.2)}$

$x = \dfrac{-19.7 \pm \sqrt{412.09}}{-0.4}$

$x = \dfrac{-19.7 \pm 20.3}{-0.4}$

$x = \dfrac{-19.7 + 20.3}{-0.4}, \; x = \dfrac{-19.7 - 20.3}{-0.4}$

$x = -1.5, \; x = 100$

The solutions are $x = 4$, $x = -1.5$, and $x = 100$.

d. Since negative solutions do not make sense in the context of the question, break-even occurs when 400 units are produced or when 10,000 units are produced.

43. a. Since the degree of the numerator is less than the degree of the denominator, the horizontal asymptote is $C = 0$.

b. As the time increases, the concentration of the drug approaches zero percent.

c.

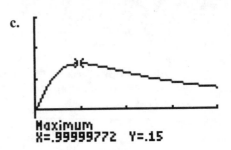

[0, 4] by [0, 0.3]

The maximum drug concentration is 15% occurring after one hour.

44. a. $\overline{C}(0)$ does not exist. If no units are produced, an average cost per unit can not be calculated.

b. Since the degree of the numerator equals the degree of the denominator, the horizontal asymptote is $\overline{C}(x) = \dfrac{50}{1} = 50$.

As the number of units produced increases without bound, the average cost per unit approaches $50.

c. The function decreases as x increases.

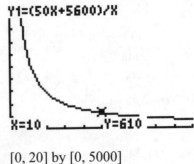

[0, 20] by [0, 5000]

45. a. `Y1=(30X^2+12000)/X`

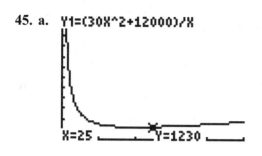

[0, 50] by [0, 12,000]

b.

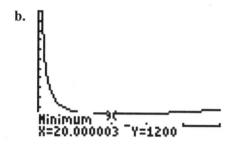

[0, 50] by [0, 12,000]

The minimum average cost is $1200 occurring when 20 units are produced.

b.

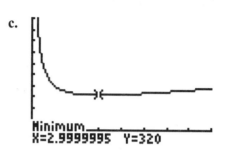

Using 5 plates creates a cost of $336.

c.

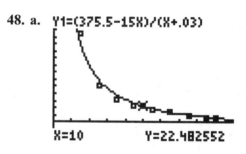

[0, 8] by [−100, 1000]

Using 3 plates creates a minimum cost of $320.

46. `Y1=30000/(X+20)`

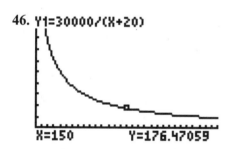

[0, 300] by [0, 1000]

47. a. $C(x) = 200 + 20x + \dfrac{180}{x}$

$C(x) = \dfrac{200x}{x} + \dfrac{20x^2}{x} + \dfrac{180}{x}$

$C(x) = \dfrac{20x^2 + 200x + 180}{x}$

48. a. `Y1=(375.5-15X)/(X+.03)`

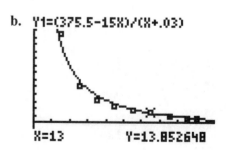

[0, 20] by [−25, 150]

b. `Y1=(375.5-15X)/(X+.03)`

[0, 20] by [−25, 150]

During the 1993-94 school year, the model predicts 13.85 students per computer. The prediction is close to the actual value of 14 students per computer, as displayed in the table.

1994 inclusive, the homicide rate is at least 8.13 per 100,000 people.

49. Applying the intersection of graphs method:

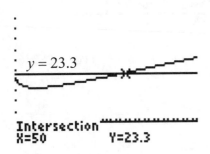

[4, 80] by [−10, 50]

Note that $S \geq 23.3$ occurs on the interval $[50, \infty)$ or when $x \geq 50$. Fifty or more hours of training results in sales greater than $23,300.

50. Applying the intersection of graphs method:

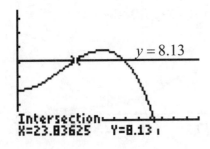

[0, 75] by [−2, 15]

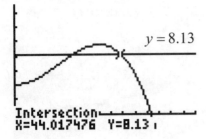

[0, 75] by [−2, 15]

Note that $y \geq 8.13$ when $23.836 \leq x \leq 44.017$. Between 1974 and

51. Applying the intersection of graphs method:

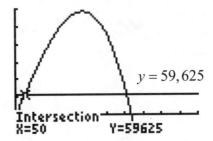

[0, 1000] by [−50,000, 300,000]

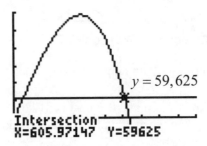

[0, 1000] by [−50,000, 300,000]

Note that $R \geq 59,625$ when $x \geq 50$ and $x \leq 605.97$. Selling 50 or more units but no more than 605 units creates a revenue stream of at least $59,625.

52. Applying the intersection of graphs method

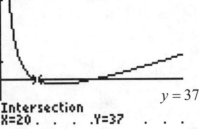

[0, 100] by [30, 50]

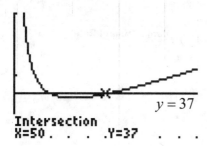

[0, 100] by [30, 50]

$\overline{C} \le 37$ when $20 \le x \le 50$. The average cost is at most $37 when between 20 and 50 units inclusive are produced.

53. Applying the intersection of graphs method:

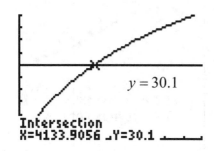

[0, 10,000] by [0, 50]

Note that $p \ge 30.1$ when $x \ge 4133.91$. To remove at least 30.1% of the particulate pollution will cost at least $4133.91.

Group Activity/Extended Applications

Unemployment Rates

1.

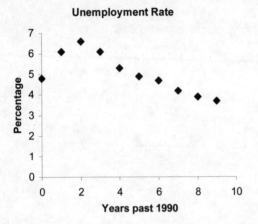

2.

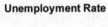

$$y = 0.0227x^3 - 0.3555x^2 + 1.2352x + 5.0101$$

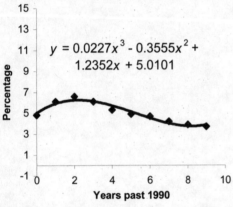

3. See Exercise 2 above.

4.

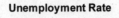

5.

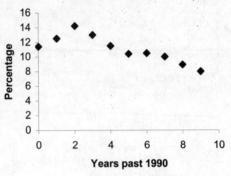

$$y = 0.0282x^3 - 0.4685x^2 + 1.5711x + 11.6078$$

6. See Exercise 5 above.

7.

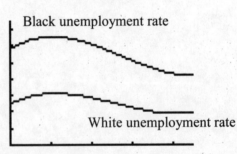

[0, 9] by [0, 15]

8. Black unemployment rates are much higher than white unemployment rates throughout the 1990's.

9.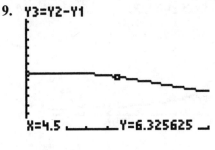

[0, 9] by [0, 15]

The unemployment rate gap seems to be decreasing.

13. No. The graph suggests that the unemployment rates move together in a more or less linear fashion. That is, as one rate rises, the other rate rises. Based on the data analysis, affirmative action plans might be useful in helping to narrow the unemployment rate gap.

10.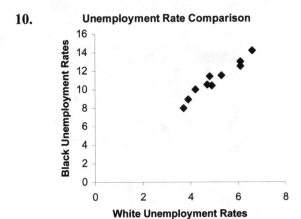

11.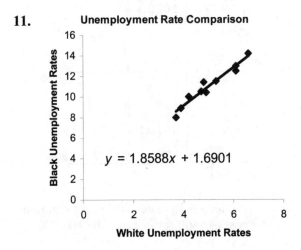

12. See Exercise 11 above.

Printing

A. 1. Assuming the printer uses 10 plates, then $1000 \cdot 10 = 10,000$ impressions can be made per hour. If 10,000 impressions are made per hour, it will take $\dfrac{100,000}{10,000} = 10$ hours to complete all the invitations.

2. Since it costs $128 per hour to run the press, the cost of using 10 plates is $10 \cdot 128 = \$1280$.

3. The 10 plates cost $10 \cdot 8 = \$80$.

4. The total cost of finishing the job is $1280 + 80 = \$1360$

B. 1. Let $x =$ number of plates. Then, the cost of the plates is $8x$.

2. Using x plates implies x invitations can be made per impression.

3. $1000x$ invitations per hour
Creating all 100,000 invitations would require $\dfrac{100,000}{1000x} = \dfrac{100}{x}$ hours.

4. $C(x) = 8x + 128\left(\dfrac{100}{x}\right)$

$C(x) = 8x + \left(\dfrac{12,800}{x}\right)$, where

x represents the number of plates and $C(x)$ represents the cost of the 100,000 invitations in dollars.

5.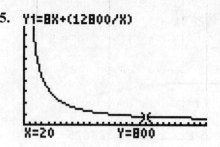

[0, 30] by [0, 10,000]

Producing 20 plates minimizes the cost, since the number of plates, x, that can be produced is between 1 and 20 inclusive.

6. Considering a table of values yields

X	Y₁
1	12808
2	6416
3	4290.7
4	3232
5	2600
6	2181.3
7	1884.6

X=1

X	Y₁
8	1664
9	1494.2
10	1360
11	1251.6
12	1162.7
13	1088.6
14	1026.3

X=8

X	Y₁
14	1026.3
15	973.33
16	928
17	888.94
18	855.11
19	825.68
20	800

X=20

Producing 20 plates creates a minimum cost of $800 for printing the 100,000 invitations.

CHAPTER 5
Systems of Equations and Matrices

Chapter 5 Algebra Toolbox

1. The constant of proportionality is 12.

2. $y = 10x$

   $x = \dfrac{y}{10}$

   $x = \dfrac{1}{10}y$

   The constant of proportionality is $\dfrac{1}{10}$.

3. No. $\dfrac{5}{6} \neq \dfrac{10}{18}$.

4. Yes. $\dfrac{8}{3} = \dfrac{24}{9}$. Note that $8 \cdot 9 = 3 \cdot 24$.

5. Does $\dfrac{4}{6} = \dfrac{2}{3}$?

   $4 \cdot 3 = 6 \cdot 2$

   $12 = 12$

   Yes. The pairs are proportional.

6. Does $\dfrac{5.1}{2.3} = \dfrac{51}{23}$?

   $5.1 \cdot 23 = 2.3 \cdot 51$

   $117.3 = 117.3$

   Yes. The pairs are proportional.

7. $y = kx$

   $18 = k(2)$

   $2k = 18$

   $k = 9$

   $y = 9x$

   Let $x = 11$.

   $y = 9(11)$

   $y = 99$

8. $x = ky$

   $3 = k(18)$

   $18k = 3$

   $k = \dfrac{3}{18} = \dfrac{1}{6}$

   $x = \dfrac{1}{6}y$

   Let $x = 13$.

   $13 = \dfrac{1}{6}y$

   $6(13) = 6\left(\dfrac{1}{6}y\right)$

   $y = 99$

9. If the pairs of triples are proportional, then there exists a number k such that $y = kx$ for all pairs (x, y).

   $y = kx$

   $20 = k(8)$

   $k = \dfrac{20}{8} = 2.5$

   $y = 2.5x$

   Checking the other pairs:

   $12.5 = 2.5(6)$

   $12.5 \neq 15$

   The pairs of triples are not proportional.

10. If the pairs of triples are proportional, then there exists a number k such that $y = kx$ for all pairs (x, y).

$$y = kx$$
$$8.2 = k(4.1)$$
$$k = \frac{8.2}{4.1} = 2$$
$$y = 2x$$

Checking the other pairs:
$$13.6 = 6.8(2)$$
$$13.6 = 13.6$$
and
$$18.6 = 2(9.3)$$
$$18.6 = 18.6$$

The pairs of triples are proportional.

11. If the pairs of triples are proportional, then there exists a number k such that $y = kx$ for all pairs (x, y).

$$y = kx$$
$$-3.4 = k(-1)$$
$$k = \frac{-3.4}{-1} = 3.4$$
$$y = 3.4x$$

Checking the other pairs yields
$$10.2 = 3.4(6)$$
$$10.2 \neq 20.4$$

The pairs of triples are not proportional.

12. If the pairs of triples are proportional, then there exists a number k such that $y = kx$ for all pairs (x, y).

$$y = kx$$
$$2 = k(5)$$
$$k = \frac{2}{5} = 0.4$$
$$y = 0.4x$$

Checking the other pairs:
$$-3 = 0.4(-12)$$
$$-3 \neq -4.8$$

The pairs of triples are not proportional.

13. Substituting yields:

$$x - 2y + z = 8$$
$$1 - 2(-2) + 3 = 8$$
$$8 = 8$$
and
$$2x - y + 2z = 10$$
$$2(1) - (-2) + 2(3) = 10$$
$$2 + 2 + 6 = 10$$
$$10 = 10$$
and
$$3x - 2y + z = 5$$
$$3(1) - 2(-2) + 3 = 5$$
$$10 \neq 5$$

The given values do not satisfy the system of equations.

14. Substituting yields:

$$x - 2y + 2z = -9$$
$$-1 - 2(4) + 2(0) = -9$$
$$-9 = -9$$

and

$$2x - 3y + z = -14$$
$$2(-1) - 3(4) + 0 = -14$$
$$-14 = -14$$

and

$$x + 2y - 3z = 7$$
$$-1 + 2(4) - 3(0) = 7$$
$$7 = 7$$

The given values satisfy the system of equations.

15. Substituting yields:

$$x + y - z = 4$$
$$5 + (-2) - (-1) = 4$$
$$4 = 4$$

and

$$2x + 3y - z = 5$$
$$2(5) + 3(-2) - (-1) = 5$$
$$5 = 5$$

and

$$3x - 2y + 5z = 14$$
$$3(5) - 2(-2) + 5(1) = 14$$
$$15 + 4 + 5 = 5$$
$$24 \neq 5$$

The given values do not satisfy the system of equations.

16. Substituting yields:

$$x - 2y - 3z = 2$$
$$2 - 2(3) - 3(-2) = 2$$
$$2 = 2$$

and

$$2x + 3y + 3z = 19$$
$$2(2) + 3(3) + 3(-2) = 19$$
$$4 + 9 - 6 = 5$$
$$7 \neq 19$$

The given values do not satisfy the system of equations.

17. Since the coefficients of the variables are proportional, but the constants are not, the planes are parallel.

18. Since the coefficients of the variables are proportional and the constants are in the same proportion, the planes are the same.

19. Since the coefficients of the variables are not proportional, the planes are different. They intersect along a line. The planes are neither the same nor parallel.

20. Since the coefficients of the variables are proportional and the constants are in the same proportion, the planes are the same.

Section 5.1 Skills Check

1. Since z is isolated, back substitution yields:

$$y + 3(3) = 11$$
$$y + 9 = 11$$
$$y = 2$$

and

$$x + 2y - z = 3$$
$$x + 2(2) - 3 = 3$$
$$x + 1 = 3$$
$$x = 2$$

The solutions are $x = 2$, $y = 2$, $z = 3$.

3. Since z is isolated, back substitution yields:

$$y + 3(-2) = 3$$
$$y - 6 = 3$$
$$y = 9$$

and

$$x + 2y - z = 6$$
$$x + 2(9) - (-2) = 6$$
$$x + 18 + 2 = 6$$
$$x + 20 = 6$$
$$x = -14$$

The solutions are $x = -14$, $y = 9$, $z = -2$.

5.
$$\begin{cases} x - y - 4z = 0 \\ y + 2z = 4 \\ 3y + 7z = 22 \end{cases} \xrightarrow{-3Eq2 + Eq3 \to Eq3}$$

$$\begin{cases} x - y - 4z = 0 \\ y + 2z = 4 \\ z = 10 \end{cases}$$

Since z is isolated, back substitution yields:

$$y + 2(10) = 4$$
$$y + 20 = 4$$
$$y = -16$$

and

$$x - y - 4z = 0$$
$$x - (-16) - 4(10) = 0$$
$$x + 16 - 40 = 0$$
$$x - 24 = 0$$
$$x = 24$$

The solutions are $x = 24$, $y = -16$, $z = 10$.

7.
$$\begin{cases} x - 2y + 3z = 0 \\ y - 2z = -1 \\ y + 5z = 6 \end{cases} \xrightarrow{-1Eq2 + Eq3 \to Eq3}$$

$$\begin{cases} x - 2y + 3z = 0 \\ y - 2z = -1 \\ 7z = 7 \end{cases} \xrightarrow{\frac{1}{7}Eq3 \to Eq3}$$

$$\begin{cases} x - 2y + 3z = 0 \\ y - 2z = -1 \\ z = 1 \end{cases}$$

Since z is isolated, back substitution yields:

$$y - 2(1) = -1$$
$$y - 2 = -1$$
$$y = 1$$

and

$$x - 2y + 3z = 0$$
$$x - 2(1) + 3(1) = 0$$
$$x - 2 + 3 = 0$$
$$x + 1 = 0$$
$$x = -1$$

The solutions are $x = -1$, $y = 1$, $z = 1$.

9. $\begin{cases} x+2y-2z=0 \\ x-y+4z=3 \\ x+2y+2z=3 \end{cases} \xrightarrow[\substack{-1Eq1+Eq2\rightarrow Eq2 \\ -1Eq1+Eq3\rightarrow Eq3}]{}$

$\begin{cases} x+2y-2z=0 \\ -3y+6z=3 \\ 4z=3 \end{cases} \xrightarrow[\frac{1}{4}Eq3\rightarrow Eq3]{}$

$\begin{cases} x+2y-2z=0 \\ -3y+6z=3 \\ z=\dfrac{3}{4} \end{cases}$

Since z is isolated, back substitution yields:

$$-3y+6\left(\frac{3}{4}\right)=3$$

$$-3y+\frac{9}{2}=3$$

$$-3y=-\frac{3}{2}$$

$$y=\frac{1}{2}$$

and

$$x+2y-2z=0$$

$$x+2\left(\frac{1}{2}\right)-2\left(\frac{3}{4}\right)=0$$

$$x+1-\frac{3}{2}=0$$

$$x-\frac{1}{2}=0$$

$$x=\frac{1}{2}$$

The solutions are $x=\dfrac{1}{2},\ y=\dfrac{1}{2},\ z=\dfrac{3}{4}$.

11. $\begin{cases} x+3y-z=0 \\ x-2y+z=8 \\ x-6y+2z=6 \end{cases} \xrightarrow[\substack{-1Eq1+Eq2\rightarrow Eq2 \\ -1Eq1_1+Eq3\rightarrow Eq3}]{}$

$\begin{cases} x+3y-z=0 \\ -5y+2z=8 \\ -9y+3z=6 \end{cases} \xrightarrow[9Eq2-5Eq3\rightarrow Eq3]{}$

$\begin{cases} x+3y-z=0 \\ -5y+2z=8 \\ 3z=42 \end{cases} \xrightarrow[\frac{1}{3}Eq3\rightarrow Eq3]{}$

$\begin{cases} x+3y-z=0 \\ -5y+2z=8 \\ z=14 \end{cases}$

Since z is isolated, back substitution yields:

$$-5y+2(14)=8$$

$$-5y+28=8$$

$$-5y=-20$$

$$y=4$$

and

$$x+3y-z=0$$

$$x+3(4)-14=0$$

$$x+12-14=0$$

$$x=2$$

The solutions are $x=2,\ y=4,\ z=14$.

13. $\begin{cases} 2x+4y-14z=0 \\ 3x+5y+z=19 \\ x+4y-z=12 \end{cases} \xrightarrow[\substack{-3Eq1+2Eq2\rightarrow Eq2 \\ Eq1-2Eq3\rightarrow Eq3}]{}$

$\begin{cases} 2x+4y-14z=0 \\ -2y+44z=38 \\ -4y-12z=-24 \end{cases} \xrightarrow[-2Eq2+Eq3\rightarrow Eq3]{}$

$\begin{cases} 2x+4y-14z=0 \\ -2y+44z=38 \\ -100z=-100 \end{cases} \xrightarrow[-\frac{1}{100}Eq3\rightarrow Eq3]{}$

$\begin{cases} 2x+4y-14z=0 \\ -2y+44z=38 \\ z=1 \end{cases}$

Since z is isolated, back substitution yields:
$$-2y + 44(1) = 38$$
$$-2y + 44 = 38$$
$$-2y = -6$$
$$y = 3$$
and
$$2x + 4y - 14z = 0$$
$$2x + 4(3) - 14(1) = 0$$
$$2x + 12 - 14 = 0$$
$$2x = 2$$
$$x = 1$$
The solutions are $x = 1$, $y = 3$, $z = 1$.

15. $\begin{cases} 3x - 4y + 6z = 10 \\ 2x - 4y - 5z = -14 \\ x + 2y - 3z = 0 \end{cases} \xrightarrow[Eq1-3Eq3\leftrightarrow Eq3]{-2Eq1+3Eq2\leftrightarrow Eq2}$

$\begin{cases} 3x - 4y + 6z = 10 \\ -4y - 27z = -62 \\ -10y + 15z = 10 \end{cases} \xrightarrow{5Eq2-2Eq3\leftrightarrow Eq3}$

$\begin{cases} 3x - 4y + 6z = 10 \\ -4y - 27z = -62 \\ -165z = -330 \end{cases} \xrightarrow{-\frac{1}{165}Eq3\leftrightarrow Eq3}$

$\begin{cases} 3x - 4y + 6z = 10 \\ -4y - 27z = -62 \\ z = 2 \end{cases}$

Since z is isolated, back substitution yields:
$$-4y - 27(2) = -62$$
$$-4y - 54 = -62$$
$$-4y = -8$$
$$y = 2$$
and
$$3x - 4y + 6z = 10$$
$$3x - 4(2) + 6(2) = 10$$
$$3x - 8 + 12 = 10$$
$$3x = 6$$
$$x = 2$$
The solutions are $x = 2$, $y = 2$, $z = 2$.

17. $\begin{cases} x - 3y + z = -2 \\ y - 2z = -4 \end{cases}$

Let z be any real number.
$$y = 2z - 4$$
and
$$x = 3y - z - 2$$
$$= 3(2z - 4) - z - 2$$
$$= 6z - 12 - z - 2$$
$$= 5z - 14$$
There are infinitely many solutions to the system all fitting the form
$x = 5z - 14$, $y = 2z - 4$, z.

19. $\begin{cases} x - y + 2z = 4 \\ y + 8z = 16 \end{cases}$

Let z be any real number.
$$y = 16 - 8z$$
and
$$x = y - 2z + 4$$
$$= (16 - 8z) - 2z + 4$$
$$= 20 - 10z$$
There are infinitely many solutions to the system all fitting the form
$x = 20 - 10z$, $y = 16 - 8z$, z.

21. $\begin{cases} x - y + 2z = -1 \\ 2x + 2y + 3z = -3 \\ x + 3y + z = -2 \end{cases} \xrightarrow[-1Eq1+Eq3\to Eq3]{-2Eq1+Eq2\to Eq2}$

$\begin{cases} x - y + 2z = -1 \\ 4y - z = -1 \\ 4y - z = -1 \end{cases} \xrightarrow{-1Eq2+Eq3\to Eq3}$

$\begin{cases} x - y + 2z = -1 \\ 4y - z = -1 \\ 0 = 0 \end{cases}$

Let z be any real number.

$4y - z = -1$

$4y = z - 1$

$y = \dfrac{1}{4}z - \dfrac{1}{4}$

and

$x = y - 2z - 1$

$\quad = \left(\dfrac{1}{4}z - \dfrac{1}{4}\right) - 2z - 1$

$\quad = -\dfrac{7}{4}z - \dfrac{5}{4}$

There are infinitely many solutions to the system all fitting the form

$x = -\dfrac{7}{4}z - \dfrac{5}{4}, \quad y = \dfrac{1}{4}z - \dfrac{1}{4}, \quad z.$

Let z be any real number.

$y = -2z - 2$

and

$x = -y + 8z + 9$

$\quad = -(-2z - 2) + 8z + 9$

$\quad = 10z + 11$

There are infinitely many solutions to the system all fitting the form

$x = 10z + 11, \quad y = -2z - 2, \quad z.$

23. $\begin{cases} 3x - 4y - 6z = 10 \\ 2x - 4y - 5z = -14 \\ x \qquad - z = 0 \end{cases} \xrightarrow[\substack{Eq2 - 2Eq3 \to Eq2 \\ Eq1 - 3Eq3 \to Eq3}]{}$

$\begin{cases} 3x - 4y - 6z = 10 \\ \quad -4y - 3z = -14 \\ \quad -4y - 3z = 10 \end{cases} \xrightarrow{-1Eq2 + Eq3 \to Eq3}$

$\begin{cases} 3x - 4y - 6z = 10 \\ \quad -4y - 3z = -14 \\ \qquad\qquad 0 = 24 \end{cases}$

Since $0 \neq 24$, the system is inconsistent and has no solution.

25. $\begin{cases} 2x + 3y - 14z = 16 \\ 4x + 5y - 30z = 34 \\ x + y - 8z = 9 \end{cases} \xrightarrow[\substack{Eq2 - 4Eq3 \to Eq2 \\ Eq1 - 2Eq3 \to Eq3}]{}$

$\begin{cases} 2x + 3y - 14z = 16 \\ \quad y + 2z = -2 \\ \quad y + 2z = -2 \end{cases} \xrightarrow{-1Eq2 + Eq3 \to Eq3}$

$\begin{cases} 2x + 3y - 14z = 16 \\ \quad y + 2z = -2 \\ \qquad\qquad 0 = 0 \end{cases}$

Section 5.1 Exercises

27. $\begin{cases} x + y + z = 60 \\ 15{,}000x + 25{,}000y + 45{,}000z = 1{,}400{,}000 \\ 30x + 40y + 50z = 2200 \end{cases} \xrightarrow[\;-30\,Eq1+Eq3\rightarrow Eq3\;]{-15{,}000\,Eq1+Eq2\rightarrow Eq2}$

$\begin{cases} x + y + z = 60 \\ 10{,}000y + 30{,}000z = 500{,}000 \\ 10y + 20z = 400 \end{cases} \xrightarrow[\;\;]{\;Eq2-1000\,Eq3\rightarrow Eq3\;}$

$\begin{cases} x + y + z = 60 \\ 10{,}000y + 30{,}000z = 500{,}000 \\ 10{,}000z = 100{,}000 \end{cases} \xrightarrow[\;\;]{\;\frac{1}{10{,}000}Eq3\rightarrow Eq3\;}$

$\begin{cases} x + y + z = 60 \\ 10{,}000y + 30{,}000z = 500{,}000 \\ z = 10 \end{cases}$

Since z is isolated, back substitution yields:

$$10{,}000y + 30{,}000(10) = 500{,}000$$
$$10{,}000y + 300{,}000 = 500{,}000$$
$$10{,}000y = 200{,}000$$
$$y = 20$$
$$\text{and}$$
$$x + y + z = 60$$
$$x + 20 + 10 = 60$$
$$x = 30$$

The solution to the system is $x = 30$, $y = 20$, $z = 10$. The agency needs to purchase 30 compact cars, 20 midsize cars, and 10 luxury cars.

29. **a.** $x + y = 2600$

   b. $40x$

   c. $60y$

   d. $40x + 60y = 120{,}000$

e. $\begin{cases} x + y = 2600 \\ 40x + 60y = 120{,}000 \end{cases}$ $\xrightarrow{-40\,Eq1+Eq2\rightarrow Eq2}$

$\begin{cases} x + y = 2600 \\ 20y = 16{,}000 \end{cases}$ $\xrightarrow{\frac{1}{20}Eq2\rightarrow Eq2}$

$\begin{cases} x + y = 2600 \\ y = 800 \end{cases}$

Since y is isolated, back substitution yields:

$x + y = 2600$

$x + 800 = 2600$

$x = 1800$

The solution to the system is $x = 1800$, $y = 800$. The concert promoter needs to sell 1800 \$40 tickets and 800 \$60 tickets.

31. a. $x + y + z = 400{,}000$

b. $7.5\%x + 8\%y + 9\%z = 33{,}700$, or rewriting

$0.075x + 0.08y + 0.09z = 33{,}700$

c. $z = x + y$, or rewriting

$x + y - z = 0$

d. $\begin{cases} x + y + z = 400{,}000 \\ 0.075x + 0.08y + 0.09z = 33{,}700 \\ x + y - z = 0 \end{cases}$ $\begin{array}{l} \scriptstyle -0.075\,Eq1+Eq2\rightarrow Eq2 \\ \scriptstyle -1\,Eq1+Eq3\rightarrow Eq3 \end{array} \longrightarrow$

$\begin{cases} x + y + z = 400{,}000 \\ 0.005y + 0.015z = 3700 \\ -2z = -400{,}000 \end{cases}$ $\xrightarrow{-\frac{1}{2}Eq3\rightarrow Eq3}$

$\begin{cases} x + y + z = 400{,}000 \\ 0.005y + 0.015z = 3700 \\ z = 200{,}000 \end{cases}$

Since z is isolated, back substitution yields:

$$0.005y + 0.015(200,000) = 3700$$
$$0.005y + 3000 = 3700$$
$$0.005y = 700$$
$$y = 140,000$$

and

$$x + y + z = 400,000$$
$$x + 140,000 + 200,000 = 400,000$$
$$x = 60,000$$

The solution to the system is $x = 60,000$, $y = 140,000$, $z = 200,000$. $60,000 is invested in the property returning 7.5%, $140,000 is invested in the property returning 8%, and $200,000 is invested in the property returning 9%.

33. Let A = the number of units of product A, B = the number of units of product B, and C = the number of units of product C.

$$\begin{cases} 24A + 20B + 40C = 8000 \\ 40A + 30B + 60C = 12,400 \\ 150A + 180B + 200C = 52,600 \end{cases} \xrightarrow[\substack{-25Eq1+4Eq3 \to Eq3}]{-5Eq1+3Eq2 \to Eq2}$$

$$\begin{cases} 24A + 20B + 40C = 8000 \\ -10B - 20C = -2800 \\ 220B - 200C = 10,400 \end{cases} \xrightarrow{22Eq2+Eq3 \to Eq3}$$

$$\begin{cases} 24A + 20B + 40C = 8000 \\ -10B - 20C = -2800 \\ -640C = -51,200 \end{cases} \xrightarrow{-\frac{1}{640}Eq3 \to Eq3}$$

$$\begin{cases} 24A + 20B + 40C = 8000 \\ -10B - 20C = -2800 \\ C = 80 \end{cases}$$

Since C is isolated, back substitution yields:

$$-10B - 20(80) = -2800$$
$$-10B - 1600 = -2800$$
$$-10B = -1200$$
$$B = 120$$

and

$$24A + 20B + 40C = 8000$$
$$24A + 20(120) + 40(80) = 8000$$
$$24A + 2400 + 3200 = 8000$$
$$24A = 2400$$
$$A = 100$$

The solution to the system is $A = 100$, $B = 120$, $C = 80$. To meet the restrictions imposed on volume, weight, and value, 100 units of product A, 120 units of product B, and 80 of product C can be carried.

35. a. $x + y + z = 500,000$ means that the sum of the money invested in the three accounts is \$500,000. $0.08x + 0.10y + 0.14z = 49,000$ means that the sum of the interest earned on the three accounts in one year is \$49,000.

b.
$$\begin{cases} x + y + z = 500,000 \\ 0.08x + 0.10y + 0.14z = 49,000 \end{cases} \xrightarrow{-0.8\,Eq1 + Eq2 \to Eq2}$$

$$\begin{cases} x + y + z = 500,000 \\ 0.02y + 0.06z = 9,000 \end{cases}$$

Let $z = z$.

$0.02y + 0.06z = 9000$

$0.02y = 9000 - 0.06z$

$y = 450,000 - 3z$

and

$x + (450,000 - 3z) + z = 500,000$

$x - 2z + 450,000 = 500,000$

$x = 2z + 50,000$

The solution is $x = 2z + 50,000$, $y = 450,000 - 3z$, $z = z$.

c. $z \geq 0$ implies that $x = 2z + 50,000 \geq 0$.

For $y \geq 0$, $450,000 - 3z \geq 0$.

Solving for z,

$-3z \geq -450,000$

$z \leq \dfrac{-450,000}{-3}$

$z \leq 150,000$

For all investments to be non-negative, $0 \leq z \leq 150,000$.

37. Let $x =$ the number of Acclaim units produced, $y =$ the number of Bestfrig units produced, and $z =$ the number of Cool King units produced.

$$\begin{cases} 5x + 4y + 4.5z = 300 \\ 2x + 1.4y + 1.7z = 120 \\ 1.4x + 1.2y + 1.3z = 210 \end{cases} \xrightarrow[\;-0.7\,Eq2+Eq3\rightarrow Eq3\;]{-2\,Eq1+5\,Eq2\rightarrow Eq2}$$

$$\begin{cases} 5x + 4y + 4.5z = 300 \\ -1y - 0.5z = 0 \\ 0.22y + 0.11z = 126 \end{cases} \xrightarrow{\;0.22\,Eq2+Eq3\rightarrow Eq3\;}$$

$$\begin{cases} 5x + 4y + 4.5z = 300 \\ -1y - 0.5z = 0 \\ 0 = 126 \end{cases}$$

Since $0 \neq 126$, the system is inconsistent and has no solution. It is not possible to satisfy the given manufacturing conditions.

39. Let $x =$ grams of food I, $y =$ grams of food II, and $z =$ grams of food III.

$$\begin{cases} 12\%x + 14\%y + 8\%z = 6.88 \\ 8\%x + 6\%y + 16\%z = 6.72 \\ 10\%x + 10\%y + 12\%z = 6.80 \end{cases}$$

or

$$\begin{cases} 0.12x + 0.14y + 0.08z = 6.88 \\ 0.08x + 0.06y + 0.16z = 6.72 \\ 0.10x + 0.10y + 0.12z = 6.80 \end{cases}$$

$$\begin{cases} 0.12x + 0.14y + 0.08z = 6.88 \\ 0.08x + 0.06y + 0.16z = 6.72 \\ 0.10x + 0.10y + 0.12z = 6.80 \end{cases} \quad \xrightarrow[\; -5Eq1+6Eq3 \to Eq3\;]{-2Eq1+3Eq2 \to Eq2}$$

$$\begin{cases} 0.12x + 0.14y + 0.08z = 6.88 \\ -0.10y + 0.32z = 6.40 \\ -0.10y + 0.32z = 6.40 \end{cases} \quad \xrightarrow{\;-1Eq2+Eq3 \to Eq3\;}$$

$$\begin{cases} 0.12x + 0.14y + 0.08z = 6.88 \\ -0.10y + 0.32z = 6.40 \\ 0 = 0 \end{cases}$$

Let $z = z$.

$-0.10y + 0.32z = 6.40$

$-0.10y = 6.40 - 0.32z$

$y = 3.2z - 64$

and

$0.12x + 0.14y + 0.08z = 6.88$

$0.12x + 0.14(3.2z - 64) + 0.08z = 6.88$

$0.12x + 0.448z - 8.96 + 0.8z = 6.88$

$0.12x + 0.528z - 8.96 = 6.88$

$0.12x = 15.84 - 0.528z$

$x = \dfrac{15.84 - 0.528z}{0.12}$

$x = 132 - 4.4z$

The solution is $x = 132 - 4.4z$, $y = 3.2z - 64$, $z = z$.

Note that $z \geq 0$. $y \geq 0$ implies that $3.2z - 64 \geq 0$.

$\qquad 3.2z \geq 64$

$\qquad\quad z \geq \dfrac{64}{3.2}$

$\qquad\quad z \geq 20$

$x \geq 0$ implies that $132 - 4.4z \geq 0$.

$\qquad -4.4z \geq -132$

$\qquad\quad z \leq \dfrac{-132}{-4.4}$

$\qquad\quad z \leq 30$

Therefore, $20 \leq z \leq 30$.

Section 5.2 Skills Check

1. $\begin{bmatrix} 1 & 1 & -1 & | & 4 \\ 1 & -2 & -1 & | & -2 \\ 2 & 2 & 1 & | & 11 \end{bmatrix}$

3. $\begin{bmatrix} 5 & -3 & 2 & | & 12 \\ 3 & 6 & -9 & | & 4 \\ 2 & 3 & -4 & | & 9 \end{bmatrix}$

5. Since the matrix is in reduced row-echelon form, the solution is $x = -1$, $y = 4$, $z = -2$.

7. $\begin{bmatrix} 1 & 1 & -1 & | & 4 \\ 1 & -2 & -1 & | & -2 \\ 2 & 2 & 1 & | & 11 \end{bmatrix} \xrightarrow{\substack{-1R_1+R_2 \to R_2 \\ -2R_1+R_3 \to R_3}}$

$\begin{bmatrix} 1 & 1 & -1 & | & 4 \\ 0 & -3 & 0 & | & -6 \\ 0 & 0 & 3 & | & 3 \end{bmatrix} \xrightarrow{\substack{\left(-\frac{1}{3}\right)R_2 \to R_2 \\ \left(\frac{1}{3}\right)R_3 \to R_3}}$

$\begin{bmatrix} 1 & 1 & -1 & | & 4 \\ 0 & 1 & 0 & | & 2 \\ 0 & 0 & 1 & | & 1 \end{bmatrix}$

Back substituting to find the solutions

$\begin{cases} x + y - z = 4 \\ \quad\quad y = 2 \\ \quad\quad\quad z = 1 \end{cases}$

$x + 2 - 1 = 4$

$x + 1 = 4$

$x = 3$

The solutions are $x = 3$, $y = 2$, $z = 1$.

9. $\begin{bmatrix} 2 & -3 & 4 & | & 13 \\ 1 & -2 & 1 & | & 3 \\ 2 & -3 & 1 & | & 4 \end{bmatrix} \xrightarrow{R_2 \leftrightarrow R_1}$

$\begin{bmatrix} 1 & -2 & 1 & | & 3 \\ 2 & -3 & 4 & | & 13 \\ 2 & -3 & 1 & | & 4 \end{bmatrix} \xrightarrow{\substack{-2R_1+R_2 \to R_2 \\ -2R_1+R_3 \to R_3}}$

$\begin{bmatrix} 1 & -2 & 1 & | & 3 \\ 0 & 1 & 2 & | & 7 \\ 0 & 1 & -1 & | & -2 \end{bmatrix} \xrightarrow{-1R_2+R_3 \to R_3}$

$\begin{bmatrix} 1 & -2 & 1 & | & 3 \\ 0 & 1 & 2 & | & 7 \\ 0 & 0 & -3 & | & -9 \end{bmatrix} \xrightarrow{\left(-\frac{1}{3}\right)R_3 \to R_3}$

$\begin{bmatrix} 1 & -2 & 1 & | & 3 \\ 0 & 1 & 2 & | & 7 \\ 0 & 0 & 1 & | & 3 \end{bmatrix}$

Back substituting to find the solutions

$\begin{cases} x - 2y + z = 3 \\ \quad\quad y + 2z = 7 \\ \quad\quad\quad z = 3 \end{cases}$

$y + 2(3) = 7$

$y + 6 = 7$

$y = 1$

and

$x - 2(1) + (3) = 3$

$x + 1 = 3$

$x = 2$

The solutions are $x = 2$, $y = 1$, $z = 3$.

11. Since the third row of the given augmented matrix implies $0 = 1$, the system is inconsistent and has no solution.

13. Since the third row of the augmented matrix states $0 = 0$, the system has infinitely many solutions. Let z be any real number. Then,

$$y - 5z = 5$$
$$y = 5z + 5$$

and

$$x + 3z = 2$$
$$x = 2 - 3z$$

There are infinitely many solutions to the system of the form

$$x = 2 - 3z, \quad y = 5z + 5, \quad z.$$

15. $\begin{bmatrix} 1 & 1 & -1 & | & 0 \\ 1 & -2 & -1 & | & 6 \\ 2 & 2 & 1 & | & 3 \end{bmatrix} \xrightarrow[\substack{-1R_1 + R_2 \to R_2 \\ -2R_1 + R_3 \to R_3}]{}$

$\begin{bmatrix} 1 & 1 & -1 & | & 0 \\ 0 & -3 & 0 & | & 6 \\ 0 & 0 & 3 & | & 3 \end{bmatrix} \xrightarrow[\substack{\left(-\frac{1}{3}\right)R_2 \to R_2 \\ \left(\frac{1}{3}\right)R_3 \to R_3}]{}$

$\begin{bmatrix} 1 & 1 & -1 & | & 0 \\ 0 & 1 & 0 & | & -2 \\ 0 & 0 & 1 & | & 1 \end{bmatrix}$

Back substituting to find the solutions

$$\begin{cases} x + y - z = 0 \\ \qquad y = -2 \\ \qquad z = 1 \end{cases}$$

$$x - 2 - 1 = 0$$
$$x - 3 = 0$$
$$x = 3$$

The solutions are $x = 3$, $y = -2$, $z = 1$.

17. $\begin{bmatrix} 3 & -2 & 5 & | & 15 \\ 1 & -2 & -2 & | & -1 \\ 2 & -2 & 0 & | & 0 \end{bmatrix} \xrightarrow{R_2 \leftrightarrow R_1}$

$\begin{bmatrix} 1 & -2 & -2 & | & -1 \\ 3 & -2 & 5 & | & 15 \\ 2 & -2 & 0 & | & 0 \end{bmatrix} \xrightarrow[\substack{-3R_1 + R_2 \to R_2 \\ -2R_1 + R_3 \to R_3}]{}$

$\begin{bmatrix} 1 & -2 & -2 & | & -1 \\ 0 & 4 & 11 & | & 18 \\ 0 & 2 & 4 & | & 2 \end{bmatrix} \xrightarrow[\substack{\left(\frac{1}{4}\right)R_2 \to R_2 \\ \left(\frac{1}{2}\right)R_3 \to R_3}]{}$

$\begin{bmatrix} 1 & -2 & -2 & | & -1 \\ 0 & 1 & \frac{11}{4} & | & \frac{9}{2} \\ 0 & 1 & 2 & | & 1 \end{bmatrix} \xrightarrow{-1R_2 + R_3 \to R_3}$

$\begin{bmatrix} 1 & -2 & -2 & | & -1 \\ 0 & 1 & \frac{11}{4} & | & \frac{9}{2} \\ 0 & 0 & \frac{-3}{4} & | & -\frac{7}{2} \end{bmatrix} \xrightarrow{\left(-\frac{4}{3}\right)R_3 \to R_3}$

$\begin{bmatrix} 1 & -2 & -2 & | & -1 \\ 0 & 1 & \frac{11}{4} & | & \frac{9}{2} \\ 0 & 0 & 1 & | & \frac{14}{3} \end{bmatrix}$

Back substituting to find the solutions

$$\begin{cases} x - 2y - 2z = -1 \\ \quad\; y + \dfrac{11}{4}z = \dfrac{9}{2} \\ \qquad\qquad z = \dfrac{14}{3} \end{cases}$$

$$y + \frac{11}{4}\left(\frac{14}{3}\right) = \frac{9}{2}$$

$$y + \frac{77}{6} = \frac{27}{6}$$

$$y = -\frac{50}{6}$$

$$y = -\frac{25}{3}$$

and

$$x - 2\left(-\frac{25}{3}\right) - 2\left(\frac{14}{3}\right) = -1$$

$$x + \frac{50}{3} - \frac{28}{3} = -1$$

$$x + \frac{22}{3} = -\frac{3}{3}$$

$$x = -\frac{25}{3}$$

The solutions are $x = -\dfrac{25}{3}$, $y = -\dfrac{25}{3}$, $z = \dfrac{14}{3}$.

19.

$$\begin{bmatrix} 2 & 3 & 4 & | & 5 \\ 6 & 7 & 8 & | & 9 \\ 2 & 1 & 1 & | & 1 \end{bmatrix} \xrightarrow[\substack{-3R_1 + R_2 \to R_2 \\ -1R_1 + R_3 \to R_3}]{}$$

$$\begin{bmatrix} 2 & 3 & 4 & | & 5 \\ 0 & -2 & -4 & | & -6 \\ 0 & -2 & -3 & | & -4 \end{bmatrix} \xrightarrow[\substack{\left(-\frac{1}{2}\right)R_2 \to R_2 \\ \left(-\frac{1}{2}\right)R_3 \to R_3}]{}$$

$$\begin{bmatrix} 2 & 3 & 4 & | & 5 \\ 0 & 1 & 2 & | & 3 \\ 0 & 1 & \frac{3}{2} & | & 2 \end{bmatrix} \xrightarrow[\substack{-1R_2 + R_3 \to R_3}]{}$$

$$\begin{bmatrix} 2 & 3 & 4 & | & 5 \\ 0 & 1 & 2 & | & 3 \\ 0 & 0 & -\frac{1}{2} & | & -1 \end{bmatrix} \xrightarrow[\substack{\left(\frac{1}{2}\right)R_1 \to R_1 \\ -2R_3 \to R_3}]{}$$

$$\begin{bmatrix} 1 & \frac{3}{2} & 2 & | & \frac{5}{2} \\ 0 & 1 & 2 & | & 3 \\ 0 & 0 & 1 & | & 2 \end{bmatrix}$$

Back substituting to find the solutions

$$\begin{cases} x+\dfrac{3}{2}y+2z=\dfrac{5}{2} \\[2mm] y+2z=3 \\[2mm] z=2 \end{cases}$$

$y+2(2)=3$

$y+4=3$

$y=-1$

and

$x+\dfrac{3}{2}(-1)+2(2)=\dfrac{5}{2}$

$x-\dfrac{3}{2}+4=\dfrac{5}{2}$

$x+\dfrac{5}{2}=\dfrac{5}{2}$

$x=0$

The solutions are $x=0,\ y=-1,\ z=2$.

21. $\begin{bmatrix} 1 & -1 & 1 & -1 & | & -2 \\ 2 & 0 & 4 & 1 & | & 5 \\ 2 & -3 & 1 & 0 & | & -5 \\ 0 & 1 & 2 & 20 & | & 4 \end{bmatrix} \xrightarrow{\substack{-2R_1+R_2\to R_2 \\ -2R_1+R_3\to R_3}}$

$\begin{bmatrix} 1 & -1 & 1 & -1 & | & -2 \\ 0 & 2 & 2 & 3 & | & 9 \\ 0 & -1 & -1 & 2 & | & -1 \\ 0 & 1 & 2 & 20 & | & 4 \end{bmatrix} \xrightarrow{\substack{2R_3+R_2\to R_2 \\ R_3+R_4\to R_4}}$

$\begin{bmatrix} 1 & -1 & 1 & -1 & | & -2 \\ 0 & 0 & 0 & 7 & | & 7 \\ 0 & -1 & -1 & 2 & | & -1 \\ 0 & 0 & 1 & 22 & | & 3 \end{bmatrix} \xrightarrow{\substack{R_3\to R_2 \\ R_2\to R_4 \\ R_4\to R_3}}$

$\begin{bmatrix} 1 & -1 & 1 & -1 & | & -2 \\ 0 & -1 & -1 & 2 & | & -1 \\ 0 & 0 & 1 & 22 & | & 3 \\ 0 & 0 & 0 & 7 & | & 7 \end{bmatrix} \xrightarrow{\substack{-1R_2\to R_2 \\ \left(\frac{1}{7}\right)R_4\to R_4}}$

$\begin{bmatrix} 1 & -1 & 1 & -1 & | & -2 \\ 0 & 1 & 1 & -2 & | & 1 \\ 0 & 0 & 1 & 22 & | & 3 \\ 0 & 0 & 0 & 1 & | & 1 \end{bmatrix}$

Back substituting to find the solutions

$$\begin{cases} x - y + z - w = -2 \\ y + z - 2w = 1 \\ z + 22w = 3 \\ w = 1 \end{cases}$$

$z + 22(1) = 3$

$z = -19$

and

$y + (-19) - 2(1) = 1$

$y - 19 - 2 = 1$

$y - 21 = 1$

$y = 22$

and

$x - (22) + (-19) - (1) = -2$

$x - 22 - 19 - 1 = -2$

$x - 42 = -2$

$x = 40$

The solutions are $x = 40$, $y = 22$,

$z = -19$, $w = 1$.

23. $\begin{bmatrix} -2 & 3 & 2 & | & 13 \\ -2 & -2 & 3 & | & 0 \\ 4 & 1 & 4 & | & 11 \end{bmatrix}$

Using the calculator to generate the reduced
row-echelon form of the matrix yields

```
rref([A])
        [[1 0 0 0]
         [0 1 0 3]
         [0 0 1 2]]
```

The solution to the system is
$x = 0, y = 3, z = 2$.

25. $\begin{bmatrix} 2 & 5 & 6 & | & 6 \\ 3 & -2 & 2 & | & 4 \\ 5 & 3 & 8 & | & 10 \end{bmatrix}$

Using the calculator to generate the reduced
row-echelon form of the matrix yields

$$\begin{bmatrix} 1 & 0 & \dfrac{22}{19} & | & \dfrac{32}{19} \\ 0 & 1 & \dfrac{14}{19} & | & \dfrac{10}{19} \\ 0 & 0 & 0 & | & 0 \end{bmatrix}$$

Since the third row of the augmented
matrix states $0 = 0$, the system has
infinitely many solutions. Let z be
any real number. Then,

$$y + \frac{14}{19}z = \frac{10}{19}$$

$$y = \frac{10}{19} - \frac{14}{19}z$$

and

$$x + \frac{22}{19}z = \frac{32}{19}$$

$$x = \frac{32}{19} - \frac{22}{19}z$$

There are infinitely many solutions
to the system of the form

$$x = \frac{32}{19} - \frac{22}{19}z, \quad y = \frac{10}{19} - \frac{14}{19}z, \ z.$$

27. $\begin{bmatrix} -1 & -5 & 3 & | & -2 \\ 3 & 7 & 2 & | & 5 \\ 4 & 12 & -1 & | & 7 \end{bmatrix}$

Using the calculator to generate the reduced
row-echelon form of the matrix yields

```
rref([A])▶Frac
...1 0 31/8   11/8...
...0 1 -11/8  1/8 ...
...0 0 0      0   ...
```

or

$$\begin{bmatrix} 1 & 0 & \dfrac{31}{8} & \bigg| & \dfrac{11}{8} \\[2mm] 0 & 1 & -\dfrac{11}{8} & \bigg| & \dfrac{1}{8} \\[2mm] 0 & 0 & 0 & \bigg| & 0 \end{bmatrix}$$

Since the third row of the augmented matrix states $0 = 0$, the system has infinitely many solutions. Let z be any real number. Then,

$$y - \frac{11}{8}z = \frac{1}{8}$$

$$y = \frac{11}{8}z + \frac{1}{8}$$

and

$$x + \frac{31}{8}z = \frac{11}{8}$$

$$x = \frac{11}{8} - \frac{31}{8}z$$

There are infinitely many solutions to the system of the form

$$x = \frac{11}{8} - \frac{31}{8}z, \;\; y = \frac{11}{8}z + \frac{1}{8}, \;\; z.$$

29. $$\begin{bmatrix} 2 & -3 & 2 & \big| & 5 \\ 4 & 1 & -3 & \big| & 6 \end{bmatrix}$$

Using the calculator to generate the reduced row-echelon form of the matrix yields

```
rref([B])▶Frac
…1 0 -1/2 23/14…
…0 1 -1    -4/7 …
```

or

$$\begin{bmatrix} 1 & 0 & -\dfrac{1}{2} & \bigg| & \dfrac{23}{14} \\[2mm] 0 & 1 & -1 & \bigg| & -\dfrac{4}{7} \end{bmatrix}$$

Since the augmented matrix contains two equations with three variables, the system has infinitely many solutions. Let z be any real number. Then,

$$y - z = -\frac{4}{7}$$

$$y = z - \frac{4}{7}$$

and

$$x - \frac{1}{2}z = \frac{23}{14}$$

$$x = \frac{1}{2}z + \frac{23}{14}$$

There are infinitely many solutions to the system of the form

$$x = \frac{1}{2}z + \frac{23}{14}, \;\; y = z - \frac{4}{7}, \;\; z.$$

31. $$\begin{bmatrix} 1 & 0 & -3 & -3 & \big| & -2 \\ 1 & 1 & 1 & 3 & \big| & 2 \\ 2 & 1 & -2 & -2 & \big| & 0 \\ 3 & 2 & -1 & 1 & \big| & 2 \end{bmatrix}$$

Using the calculator to generate the reduced row-echelon form of the matrix yields

```
rref([C])▶Frac
  [[1 0 -3 0 -2]
   [0 1 4  0 4 ]
   [0 0 0  1 0 ]
   [0 0 0  0 0 ]]
```

Since the fourth row of the augmented matrix states $0 = 0$, the system has infinitely many solutions.

$w = 0$

Let z be any real number. Then,

$y + 4z = 4$

$y = 4 - 4z$

and

$x - 3z = -2$

$x = 3z - 2$

There are infinitely many solutions to the system of the form

$x = 3z - 2, \ y = 4 - 4z, \ z, \ w = 0.$

Section 5.2 Exercises

33. Let x, y, and z represent the number of tickets in each section.

Note that $x = 2(y + z)$, which can be rewritten as $x - 2y - 2z = 0$.

$$\begin{bmatrix} 1 & 1 & 1 & | & 3600 \\ 1 & -2 & -2 & | & 0 \\ 40 & 70 & 100 & | & 192{,}000 \end{bmatrix}$$

Using the calculator to generate the reduced row-echelon form of the matrix yields

```
rref([A])▶Frac
    [[1 0 0 2400]
     [0 1 0 800 ]
     [0 0 1 400 ]]
```

The solution to the system is $x = 2400$, $y = 800$, $z = 400$. The theater owner needs to sell 2400 \$40 tickets, 800 \$70 tickets, and 400 \$100 tickets.

35. a. Let $x =$ the points per each true-false question, $y =$ the points per each multiple-choice question, and $z =$ the points per each essay question.

$15x + 10y + 5z = 100$

and

$2x = y$, or rewriting

$2x - y = 0$

and

$3x = z$, or rewriting

$3x - z = 0$

The system is

$$\begin{cases} 15x + 10y + 5z = 100 \\ 2x - y = 0 \\ 3x - z = 0 \end{cases}$$

b.
$$\begin{bmatrix} 15 & 10 & 5 & | & 100 \\ 2 & -1 & 0 & | & 0 \\ 3 & 0 & -1 & | & 0 \end{bmatrix}$$

Using the calculator to generate the reduced row-echelon form of the matrix yields

```
rref([A])
    [[1 0 0 2]
     [0 1 0 4]
     [0 0 1 6]]
```

The solution to the system is $x = 2$, $y = 4$, $z = 6$. On the exam true-false questions are worth two points, multiple-choice questions are worth four points, and essay questions are worth six points.

37. Let x = the number of Plan I units, y = the number of Plan II units, and z = the number of Plan III units.

$$\begin{cases} 4x + 8y + 14z = 42 \\ 2x + 4y + 6z = 20 \\ 6y + 6z = 18 \end{cases}$$

$$\begin{bmatrix} 4 & 8 & 14 & | & 42 \\ 2 & 4 & 6 & | & 20 \\ 0 & 6 & 6 & | & 18 \end{bmatrix}$$

Using the calculator to generate the reduced row-echelon form of the matrix yields

```
rref([A])
    [[1 0 0 3]
     [0 1 0 2]
     [0 0 1 1]]
```

The solution to the system is $x = 3$, $y = 2$, $z = 1$. The client needs to purchase 3 units of Plan I, 2 units of Plan II, and 1 unit of Plan III to achieve the investment objectives.

39. Let x = grams of Food I, y = grams of Food II, and z = grams of Food III.

$$\begin{cases} 12\%x + 15\%y + 28\%z = 3.74 \\ 8\%x + 6\%y + 16\%z = 2.04 \\ 15\%x + 2\%y + 6\%z = 1.35 \end{cases}$$

or

$$\begin{cases} 0.12x + 0.15y + 0.28z = 3.74 \\ 0.08x + 0.06y + 0.16z = 2.04 \\ 0.15x + 0.02y + 0.06z = 1.35 \end{cases}$$

$$\begin{bmatrix} 0.12 & 0.15 & 0.28 & | & 3.74 \\ 0.08 & 0.06 & 0.16 & | & 2.04 \\ 0.15 & 0.02 & 0.06 & | & 1.35 \end{bmatrix}$$

Using the calculator to generate the reduced row-echelon form of the matrix yields

```
rref([A])
    [[1 0 0 5]
     [0 1 0 6]
     [0 0 1 8]]
```

The solution to the system is $x = 5$, $y = 6$, $z = 8$. The psychologist recommends 5 grams of Food I, 6 grams of Food II, and 8 grams of Food III.

41. a. Let x = the amount invested at 8%, y = the amount invested at 10%, and z = the amount invested at 12%.

$$\begin{cases} x + y + z = 400{,}000 \\ 8\%x + 10\%y + 12\%z = 42{,}400 \end{cases}$$

or

$$\begin{cases} x + y + z = 400{,}000 \\ 0.08x + 0.10y + 0.12z = 42{,}400 \end{cases}$$

b.
$$\begin{bmatrix} 1 & 1 & 1 & | & 400{,}000 \\ 0.08 & 0.10 & 0.12 & | & 42{,}400 \end{bmatrix}$$

Using the calculator to generate the reduced row-echelon form of the matrix yields

```
rref([A])
..1 0 -1  -120000...
..0 1 2   520000 ...
```

Let $z = z$.

$y + 2z = 520,000$

$y = 520,000 - 2z$

and

$x - z = -120,000$

$x = z - 120,000$

The solution is

$x = z - 120,000,$

$y = 520,000 - 2z, \ z = z.$

The largest possible investment in the 12% also produces the largest possible investment in the 8% account. The largest investment at 12% must also keep $y \geq 0$. Therefore,

$520,000 - 2z \geq 0$

$-2z \geq -520,000$

$z \leq \dfrac{-520,000}{-2}$

$z \leq 260,000$

The maximum value of z is $260,000. Therefore, the maximum values of x and y are

$x = 260,000 - 120,000$

$= \$140,000$

and

$y = 520,000 - 2z$

$= 520,000 - 2(260,000)$

$= \$0.$

43. a.   Let $x =$ the number of portfolio I units, $y =$ the number of portfolio II units, and $z =$ the number of portfolio III units.

$$\begin{cases} 10x + 12y + 10z = 180 \\ 2x + 8y + 6z = 140 \\ 3x + 5y + 4z = 110 \end{cases}$$

b. $\begin{bmatrix} 10 & 12 & 10 & | & 180 \\ 2 & 8 & 6 & | & 140 \\ 3 & 5 & 4 & | & 110 \end{bmatrix}$

Using the calculator to generate the reduced row-echelon form of the matrix yields a last row containing all zeros along with an augmented 1. Therefore, the system is inconsistent and has no solution. The client can not achieve the desired investment results.

45. a.   Let $x =$ the number of \$40,000 cars, $y =$ the number of \$30,000 cars, and $z =$ the number of \$20,000.

$$\begin{cases} x + y + z = 4 \\ 40,000x + 30,000y + 20,000z = 100,000 \end{cases}$$

b.   Since the system has more variables than equations, the system can not have a unique solution.

c. $\begin{bmatrix} 1 & 1 & 1 & | & 4 \\ 40,000 & 30,000 & 20,000 & | & 100,000 \end{bmatrix}$

Using the calculator to generate the reduced row-echelon form of the matrix yields

```
rref([A])
     [[1 0 -1  -2]
      [0 1 2   6 ]]
```

Let $z = z$.

$y + 2z = 6$

$y = 6 - 2z$

and

$x - z = -2$

$x = z - 2$

The solution to the system is

$x = z - 2, \ y = 6 - 2z, \ z = z.$

d. The only values of z that make sense in the context of the question are $z = 2$ and $z = 3$. Other values of z create negative solutions for x and y. If $z = 2$, the young man purchases two $20,000 cars, two $30,000 cars, and zero $40,000 cars. If $z = 3$, the young man purchases three $20,000 cars, zero $30,000 cars, and one $40,000 car.

47. $\begin{cases} x_1 - x_2 = -550 \\ x_2 - x_3 = -1300 \\ x_3 - x_4 = 1200 \\ x_1 - x_4 = -650 \end{cases}$

$$\begin{bmatrix} 1 & -1 & 0 & 0 & | & -550 \\ 0 & 1 & -1 & 0 & | & -1300 \\ 0 & 0 & 1 & -1 & | & 1200 \\ 1 & 0 & 0 & -1 & | & -650 \end{bmatrix}$$

Using the calculator to generate the reduced row-echelon form of the matrix yields

```
rref([C])▶Frac
…1 0 0 -1 -650]
…0 1 0 -1 -100]
…0 0 1 -1 1200]
…0 0 0 0  0   ]]
```

Let $x_4 =$ any real number, then

$x_1 - x_4 = -650$

$x_1 = x_4 - 650$

and

$x_2 - x_4 = -100$

$x_2 = x_4 - 100$

and

$x_3 - x_4 = 1200$

$x_3 = x_4 + 1200$

The solution is

$x_1 = x_4 - 650, \quad x_2 = x_4 - 100,$

$x_3 = x_4 + 1200, \quad x_4.$

Since all the variables must remain positive in the physical context of the question, $x_4 \geq 650$.

49. a. Water flow at A is $x_1 + x_2 = 400,000$.
Water flow at B is $x_1 = 100,000 + x_4$.
Water flow at D is $x_3 + x_4 = 100,000$.

Rewriting the equations yields the following system

$\begin{cases} x_1 + x_2 = 400,000 \\ x_1 - x_4 = 100,000 \\ x_3 + x_4 = 100,000 \\ x_2 - x_3 = 200,000 \end{cases}$

The matrix is

$$\begin{bmatrix} 1 & 1 & 0 & 0 & | & 400,000 \\ 1 & 0 & 0 & -1 & | & 100,000 \\ 0 & 0 & 1 & 1 & | & 100,000 \\ 0 & 1 & -1 & 0 & | & 200,000 \end{bmatrix}$$

b. Using the calculator to generate the reduced row-echelon form of the matrix yields

```
rref([A])
[[1  0  0  -1  1000…
 [0  1  0   1  3000…
 [0  0  1   1  1000…
 [0  0  0   0   0   …
```

```
rref([A])
…0  0  -1  100000]
…1  0   1  300000]
…0  1   1  100000]
…0  0   0    0    ]]
```

The system has infinitely many solutions. Let x_4 be any number. Then,

$x_3 = 100,000 - x_4$,

$x_2 = 300,000 - x_4$, and

$x_1 = 100,000 + x_4$.

Water flow from A to B is 100,000 plus the water flow from B to D. Water flow from A to C is 300,000 minus the water flow from B to D. Water flow from C to D is 100,000 minus the water flow from B to D. Water flow is measured by the number of gallons of water moving from one intersection to another.

Section 5.3 Skills Check

1. Only matrices of the same dimensions can be added. Therefore, matrices A, D, and E can be added, since they all have 3 rows and 3 columns. Furthermore, matrices B and F can be added, since they both have 2 rows and 3 columns. Note that a matrix can always be added to itself. Therefore, the following sums can be calculated:

$$A + D, \ A + E, \ D + E$$
$$B + F$$
$$A + A, \ B + B, \ C + C, \ D + D, \ E + E, \ F + F$$

3. To add the matrices, add the corresponding entries.

$$\begin{bmatrix} 1 & 3 & -2 \\ 3 & 1 & 4 \\ -5 & 3 & 6 \end{bmatrix} + \begin{bmatrix} 2 & 3 & 1 \\ 3 & 4 & -1 \\ 2 & 5 & 1 \end{bmatrix}$$

$$\begin{bmatrix} 1+2 & 3+3 & (-2)+1 \\ 3+3 & 1+4 & 4+(-1) \\ -5+2 & 3+5 & 6+1 \end{bmatrix}$$

$$\begin{bmatrix} 3 & 6 & -1 \\ 6 & 5 & 3 \\ -3 & 8 & 7 \end{bmatrix}$$

5. $3A$

$$= \begin{bmatrix} 3(1) & 3(3) & 3(-2) \\ 3(3) & 3(1) & 3(4) \\ 3(-5) & 3(3) & 3(6) \end{bmatrix}$$

$$= \begin{bmatrix} 3 & 9 & -6 \\ 9 & 3 & 12 \\ -15 & 9 & 18 \end{bmatrix}$$

7. $2D - 4A$

$$= 2\begin{bmatrix} 2 & 3 & 1 \\ 3 & 4 & -1 \\ 2 & 5 & 1 \end{bmatrix} - 4\begin{bmatrix} 1 & 3 & -2 \\ 3 & 1 & 4 \\ -5 & 3 & 6 \end{bmatrix}$$

$$= \begin{bmatrix} 4 & 6 & 2 \\ 6 & 8 & -2 \\ 4 & 10 & 2 \end{bmatrix} + \begin{bmatrix} -4 & -12 & 8 \\ -12 & -4 & -16 \\ 20 & -12 & -24 \end{bmatrix}$$

$$= \begin{bmatrix} 4+(-4) & 6+(-12) & 2+8 \\ 6+(-12) & 8+(-4) & -2+(-16) \\ 4+20 & 10+(-12) & 2+(-24) \end{bmatrix}$$

$$= \begin{bmatrix} 0 & -6 & 10 \\ -6 & 4 & -18 \\ 24 & -2 & -22 \end{bmatrix}$$

9. **a.** AD

$$= \begin{bmatrix} 1 & 3 & -2 \\ 3 & 1 & 4 \\ -5 & 3 & 6 \end{bmatrix}\begin{bmatrix} 2 & 3 & 1 \\ 3 & 4 & -1 \\ 2 & 5 & 1 \end{bmatrix}$$

$$= \begin{bmatrix} 2+9-4 & 3+12-10 & 1-3-2 \\ 6+3+8 & 9+4+20 & 3-1+4 \\ -10+9+12 & -15+12+30 & -5-3+6 \end{bmatrix}$$

$$= \begin{bmatrix} 7 & 5 & -4 \\ 17 & 33 & 6 \\ 11 & 27 & -2 \end{bmatrix}$$

DA

$$= \begin{bmatrix} 2 & 3 & 1 \\ 3 & 4 & -1 \\ 2 & 5 & 1 \end{bmatrix} \begin{bmatrix} 1 & 3 & -2 \\ 3 & 1 & 4 \\ -5 & 3 & 6 \end{bmatrix}$$

$$= \begin{bmatrix} 2+9-5 & 6+3+3 & -4+12+6 \\ 3+12+5 & 9+4-3 & -6+16-6 \\ 2+15-5 & 6+5+3 & -4+20+6 \end{bmatrix}$$

$$= \begin{bmatrix} 6 & 12 & 14 \\ 20 & 10 & 4 \\ 12 & 14 & 22 \end{bmatrix}$$

b. The products are different.

c. The dimensions of the matrices are the same. Both matrices are 3×3.

11. a. $DE = \begin{bmatrix} 2 & 3 & 1 \\ 3 & 4 & -1 \\ 2 & 5 & 1 \end{bmatrix} \begin{bmatrix} 9 & 2 & -7 \\ -5 & 0 & 5 \\ 7 & -4 & -1 \end{bmatrix}$

$$= \begin{bmatrix} 18-15+7 & 4+0-4 & -14+15-1 \\ 27-20-7 & 6+0+4 & -21+20+1 \\ 18-25+7 & 4+0+-4 & -14+25-1 \end{bmatrix}$$

$$= \begin{bmatrix} 10 & 0 & 0 \\ 0 & 10 & 0 \\ 0 & 0 & 10 \end{bmatrix}$$

$ED = \begin{bmatrix} 9 & 2 & -7 \\ -5 & 0 & 5 \\ 7 & -4 & -1 \end{bmatrix} \begin{bmatrix} 2 & 3 & 1 \\ 3 & 4 & -1 \\ 2 & 5 & 1 \end{bmatrix}$

$$= \begin{bmatrix} 18+6-14 & 27+8-35 & 9-2-7 \\ -10+0+10 & -15+0+25 & -5+0+5 \\ 14-12-2 & 21-16-5 & 7+4-1 \end{bmatrix}$$

$$= \begin{bmatrix} 10 & 0 & 0 \\ 0 & 10 & 0 \\ 0 & 0 & 10 \end{bmatrix}$$

Note that $DE = ED$.

b. $\dfrac{1}{10}DE = \dfrac{1}{10}\begin{bmatrix} 10 & 0 & 0 \\ 0 & 10 & 0 \\ 0 & 0 & 10 \end{bmatrix} = \begin{bmatrix} \frac{1}{10}(10) & \frac{1}{10}(0) & \frac{1}{10}(0) \\ \frac{1}{10}(0) & \frac{1}{10}(10) & \frac{1}{10}(0) \\ \frac{1}{10}(0) & \frac{1}{10}(0) & \frac{1}{10}(10) \end{bmatrix} = \begin{bmatrix} 1 & 0 & 0 \\ 0 & 1 & 0 \\ 0 & 0 & 1 \end{bmatrix}$

Note that the solution is the 3×3 identity matrix.

13. $A+B$

$$=\begin{bmatrix} 1 & 5 \\ 3 & 2 \end{bmatrix}+\begin{bmatrix} 2a & 3b \\ -c & -2d \end{bmatrix}$$

$$=\begin{bmatrix} 1+2a & 5+3b \\ 3-c & 2-2d \end{bmatrix}$$

15. $3A-2B$

$$=3\begin{bmatrix} a & b \\ c & d \end{bmatrix}-2\begin{bmatrix} 1 & 2 \\ 3 & 4 \end{bmatrix}$$

$$=\begin{bmatrix} 3a & 3b \\ 3c & 3d \end{bmatrix}+\begin{bmatrix} -2 & -4 \\ -6 & -8 \end{bmatrix}$$

$$=\begin{bmatrix} 3a-2 & 3b-4 \\ 3c-6 & 3d-8 \end{bmatrix}$$

17. The new matrix will be $m \times k$.

19. a. The product of two matrices can be calculated if the number of columns of the first matrix is equal to the number of rows of the second matrix. Therefore, BA can be calculated since B has 2 columns and A has 2 rows.

b. The matrix formed by the product will by 4×3.

21. AB

$$=\begin{bmatrix} a & b & c \\ d & e & f \end{bmatrix}\begin{bmatrix} 1 & 2 \\ 3 & 4 \\ 5 & 6 \end{bmatrix}$$

$$=\begin{bmatrix} a+3b+5c & 2a+4b+6c \\ d+3e+5f & 2d+4e+6f \end{bmatrix}$$

BA

$$=\begin{bmatrix} 1 & 2 \\ 3 & 4 \\ 5 & 6 \end{bmatrix}\begin{bmatrix} a & b & c \\ d & e & f \end{bmatrix}$$

$$=\begin{bmatrix} a+2d & b+2e & c+2f \\ 3a+4d & 3b+4e & 3c+4f \\ 5a+6d & 5b+6e & 5c+6f \end{bmatrix}$$

23. AB

$$=\begin{bmatrix} 1 & 5 \\ 3 & 2 \end{bmatrix}\begin{bmatrix} 2 & 3 \\ -1 & -2 \end{bmatrix}$$

$$=\begin{bmatrix} 2-5 & 3-10 \\ 6-2 & 9-4 \end{bmatrix}$$

$$=\begin{bmatrix} -3 & -7 \\ 4 & 5 \end{bmatrix}$$

BA

$$=\begin{bmatrix} 2 & 3 \\ -1 & -2 \end{bmatrix}\begin{bmatrix} 1 & 5 \\ 3 & 2 \end{bmatrix}$$

$$=\begin{bmatrix} 2+9 & 10+6 \\ -1-6 & -5-4 \end{bmatrix}$$

$$=\begin{bmatrix} 11 & 16 \\ -7 & -9 \end{bmatrix}$$

25. *AB*

$$= \begin{bmatrix} 1 & -1 & 2 \\ 3 & 4 & 4 \end{bmatrix} \begin{bmatrix} 3 & 1 \\ 1 & 3 \\ -2 & 1 \end{bmatrix}$$

$$= \begin{bmatrix} 3-1-4 & 1-3+2 \\ 9+4-8 & 3+12+4 \end{bmatrix}$$

$$= \begin{bmatrix} -2 & 0 \\ 5 & 19 \end{bmatrix}$$

BA

$$= \begin{bmatrix} 3 & 1 \\ 1 & 3 \\ -2 & 1 \end{bmatrix} \begin{bmatrix} 1 & -1 & 2 \\ 3 & 4 & 4 \end{bmatrix}$$

$$= \begin{bmatrix} 3+3 & -3+4 & 6+4 \\ 1+9 & -1+12 & 2+12 \\ -2+3 & 2+4 & -4+4 \end{bmatrix}$$

$$= \begin{bmatrix} 6 & 1 & 10 \\ 10 & 11 & 14 \\ 1 & 6 & 0 \end{bmatrix}$$

Section 5.3 Exercises

27. a. $A = \begin{bmatrix} 252 & 74 & 14 & 7 & 65 \\ 19 & 16 & 19 & 5 & 40 \end{bmatrix}$

$B = \begin{bmatrix} 252 & 178 & 65 & 8 & 11 \\ 19 & 6 & 14 & 1 & 0 \end{bmatrix}$

b. $A + B$

$= \begin{bmatrix} 252 & 74 & 14 & 7 & 65 \\ 19 & 16 & 19 & 5 & 40 \end{bmatrix} + \begin{bmatrix} 252 & 178 & 65 & 8 & 11 \\ 19 & 6 & 14 & 1 & 0 \end{bmatrix}$

$= \begin{bmatrix} 252+252 & 74+178 & 14+65 & 7+8 & 65+11 \\ 19+19 & 16+6 & 19+14 & 5+1 & 40+0 \end{bmatrix}$

$= \begin{bmatrix} 504 & 252 & 79 & 15 & 76 \\ 38 & 22 & 33 & 6 & 40 \end{bmatrix}$

29. a. $A = \begin{bmatrix} 59{,}385 & 47{,}091 \\ 486{,}171 & 565{,}490 \\ 12{,}181 & 9880 \end{bmatrix}$

b. $B = \begin{bmatrix} 32{,}575 & 36{,}681 \\ 658{,}782 & 882{,}013 \\ 78{,}086 & 75{,}803 \end{bmatrix}$

c. $C = A - B$

$= \begin{bmatrix} 59{,}385 & 47{,}091 \\ 486{,}171 & 565{,}490 \\ 12{,}181 & 9880 \end{bmatrix} - \begin{bmatrix} 32{,}575 & 36{,}681 \\ 658{,}782 & 882{,}013 \\ 78{,}086 & 75{,}803 \end{bmatrix}$

$= \begin{bmatrix} 26{,}810 & 10{,}410 \\ -172{,}611 & -316{,}523 \\ -65{,}905 & -65{,}923 \end{bmatrix}$

d. Referring to the matrix in part c, the only positive trade balance for the U.S. is agricultural, occurring in both 1996 and 1999.

e. Referring to the matrix in part c, the largest trade deficit for the U.S. occurs in 1999 in manufactured goods category. Note that the number corresponding to manufactured goods in 1999 is the most extreme negative number in the matrix.

31. The original matrix is
$$A = \begin{bmatrix} 39{,}331 & 28{,}023 \\ 30{,}297 & 25{,}142 \\ 23{,}342 & 20{,}052 \end{bmatrix}.$$

If the median income increases by 12%, then the new matrix is $A + 12\%A$ or $(1 + 12\%)A$ or $1.12A$.

$$1.12A$$
$$= \begin{bmatrix} 1.12(39{,}331) & 1.12(28{,}023) \\ 1.12(30{,}297) & 1.12(25{,}142) \\ 1.12(23{,}342) & 1.12(20{,}052) \end{bmatrix}$$
$$= \begin{bmatrix} 44{,}050.72 & 31{,}385.76 \\ 33{,}932.64 & 28{,}159.04 \\ 26{,}143.04 & 22{,}458.24 \end{bmatrix}$$

33. To compute the required matrix, the costs of TV, radio, and newspaper advertisements need to be multiplied by the number of each type advertisement targeted to the various audiences. The matrix is represented by BA.

$$BA = \begin{bmatrix} 30 & 45 & 35 \\ 25 & 32 & 40 \\ 22 & 12 & 30 \end{bmatrix}\begin{bmatrix} 12 \\ 15 \\ 5 \end{bmatrix}$$

Using technology to calculate the product yields:

```
[B] [A]
      [[1210]
       [980 ]
       [594 ]]
```

Therefore, the cost of advertising to singles is $1210, the cost of advertising to males 35–55 is $980, and the cost of advertising to females 65+ is $594.

35. a. To calculate the total cost of various products per department, for each product multiply the quantity needed by

the unit cost and add the results. The matrix form of the multiplication corresponds to

$$\begin{bmatrix} 60 & 40 & 20 \\ 40 & 20 & 40 \end{bmatrix}\begin{bmatrix} 600 & 560 \\ 300 & 200 \\ 300 & 400 \end{bmatrix}.$$

Using technology to calculate the product yields

```
[A] [B]
 [[54000 49600]
  [42000 42400]]
```

	DeTuris	Marriott
Department A	54,000	49,600
Department B	42,000	42,400

b. To minimize the cost of the purchase, Department A needs to purchase the products from Marriott, while Department B needs to purchase the products from DeTuris.

37. $\begin{bmatrix} R \\ D \end{bmatrix} = \begin{bmatrix} 0.90 & 0.20 \\ 0.10 & 0.80 \end{bmatrix}\begin{bmatrix} a \\ b \end{bmatrix}$

Let $a = 0.50$ and $b = 0.50$.
$$\begin{bmatrix} R \\ D \end{bmatrix} = \begin{bmatrix} 0.90 & 0.20 \\ 0.10 & 0.80 \end{bmatrix}\begin{bmatrix} 0.50 \\ 0.50 \end{bmatrix}$$

Using technology to calculate the product yields

```
[A] [B]
       [[.55]
        [.45]]
```

$$\begin{bmatrix} R \\ D \end{bmatrix} = \begin{bmatrix} 0.55 \\ 0.45 \end{bmatrix} = \begin{bmatrix} 55\% \\ 45\% \end{bmatrix}$$

Based on the model, in the next election Republicans will receive 55% of the vote, while Democrats will receive 45% of the vote.

39. a. $\begin{bmatrix} 348 & 648 & 560 & 691 & 751 \\ 321 & 477 & 486 & 502 & 467 \end{bmatrix}$

b. To generate a new matrix representing an increase in weekly earnings for men of 10% and an increase in weekly earnings for women of 25%, the matrix from part a) needs to be multiplied by the matrix $\begin{bmatrix} 1.10 & 0 \\ 0 & 1.25 \end{bmatrix}$.

$$\begin{bmatrix} 1.10 & 0 \\ 0 & 1.25 \end{bmatrix}\begin{bmatrix} 348 & 648 & 560 & 691 & 751 \\ 321 & 477 & 486 & 502 & 467 \end{bmatrix}$$

Using technology to calculate the product yields:

$$[A][B] = \begin{bmatrix} 382.80 & 712.80 & 616 & 760.10 & 826.10 \\ 401.25 & 596.25 & 607.50 & 627.50 & 583.75 \end{bmatrix}$$

Section 5.4 Skills Check

1. a. AB

$$= \begin{bmatrix} 3 & 1 \\ 4 & 2 \end{bmatrix} \begin{bmatrix} 1 & -0.5 \\ -2 & 1.5 \end{bmatrix}$$

$$= \begin{bmatrix} 3-2 & -1.5+1.5 \\ 4-4 & -2+3 \end{bmatrix}$$

$$= \begin{bmatrix} 1 & 0 \\ 0 & 1 \end{bmatrix}$$

BA

$$= \begin{bmatrix} 1 & -0.5 \\ -2 & 1.5 \end{bmatrix} \begin{bmatrix} 3 & 1 \\ 4 & 2 \end{bmatrix}$$

$$= \begin{bmatrix} 3-2 & 1-1 \\ -6+6 & -2+3 \end{bmatrix}$$

$$= \begin{bmatrix} 1 & 0 \\ 0 & 1 \end{bmatrix}$$

b. Since $AB = BA = I$, A and B are inverses of one another.

3. Using technology to calculate AB yields

```
[A] [B]
      [[1 0 0]
       [0 1 0]
       [0 0 1]]
```

Likewise, using technology to calculate BA yields

```
[B] [A]
      [[1 0 0]
       [0 1 0]
       [0 0 1]]
```

Since $AB = BA = I$, A and B are inverses of one another.

5. $\begin{bmatrix} A & | & I \end{bmatrix}$

$$\begin{bmatrix} 1 & 3 & | & 1 & 0 \\ 2 & 7 & | & 0 & 1 \end{bmatrix} \xrightarrow{-2R_1 + R_2 \to R_2}$$

$$\begin{bmatrix} 1 & 3 & | & 1 & 0 \\ 0 & 1 & | & -2 & 1 \end{bmatrix} \xrightarrow{-3R_2 + R_1 \to R_1}$$

$$\begin{bmatrix} 1 & 0 & | & 7 & -3 \\ 0 & 1 & | & -2 & 1 \end{bmatrix}$$

$$\begin{bmatrix} I & | & A^{-1} \end{bmatrix}$$

$$A^{-1} = \begin{bmatrix} 7 & -3 \\ -2 & 1 \end{bmatrix}$$

7. Using technology to calculate A^{-1} yields

```
[A]-1▶Frac
[[-1/6 -1/3 1 ]
 [-1/3 1/3  0 ]
 [1/3  2/3  -1]]
```

9. Using technology to calculate A^{-1} yields

```
[A]-1▶Frac
[[-1/3 -1 1/3]
 [1    1  0 ]
 [-1/3 0  1/3]]
```

11. Using technology to calculate A^{-1} yields

```
[A]-1
[[.9  .2  -.7]
 [-.5 0   .5 ]
 [.7  -.4 -.1]]
```

13. Using technology to calculate C^{-1} yields

```
[C]-1
      [[1 0 1 1]
       [0 1 1 1]
       [0 0 1 0]
       [0 0 0 1]]
```

15. $AX = \begin{bmatrix} 2 \\ 4 \end{bmatrix}$

$A^{-1}(AX) = A^{-1}\left(\begin{bmatrix} 2 \\ 4 \end{bmatrix} \right)$

$IX = \begin{bmatrix} 1 & 2 \\ 4 & 3 \end{bmatrix} \begin{bmatrix} 2 \\ 4 \end{bmatrix}$

$X = \begin{bmatrix} 2+8 \\ 8+12 \end{bmatrix}$

$X = \begin{bmatrix} 10 \\ 20 \end{bmatrix}$

17. $\begin{bmatrix} -1 & 1 & 0 \\ -2 & 3 & -2 \\ 2 & -2 & 1 \end{bmatrix} \begin{bmatrix} x \\ y \\ z \end{bmatrix} = \begin{bmatrix} 3 \\ 5 \\ 8 \end{bmatrix}$

Let $A = \begin{bmatrix} -1 & 1 & 0 \\ -2 & 3 & -2 \\ 2 & -2 & 1 \end{bmatrix}$

Applying technology to calculate A^{-1}

$A^{-1} = \begin{bmatrix} 1 & 1 & 2 \\ 2 & 1 & 2 \\ 2 & 0 & 1 \end{bmatrix}$

Solving for X

$A^{-1}(AX) = A^{-1}\left(\begin{bmatrix} 3 \\ 5 \\ 8 \end{bmatrix} \right)$

$IX = \begin{bmatrix} 1 & 1 & 2 \\ 2 & 1 & 2 \\ 2 & 0 & 1 \end{bmatrix} \begin{bmatrix} 3 \\ 5 \\ 8 \end{bmatrix}$

$X = \begin{bmatrix} x \\ y \\ z \end{bmatrix} = \begin{bmatrix} 3+5+16 \\ 6+5+16 \\ 6+0+8 \end{bmatrix}$

$\begin{bmatrix} x \\ y \\ z \end{bmatrix} = \begin{bmatrix} 24 \\ 27 \\ 14 \end{bmatrix}$

$x = 24, \ y = 27, \ z = 14$

19. $\begin{bmatrix} 4 & -3 & 1 \\ -6 & 5 & -2 \\ 1 & -1 & 1 \end{bmatrix} \begin{bmatrix} x \\ y \\ z \end{bmatrix} = \begin{bmatrix} 2 \\ -3 \\ 1 \end{bmatrix}$

Let $A = \begin{bmatrix} 4 & -3 & 1 \\ -6 & 5 & -2 \\ 1 & -1 & 1 \end{bmatrix}$

Applying technology to calculate A^{-1}

$A^{-1} = \begin{bmatrix} 3 & 2 & 1 \\ 4 & 3 & 2 \\ 1 & 1 & 2 \end{bmatrix}$

Solving for X

$A^{-1}(AX) = A^{-1}\left(\begin{bmatrix} 2 \\ -3 \\ 1 \end{bmatrix} \right)$

$IX = \begin{bmatrix} 3 & 2 & 1 \\ 4 & 3 & 2 \\ 1 & 1 & 2 \end{bmatrix} \begin{bmatrix} 2 \\ -3 \\ 1 \end{bmatrix}$

$X = \begin{bmatrix} x \\ y \\ z \end{bmatrix} = \begin{bmatrix} 6-6+1 \\ 8-9+2 \\ 2-3+2 \end{bmatrix}$

$\begin{bmatrix} x \\ y \\ z \end{bmatrix} = \begin{bmatrix} 1 \\ 1 \\ 1 \end{bmatrix}$

$x = 1, \ y = 1, \ z = 1$

21. $\begin{bmatrix} 2 & 1 & 1 \\ 1 & 4 & 2 \\ 2 & 1 & 2 \end{bmatrix}\begin{bmatrix} x \\ y \\ z \end{bmatrix} = \begin{bmatrix} 4 \\ 4 \\ 3 \end{bmatrix}$

Let $A = \begin{bmatrix} 2 & 1 & 1 \\ 1 & 4 & 2 \\ 2 & 1 & 2 \end{bmatrix}$

Applying technology to calculate A^{-1}

$A^{-1} = \begin{bmatrix} \dfrac{6}{7} & -\dfrac{1}{7} & -\dfrac{2}{7} \\ \dfrac{2}{7} & \dfrac{2}{7} & -\dfrac{3}{7} \\ -1 & 0 & 1 \end{bmatrix}$

$A^{-1}\left(AX\right) = A^{-1}\left(\begin{bmatrix} 4 \\ 4 \\ 3 \end{bmatrix}\right)$

$IX = \begin{bmatrix} \dfrac{6}{7} & -\dfrac{1}{7} & -\dfrac{2}{7} \\ \dfrac{2}{7} & \dfrac{2}{7} & -\dfrac{3}{7} \\ -1 & 0 & 1 \end{bmatrix}\begin{bmatrix} 4 \\ 4 \\ 3 \end{bmatrix}$

Solving for X

$X = \begin{bmatrix} x \\ y \\ z \end{bmatrix} = \begin{bmatrix} \dfrac{24}{7} - \dfrac{4}{7} - \dfrac{6}{7} \\ \dfrac{8}{7} + \dfrac{8}{7} - \dfrac{9}{7} \\ -4 + 0 + 3 \end{bmatrix}$

$\begin{bmatrix} x \\ y \\ z \end{bmatrix} = \begin{bmatrix} \dfrac{14}{7} \\ \dfrac{7}{7} \\ -1 \end{bmatrix}$

$\begin{bmatrix} x \\ y \\ z \end{bmatrix} = \begin{bmatrix} 2 \\ 1 \\ -1 \end{bmatrix}$

$x = 2, \ y = 1, \ z = -1$

23. $\begin{bmatrix} 1 & 0 & 1 & 1 \\ 0 & 1 & 1 & 1 \\ 2 & 5 & 1 & 0 \\ 0 & 3 & 0 & 1 \end{bmatrix}\begin{bmatrix} x_1 \\ x_2 \\ x_3 \\ x_4 \end{bmatrix} = \begin{bmatrix} 90 \\ 72 \\ 108 \\ 144 \end{bmatrix}$

Let $A = \begin{bmatrix} 1 & 0 & 1 & 1 \\ 0 & 1 & 1 & 1 \\ 2 & 5 & 1 & 0 \\ 0 & 3 & 0 & 1 \end{bmatrix}$.

Applying technology to calculate A^{-1}

$A^{-1} = \begin{bmatrix} \dfrac{7}{9} & -\dfrac{8}{9} & \dfrac{1}{9} & \dfrac{1}{9} \\ -\dfrac{2}{9} & \dfrac{1}{9} & \dfrac{1}{9} & \dfrac{1}{9} \\ -\dfrac{4}{9} & \dfrac{11}{9} & \dfrac{2}{9} & -\dfrac{7}{9} \\ \dfrac{2}{3} & -\dfrac{1}{3} & -\dfrac{1}{3} & \dfrac{2}{3} \end{bmatrix}$

Solving for X

$A^{-1}\left(AX\right) = A^{-1}\left(\begin{bmatrix} 90 \\ 72 \\ 108 \\ 144 \end{bmatrix}\right)$

$IX = \begin{bmatrix} \dfrac{7}{9} & -\dfrac{8}{9} & \dfrac{1}{9} & \dfrac{1}{9} \\ -\dfrac{2}{9} & \dfrac{1}{9} & \dfrac{1}{9} & \dfrac{1}{9} \\ -\dfrac{4}{9} & \dfrac{11}{9} & \dfrac{2}{9} & -\dfrac{7}{9} \\ \dfrac{2}{3} & -\dfrac{1}{3} & -\dfrac{1}{3} & \dfrac{2}{3} \end{bmatrix}\begin{bmatrix} 90 \\ 72 \\ 108 \\ 144 \end{bmatrix}$

$X = \begin{bmatrix} x_1 \\ x_2 \\ x_3 \\ x_4 \end{bmatrix} = \begin{bmatrix} 34 \\ 16 \\ -40 \\ 96 \end{bmatrix}$

$x_1 = 34, \ x_2 = 16, \ x_3 = -40, \ x_4 = 96$

Section 5.4 Exercises

25. $\begin{bmatrix} X \\ Y \end{bmatrix} = \begin{bmatrix} \dfrac{3}{5} & \dfrac{1}{3} \\ \dfrac{2}{5} & \dfrac{2}{3} \end{bmatrix} \begin{bmatrix} x \\ y \end{bmatrix}$

Let $X = 150,000$ and $Y = 120,000$.

$\begin{bmatrix} 150,000 \\ 120,000 \end{bmatrix} = \begin{bmatrix} \dfrac{3}{5} & \dfrac{1}{3} \\ \dfrac{2}{5} & \dfrac{2}{3} \end{bmatrix} \begin{bmatrix} x \\ y \end{bmatrix}$

Let $A = \begin{bmatrix} \dfrac{3}{5} & \dfrac{1}{3} \\ \dfrac{2}{5} & \dfrac{2}{3} \end{bmatrix}$. Using technology to calculate A^{-1} yields $A^{-1} = \begin{bmatrix} \dfrac{5}{2} & -\dfrac{5}{4} \\ -\dfrac{3}{2} & \dfrac{9}{4} \end{bmatrix}$.

Multiplying both sides by A^{-1} yields

$\begin{bmatrix} \dfrac{5}{2} & -\dfrac{5}{4} \\ -\dfrac{3}{2} & \dfrac{9}{4} \end{bmatrix} \begin{bmatrix} 150,000 \\ 120,000 \end{bmatrix} = \begin{bmatrix} \dfrac{5}{2} & -\dfrac{5}{4} \\ -\dfrac{3}{2} & \dfrac{9}{4} \end{bmatrix} \begin{bmatrix} \dfrac{3}{5} & \dfrac{1}{3} \\ \dfrac{2}{5} & \dfrac{2}{3} \end{bmatrix} \begin{bmatrix} x \\ y \end{bmatrix}$.

Using technology to carry out the multiplication:

$\begin{bmatrix} 225,000 \\ 45,000 \end{bmatrix} = \begin{bmatrix} 1 & 0 \\ 0 & 1 \end{bmatrix} \begin{bmatrix} x \\ y \end{bmatrix}$

$\begin{bmatrix} x \\ y \end{bmatrix} = \begin{bmatrix} 225,000 \\ 45,000 \end{bmatrix}$

Last year Company X had 225,000 customers, and Company Y had 45,000 customers.

27. $\begin{bmatrix} R \\ D \end{bmatrix} = \begin{bmatrix} 0.90 & 0.20 \\ 0.10 & 0.80 \end{bmatrix} \begin{bmatrix} r \\ d \end{bmatrix}$

Let $R = 0.55$ and $D = 0.45$.

$\begin{bmatrix} 0.55 \\ 0.45 \end{bmatrix} = \begin{bmatrix} 0.90 & 0.20 \\ 0.10 & 0.80 \end{bmatrix} \begin{bmatrix} r \\ d \end{bmatrix}$

Let $A = \begin{bmatrix} 0.90 & 0.20 \\ 0.10 & 0.80 \end{bmatrix}$.

Using technology to calculate A^{-1} yields

```
[A]⁻¹▶Frac
    [[8/7   -2/7]
     [-1/7  9/7 ]]
```

$$A^{-1} = \begin{bmatrix} \dfrac{8}{7} & -\dfrac{2}{7} \\[2ex] -\dfrac{1}{7} & \dfrac{9}{7} \end{bmatrix}$$

Multiplying both sides by A^{-1} yields $\begin{bmatrix} \dfrac{8}{7} & -\dfrac{2}{7} \\[2ex] -\dfrac{1}{7} & \dfrac{9}{7} \end{bmatrix}\begin{bmatrix} 0.55 \\ 0.45 \end{bmatrix} = \begin{bmatrix} \dfrac{8}{7} & -\dfrac{2}{7} \\[2ex] -\dfrac{1}{7} & \dfrac{9}{7} \end{bmatrix}\begin{bmatrix} 0.90 & 0.20 \\ 0.10 & 0.80 \end{bmatrix}\begin{bmatrix} r \\ d \end{bmatrix}.$

Using technology to carry out the multiplication yields:

$$\begin{bmatrix} 0.5 \\ 0.5 \end{bmatrix} = \begin{bmatrix} 1 & 0 \\ 0 & 1 \end{bmatrix}\begin{bmatrix} r \\ d \end{bmatrix}$$

$$\begin{bmatrix} r \\ d \end{bmatrix} = \begin{bmatrix} 0.5 \\ 0.5 \end{bmatrix}$$

In last year's election 50% of voters were Republicans and 50% were Democrats.

29. a.   Let x represent the largest loan, y represent the medium size loan, and z represent the smallest loan.

$$\begin{cases} x + y + z = 400{,}000 \\ x = (y + z) + 100{,}000 \\ z = \dfrac{1}{2}y \end{cases}$$

or

$$\begin{cases} x + y + z = 400{,}000 \\ x - y - z = 100{,}000 \\ -\dfrac{1}{2}y + z = 0 \end{cases}$$

$$\begin{bmatrix} 1 & 1 & 1 \\ 1 & -1 & -1 \\ 0 & -\dfrac{1}{2} & 1 \end{bmatrix}\begin{bmatrix} x \\ y \\ z \end{bmatrix} = \begin{bmatrix} 400{,}000 \\ 100{,}000 \\ 0 \end{bmatrix}$$

b. $\begin{bmatrix} 1 & 1 & 1 \\ 1 & -1 & -1 \\ 0 & -\dfrac{1}{2} & 1 \end{bmatrix} \begin{bmatrix} x \\ y \\ z \end{bmatrix} = \begin{bmatrix} 400,000 \\ 100,000 \\ 0 \end{bmatrix}$

Let $A = \begin{bmatrix} 1 & 1 & 1 \\ 1 & -1 & -1 \\ 0 & -\dfrac{1}{2} & 1 \end{bmatrix}$. Using technology to calculate A^{-1} yields $A^{-1} = \begin{bmatrix} \dfrac{1}{2} & \dfrac{1}{2} & 0 \\ \dfrac{1}{3} & -\dfrac{1}{3} & -\dfrac{2}{3} \\ \dfrac{1}{6} & -\dfrac{1}{6} & \dfrac{2}{3} \end{bmatrix}$.

$\begin{bmatrix} \dfrac{1}{2} & \dfrac{1}{2} & 0 \\ \dfrac{1}{3} & -\dfrac{1}{3} & -\dfrac{2}{3} \\ \dfrac{1}{6} & -\dfrac{1}{6} & \dfrac{2}{3} \end{bmatrix} \begin{bmatrix} 1 & 1 & 1 \\ 1 & -1 & -1 \\ 0 & -\dfrac{1}{2} & 1 \end{bmatrix} \begin{bmatrix} x \\ y \\ z \end{bmatrix} = \begin{bmatrix} \dfrac{1}{2} & \dfrac{1}{2} & 0 \\ \dfrac{1}{3} & -\dfrac{1}{3} & -\dfrac{2}{3} \\ \dfrac{1}{6} & -\dfrac{1}{6} & \dfrac{2}{3} \end{bmatrix} \begin{bmatrix} 400,000 \\ 100,000 \\ 0 \end{bmatrix}$

Using technology to carry out the multiplication:

$\begin{bmatrix} 1 & 0 & 0 \\ 0 & 1 & 0 \\ 0 & 0 & 1 \end{bmatrix} \begin{bmatrix} x \\ y \\ z \end{bmatrix} = \begin{bmatrix} 250,000 \\ 100,000 \\ 50,000 \end{bmatrix}$

$\begin{bmatrix} x \\ y \\ z \end{bmatrix} = \begin{bmatrix} 250,000 \\ 100,000 \\ 50,000 \end{bmatrix}$

$x = 250,000, \ y = 100,000, \ z = 50,000$

The largest loan is \$250,000. The next medium size is \$100,000. The smallest loan is \$50,000.

31. Let x represent the amount invested at 6%, y represent the amount invested at 8%, and z represent the amount invested at 10%.

$$\begin{cases} x + y + z = 400,000 \\ 2x = y \\ 0.06x + 0.08y + 0.10z = 36,000 \end{cases} \quad \text{or} \quad \begin{cases} x + y + z = 400,000 \\ 2x - y = 0 \\ 0.06x + 0.08y + 0.10z = 36,000 \end{cases}$$

$$\begin{bmatrix} 1 & 1 & 1 \\ 2 & -1 & 0 \\ 0.06 & 0.08 & 0.10 \end{bmatrix} \begin{bmatrix} x \\ y \\ z \end{bmatrix} = \begin{bmatrix} 400,000 \\ 0 \\ 36,000 \end{bmatrix}$$

Let $A = \begin{bmatrix} 1 & 1 & 1 \\ 2 & -1 & 0 \\ 0.06 & 0.08 & 0.10 \end{bmatrix}$. Using technology to calculate A^{-1} yields $A^{-1} = \begin{bmatrix} \dfrac{5}{4} & \dfrac{1}{4} & -\dfrac{25}{2} \\ \dfrac{5}{2} & -\dfrac{1}{2} & -25 \\ -\dfrac{11}{4} & \dfrac{1}{4} & \dfrac{75}{2} \end{bmatrix}$.

$$\begin{bmatrix} \dfrac{5}{4} & \dfrac{1}{4} & -\dfrac{25}{2} \\ \dfrac{5}{2} & -\dfrac{1}{2} & -25 \\ -\dfrac{11}{4} & \dfrac{1}{4} & \dfrac{75}{2} \end{bmatrix} \begin{bmatrix} 1 & 1 & 1 \\ 2 & -1 & 0 \\ 0.06 & 0.08 & 0.10 \end{bmatrix} \begin{bmatrix} x \\ y \\ z \end{bmatrix} = \begin{bmatrix} \dfrac{5}{4} & \dfrac{1}{4} & -\dfrac{25}{2} \\ \dfrac{5}{2} & -\dfrac{1}{2} & -25 \\ -\dfrac{11}{4} & \dfrac{1}{4} & \dfrac{75}{2} \end{bmatrix} \begin{bmatrix} 400,000 \\ 0 \\ 36,000 \end{bmatrix}$$

Using technology to carry out the multiplication yields

$$\begin{bmatrix} 1 & 0 & 0 \\ 0 & 1 & 0 \\ 0 & 0 & 1 \end{bmatrix} \begin{bmatrix} x \\ y \\ z \end{bmatrix} = \begin{bmatrix} 50,000 \\ 100,000 \\ 250,000 \end{bmatrix}$$

$$\begin{bmatrix} x \\ y \\ z \end{bmatrix} = \begin{bmatrix} 50,000 \\ 100,000 \\ 250,000 \end{bmatrix}$$

$x = 50,000$, $y = 100,000$, $z = 250,000$

$50,000 is invested in the 6% account, $100,000 is invested in the 8% account, and $250,000 is invested in the 10% account.

33. Let x represent the percentage of venture capital from business loans, y represent thepercentage of venture capital from auto loans, and z represent the percentage of venture capital from home loans.

$$\begin{cases} 532x + 58y + 682z = 483.94 \\ 562x + 62y + 695z = 503.28 \\ 578x + 69y + 722z = 521.33 \end{cases}$$

$$\begin{bmatrix} 532 & 58 & 682 \\ 562 & 62 & 695 \\ 578 & 69 & 722 \end{bmatrix} \begin{bmatrix} x \\ y \\ z \end{bmatrix} = \begin{bmatrix} 483.94 \\ 503.28 \\ 521.33 \end{bmatrix}$$

Let $A = \begin{bmatrix} 532 & 58 & 682 \\ 562 & 62 & 695 \\ 578 & 69 & 722 \end{bmatrix}$. Using technology to calculate A^{-1} and applying A^{-1} to both sides

of the equation yields $A^{-1} \begin{bmatrix} 532 & 58 & 682 \\ 562 & 62 & 695 \\ 578 & 69 & 722 \end{bmatrix} \begin{bmatrix} x \\ y \\ z \end{bmatrix} = A^{-1} \begin{bmatrix} 483.94 \\ 503.28 \\ 521.33 \end{bmatrix}$.

Using technology to carry out the multiplication yields

```
[B]                          [A]-1[B]
   [[483.94]                    [[.47]
    [503.28]                     [.27]
    [521.33]]                    [.32]]
```

$$\begin{bmatrix} 1 & 0 & 0 \\ 0 & 1 & 0 \\ 0 & 0 & 1 \end{bmatrix} \begin{bmatrix} x \\ y \\ z \end{bmatrix} = \begin{bmatrix} 0.47 \\ 0.27 \\ 0.32 \end{bmatrix}$$

$$\begin{bmatrix} x \\ y \\ z \end{bmatrix} = \begin{bmatrix} 0.47 \\ 0.27 \\ 0.32 \end{bmatrix}$$

$x = 0.47, \ y = 0.27, \ z = 0.32$

The percentage of venture capital from business loans is 47%, the percentage of venture capital from auto loans is 27%, and the percentage of venture capital from home loans is 32%.

35. a. "Just do it" is represented by the 10, 21, 19, 20, 27, 4, 15, 27, 9, and 20. Recall that spaces are coded as 27.

b. Multiplying by the encoding matrix and applying technology to generate the solutions yields:

$$\begin{bmatrix} 4 & 4 \\ 1 & 2 \end{bmatrix}\begin{bmatrix} 10 \\ 21 \end{bmatrix} = \begin{bmatrix} 124 \\ 52 \end{bmatrix}$$

$$\begin{bmatrix} 4 & 4 \\ 1 & 2 \end{bmatrix}\begin{bmatrix} 19 \\ 20 \end{bmatrix} = \begin{bmatrix} 156 \\ 59 \end{bmatrix}$$

$$\begin{bmatrix} 4 & 4 \\ 1 & 2 \end{bmatrix}\begin{bmatrix} 27 \\ 4 \end{bmatrix} = \begin{bmatrix} 124 \\ 35 \end{bmatrix}$$

$$\begin{bmatrix} 4 & 4 \\ 1 & 2 \end{bmatrix}\begin{bmatrix} 15 \\ 27 \end{bmatrix} = \begin{bmatrix} 168 \\ 69 \end{bmatrix}$$

$$\begin{bmatrix} 4 & 4 \\ 1 & 2 \end{bmatrix}\begin{bmatrix} 9 \\ 20 \end{bmatrix} = \begin{bmatrix} 116 \\ 49 \end{bmatrix}$$

The pairs of coded numbers are 124, 52, 156, 59, 124, 35, 168, 69, 116, and 49.

37 "Neatness counts" is represented by the 14, 5, 1, 20, 14, 5, 19, 19, 27, 3, 15, 21, 14, 20, and 19. Recall that spaces are coded as 27.

Multiplying by the encoding matrix and applying technology to generate the solutions yields:

$$\begin{bmatrix} 4 & 4 & 4 \\ 1 & 2 & 3 \\ 2 & 4 & 2 \end{bmatrix}\begin{bmatrix} 14 \\ 5 \\ 1 \end{bmatrix} = \begin{bmatrix} 80 \\ 27 \\ 50 \end{bmatrix}$$

$$\begin{bmatrix} 4 & 4 & 4 \\ 1 & 2 & 3 \\ 2 & 4 & 2 \end{bmatrix}\begin{bmatrix} 20 \\ 14 \\ 5 \end{bmatrix} = \begin{bmatrix} 156 \\ 63 \\ 106 \end{bmatrix}$$

$$\begin{bmatrix} 4 & 4 & 4 \\ 1 & 2 & 3 \\ 2 & 4 & 2 \end{bmatrix}\begin{bmatrix} 19 \\ 19 \\ 27 \end{bmatrix} = \begin{bmatrix} 260 \\ 138 \\ 168 \end{bmatrix}$$

$$\begin{bmatrix} 4 & 4 & 4 \\ 1 & 2 & 3 \\ 2 & 4 & 2 \end{bmatrix}\begin{bmatrix} 3 \\ 15 \\ 21 \end{bmatrix} = \begin{bmatrix} 156 \\ 96 \\ 108 \end{bmatrix}$$

$$\begin{bmatrix} 4 & 4 & 4 \\ 1 & 2 & 3 \\ 2 & 4 & 2 \end{bmatrix}\begin{bmatrix} 14 \\ 20 \\ 19 \end{bmatrix} = \begin{bmatrix} 212 \\ 111 \\ 146 \end{bmatrix}$$

The triples of coded numbers are 80, 27, 50, 156, 63, 106, 260, 138, 168, 156, 96, 108, 212, 111, and 146.

39. `[A]⁻¹▸Frac`
`[[1  1]`
`[2  3]]`

Multiplying pairs of codes by the inverse of the encoding matrix decodes the message. Multiplying by A^{-1} and using technology to simplify yields:

$$\begin{bmatrix} 1 & 1 \\ 2 & 3 \end{bmatrix}\begin{bmatrix} 51 \\ -29 \end{bmatrix} = \begin{bmatrix} 22 \\ 15 \end{bmatrix}$$

$$\begin{bmatrix} 1 & 1 \\ 2 & 3 \end{bmatrix}\begin{bmatrix} 55 \\ -35 \end{bmatrix} = \begin{bmatrix} 20 \\ 5 \end{bmatrix}$$

$$\begin{bmatrix} 1 & 1 \\ 2 & 3 \end{bmatrix}\begin{bmatrix} 76 \\ -49 \end{bmatrix} = \begin{bmatrix} 27 \\ 5 \end{bmatrix}$$

$$\begin{bmatrix} 1 & 1 \\ 2 & 3 \end{bmatrix}\begin{bmatrix} -15 \\ 16 \end{bmatrix} = \begin{bmatrix} 1 \\ 18 \end{bmatrix}$$

$$\begin{bmatrix} 1 & 1 \\ 2 & 3 \end{bmatrix}\begin{bmatrix} 11 \\ 1 \end{bmatrix} = \begin{bmatrix} 12 \\ 25 \end{bmatrix}$$

The decoded message is "Vote early."

41. `[A]⁻¹▸Frac`
`[[1  1  1]`
`[1  2  1]`
`[2  1  1]]`

Multiplying triples of codes by the inverse of the encoding matrix decodes the message. Multiplying by A^{-1} and using technology to simplify yields:

$$\begin{bmatrix} 1 & 1 & 1 \\ 1 & 2 & 1 \\ 2 & 1 & 1 \end{bmatrix} \begin{bmatrix} 1 \\ -4 \\ 16 \end{bmatrix} = \begin{bmatrix} 13 \\ 9 \\ 14 \end{bmatrix}$$

$$\begin{bmatrix} 1 & 1 & 1 \\ 1 & 2 & 1 \\ 2 & 1 & 1 \end{bmatrix} \begin{bmatrix} 21 \\ 23 \\ -40 \end{bmatrix} = \begin{bmatrix} 4 \\ 27 \\ 25 \end{bmatrix}$$

$$\begin{bmatrix} 1 & 1 & 1 \\ 1 & 2 & 1 \\ 2 & 1 & 1 \end{bmatrix} \begin{bmatrix} 3 \\ 6 \\ 6 \end{bmatrix} = \begin{bmatrix} 15 \\ 21 \\ 18 \end{bmatrix}$$

$$\begin{bmatrix} 1 & 1 & 1 \\ 1 & 2 & 1 \\ 2 & 1 & 1 \end{bmatrix} \begin{bmatrix} -26 \\ -14 \\ 67 \end{bmatrix} = \begin{bmatrix} 27 \\ 13 \\ 1 \end{bmatrix}$$

$$\begin{bmatrix} 1 & 1 & 1 \\ 1 & 2 & 1 \\ 2 & 1 & 1 \end{bmatrix} \begin{bmatrix} -9 \\ 0 \\ 23 \end{bmatrix} = \begin{bmatrix} 14 \\ 14 \\ 5 \end{bmatrix}$$

$$\begin{bmatrix} 1 & 1 & 1 \\ 1 & 2 & 1 \\ 2 & 1 & 1 \end{bmatrix} \begin{bmatrix} 9 \\ 1 \\ 8 \end{bmatrix} = \begin{bmatrix} 18 \\ 19 \\ 27 \end{bmatrix}$$

The decoded message is "Mind your manners."

43.
```
[A]⁻¹▶Frac
...14/3  -8/3  -5/3...
...-1/3  1/3   1/3 ...
...3    -2    -1  ...
```

Multiplying triples of codes by the inverse of the encoding matrix decodes the message.

Multiplying by A^{-1} and using technology to simplify yields:

$$\begin{bmatrix} -\dfrac{14}{3} & -\dfrac{8}{3} & -\dfrac{5}{3} \\ -\dfrac{1}{3} & \dfrac{1}{3} & \dfrac{1}{3} \\ 3 & -2 & -1 \end{bmatrix} \begin{bmatrix} 29 \\ -1 \\ 75 \end{bmatrix} = \begin{bmatrix} 13 \\ 15 \\ 14 \end{bmatrix}$$

$$\begin{bmatrix} -\dfrac{14}{3} & -\dfrac{8}{3} & -\dfrac{5}{3} \\ -\dfrac{1}{3} & \dfrac{1}{3} & \dfrac{1}{3} \\ 3 & -2 & -1 \end{bmatrix} \begin{bmatrix} -19 \\ -66 \\ 50 \end{bmatrix} = \begin{bmatrix} 4 \\ 1 \\ 25 \end{bmatrix}$$

$$\begin{bmatrix} -\dfrac{14}{3} & -\dfrac{8}{3} & -\dfrac{5}{3} \\ -\dfrac{1}{3} & \dfrac{1}{3} & \dfrac{1}{3} \\ 3 & -2 & -1 \end{bmatrix} \begin{bmatrix} 46 \\ 41 \\ 47 \end{bmatrix} = \begin{bmatrix} 27 \\ 14 \\ 9 \end{bmatrix}$$

$$\begin{bmatrix} -\dfrac{14}{3} & -\dfrac{8}{3} & -\dfrac{5}{3} \\ -\dfrac{1}{3} & \dfrac{1}{3} & \dfrac{1}{3} \\ 3 & -2 & -1 \end{bmatrix} \begin{bmatrix} 3 \\ -38 \\ 65 \end{bmatrix} = \begin{bmatrix} 7 \\ 8 \\ 20 \end{bmatrix}$$

The decoded message is "Monday night."

45. Answers will vary.

Section 5.5 Skills Check

1. Isolating y in the first equation yields
   $y = x^2$.
   Substituting into the other equation yields
   $3x + y = 0$
   $3x + x^2 = 0$
   $x(3 + x) = 0$
   $x = 0, \; x = -3$
   Back substituting to calculate y
   $x = 0 \Rightarrow y = (0)^2 = 0$
   $x = -3 \Rightarrow y = (-3)^2 = 9$
   The solutions to the system are
   $(0,0)$ and $(-3,9)$.

3. Isolating y in the second equation yields
   $2x + 3y = 4$
   $3y = 4 - 2x$
   $y = \dfrac{4 - 2x}{3}$
   Substituting into the other equation yields
   $x^2 - 3y = 4$
   $x^2 - 3\left(\dfrac{4 - 2x}{3}\right) = 4$
   $x^2 - (4 - 2x) = 4$
   $x^2 - 4 + 2x = 4$
   $x^2 + 2x - 8 = 0$
   $(x - 2)(x + 4) = 0$
   $x = 2, \; x = -4$
   Back substituting to calculate y
   $x = 2 \Rightarrow y = \dfrac{4 - 2(2)}{3} = 0$
   $x = -4 \Rightarrow y = \dfrac{4 - 2(-4)}{3} = 4$
   The solutions to the system are
   $(2,0)$ and $(-4,4)$.

5. Substituting $y = 2x$ into the other equation
   yields
   $x^2 + y^2 = 80$
   $x^2 + (2x)^2 = 80$
   $x^2 + 4x^2 = 80$
   $5x^2 = 80$
   $x^2 = 16$
   $x = \pm\sqrt{16} = \pm 4$
   $x = 4, \; x = -4$
   Back substituting to calculate y
   $x = 4 \Rightarrow y = 2(4) = 8$
   $x = -4 \Rightarrow y = 2(-4) = -8$
   The solutions to the system are
   $(4,8)$ and $(-4,-8)$.

7. Isolating y in the first equation yields
   $y = 8 - x$.
   Substituting into the other equation yields
   $xy = 12$
   $x(8 - x) = 12$
   $8x - x^2 = 12$
   $x^2 - 8x + 12 = 0$
   $(x - 6)(x - 2) = 0$
   $x = 6, \; x = 2$
   Back substituting to calculate y
   $x = 6 \Rightarrow y = 8 - (6) = 2$
   $x = 2 \Rightarrow y = 8 - (2) = 6$
   The solutions to the system are
   $(6,2)$ and $(2,6)$.

9. Isolating y in the second equation yields
   $2x - y + 4 = 0$ or $y = 2x + 4$.

Substituting into the other equation yields

$x^2 + 5x - y = 6$

$x^2 + 5x - (2x + 4) = 6$

$x^2 + 5x - 2x - 4 = 6$

$x^2 + 3x - 10 = 0$

$(x - 2)(x + 5) = 0$

$x = 2, \; x = -5$

Back substituting to calculate y

$x = 2 \Rightarrow y = 2(2) + 4 = 8$

$x = -5 \Rightarrow y = 2(-5) + 4 = -6$

The solutions to the system are

$(2, 8)$ and $(-5, -6)$.

$2x^2 + 4x - 2y = 24$

$2x^2 + 4x - 2(2x + 37) = 24$

$2x^2 + 4x - 4x - 74 = 24$

$2x^2 - 98 = 0$

$2(x^2 - 49) = 0$

$2(x - 7)(x + 7) = 0$

$x = 7, \; x = -7$

Back substituting to calculate y

$x = -7 \Rightarrow y = 2(-7) + 37 = 23$

$x = 7 \Rightarrow y = 2(7) + 37 = 51$

The solutions to the system are

$(-7, 23)$ and $(7, 51)$.

11. Isolating y in the second equation yields

$2y - x = 61$ or $y = \dfrac{61 + x}{2}$.

Substituting into the other equation yields

$2x^2 - 2y + 7x = 19$

$2x^2 - 2\left(\dfrac{61 + x}{2}\right) + 7x = 19$

$2x^2 - 61 - x + 7x = 19$

$2x^2 + 6x - 80 = 0$

$2(x^2 + 3x - 40) = 0$

$2(x - 5)(x + 8) = 0$

$x = 5, \; x = -8$

Back substituting to calculate y

$x = 5 \Rightarrow y = \dfrac{61 + (5)}{2} = 33$

$x = -8 \Rightarrow y = \dfrac{61 + (-8)}{2} = \dfrac{53}{2}$

The solutions to the system are

$(5, 33)$ and $\left(-8, \dfrac{53}{2}\right)$.

12. Isolating y in the second equation yields

$2x - y + 37 = 0$ or $y = 2x + 37$.

Substituting into the other equation yields

13. Isolating y in both equations:

$4y = 28 - x^2$

$y = \dfrac{28 - x^2}{4}$

and

$y = 1 + \sqrt{x}$

Solving graphically by applying the intersection of graphs method:

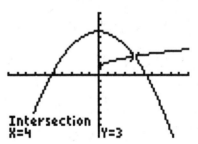

$[-10, 10]$ by $[-10, 10]$

The solution to the system is $(4, 3)$.

15. Isolating y in the second equation:

$4y = x + 8$

$y = \dfrac{x + 8}{4}$

Substituting into the other equation:
$$4xy + x = 10$$

$$4x\left(\frac{x+8}{4}\right) + x = 10$$

$$x^2 + 8x + x = 10$$

$$x^2 + 9x - 10 = 0$$

$$(x+10)(x-1) = 0$$

$$x = -10, \ x = 1$$

Back substituting to calculate y

$$x = -10 \Rightarrow y = \frac{(-10)+8}{4} = -\frac{1}{2}$$

$$x = 1 \Rightarrow y = \frac{(1)+8}{4} = \frac{9}{4}$$

The solutions to the system are
$$\left(-10, -\frac{1}{2}\right) \text{ and } \left(1, \frac{9}{4}\right).$$

17. Solving graphically by applying the intersection of graphs method yields:

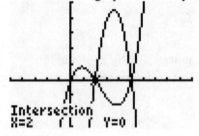

$[-5, 10]$ by $[-15, 25]$

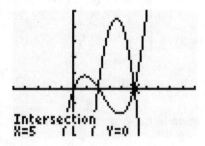

$[-5, 10]$ by $[-15, 25]$

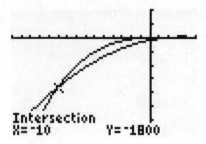

$[-15, 5]$ by $[-3500, 1000]$

The solutions to the system are
$(-10, -1800), (2, 0), \text{ and } (5, 0).$

19. Isolating y in both equations:
$$y = 15 - x^2 - x^3$$
and
$$y = 11 - 2x^2$$

Solving graphically by applying the intersection of graphs method:

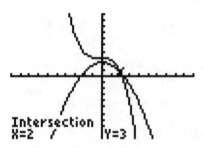

$[-10, 10]$ by $[-50, 50]$

The solution to the system is $(2, 3)$.

21. a. $y = 11 - 2x^2$

Substituting into the other equation
$$x^3 + x^2 + (11 - 2x^2) = 15$$

$$x^3 - x^2 - 4 = 0$$

b.
$$2\overline{)\begin{array}{cccc} 1 & -1 & 0 & -4 \\ & 2 & 2 & 4 \\ \hline 1 & 1 & 2 & 0 \end{array}}$$

The new polynomial equation is $x^2 + x + 2$. It has no real number solutions.

Therefore, the only solution to the system is $(2,3)$.

23. Isolating y in both equations:
$$4y = 17 - x^2$$
$$y = \frac{17 - x^2}{4}$$
and
$$x^2 y = 3 + 5x$$
$$y = \frac{3 + 5x}{x^2}$$

Solving graphically by applying the intersection of graphs method yields:

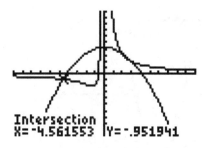

Intersection
X=-4.561553 Y=-.951941

[−10, 10] by [−10, 10]

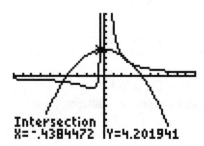

Intersection
X=-.4384472 Y=4.201941

[−10, 10] by [−10, 10]

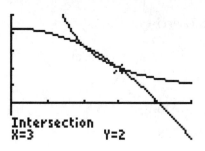

Intersection
X=3   Y=2

[0, 5] by [−2, 5]

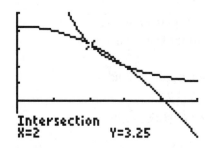

Intersection
X=2   Y=3.25

[0, 5] by [−2, 5]

Pick any two of the following four solutions to the system:

Exact solutions

$(2, 3.25), (3, 2)$

Approximate solutions

$(-0.438, 4.20), (-4.562, -0.952)$

Section 5.5 Exercises

25. Equilibrium occurs when demand equals supply.

$$q^2 + 2q + 122 = 650 - 30q$$
$$q^2 + 32q - 528 = 0$$
$$(q + 44)(q - 12) = 0$$
$$q = -44, \quad q = 12$$

Since $q \geq 0$ in the physical context of the question, $q = 12$.

Back substituting to find p
$$p = 650 - 30q$$
$$p = 650 - 30(12) = 290$$

Equilibrium occurs when the price is $290 and the demand is 1200 units.

27. Equilibrium occurs when demand equals supply.

$$0.1q^2 + 50q + 1027.50 = 6000 - 20q$$
$$0.1q^2 + 70q - 4972.50 = 0$$
$$q^2 + 700q - 49,725 = 0$$

$$q = \frac{-b \pm \sqrt{b^2 - 4ac}}{2a}$$

$$q = \frac{-700 \pm \sqrt{(700)^2 - 4(1)(-49,725)}}{2(1)}$$

$$q = \frac{-700 \pm \sqrt{490,000 + 198,900}}{2}$$

$$q = \frac{-700 \pm 830}{2}$$

$$q = \frac{-700 - 830}{2} = -765,$$

$$q = \frac{-700 + 830}{2} = 65$$

Since $q \geq 0$ in the physical context of the question, $q = 65$.

Back substituting to find p

$$p = 6000 - 20q$$
$$p = 6000 - 20(65) = 4700$$

Equilibrium occurs when the price is $4700 and the demand is 6500 units.

29. Break-even occurs when cost equals revenue.

$$C(x) = R(x)$$
$$2000x + 18,000 + 60x^2 = 4620x - 12x^2 - x^3$$
$$x^3 + 72x^2 - 2620x + 18,000 = 0$$

Solving graphically by applying the x-intercept method:

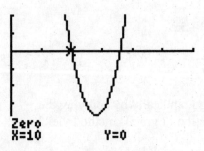

[0, 30] by [−2500, 1000]

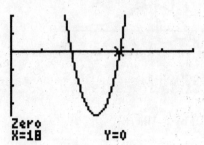

[0, 30] by [−2500, 1000]

Since $x \geq 0$ in the physical context of the question, negative solutions are not relevant.

Break-even occurs when the number of thousands of units is 10 or 18. Producing and selling 10,000 units or 18,000 units results in revenue equaling cost.

31. Break-even occurs when cost equals revenue. $C(x) = R(x)$

Solving graphically by applying the intersection of graphs method yields:

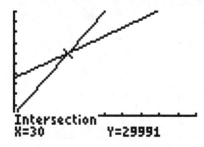

Intersection
X=30   Y=29991

[0, 100] by [−10,000, 50,000]

Using the TI-83 ZoomFit feature:

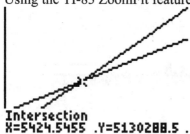

Intersection
X=5424.5455 .Y=5130288.5 .

[5000, 6000] by [4,517,901, 6,017,901]

Since $x \geq 0$ in the physical context of the question, negative solutions are not relevant.

Break-even occurs when the number of units is 30 or approximately 5425. Producing and selling 30 units or approximately 5425 units results in revenue equaling cost.

33. Let $x =$ length and $y =$ width.

$$\begin{cases} xy = 180 \\ 2(x-4)(y-4) = 176 \end{cases}$$

Solving the first equation for y

$$y = \frac{180}{x}$$

Substituting

$$2(x-4)(y-4) = 176$$

$$2(x-4)\left(\left[\frac{180}{x}\right] - 4\right) = 176$$

$$(2x-8)\left(\frac{180}{x} - 4\right) = 176$$

$$360 - 8x - \frac{1440}{x} + 32 = 176$$

$$x\left[360 - 8x - \frac{1440}{x} + 32\right] = x[176]$$

$$360x - 8x^2 - 1440 + 32x - 176x = 0$$

$$-8x^2 + 216x - 1440 = 0$$

$$-8(x^2 - 27x + 180) = 0$$

$$-8(x-15)(x-12) = 0$$

$$x = 15,\ x = 12$$

The dimensions are 15 feet by 12 feet.

35. Let $x =$ length of the shorter side and $y =$ length of the longer side.
Assuming that the box is open,

$$V = x \cdot x \cdot y = x^2 y.$$

$$\text{Surface Area} = xy + 2xy + 2x^2$$

$$= 3xy + 2x^2$$

$$\begin{cases} x^2 y = 2000 \\ 3xy + 2x^2 = 800 \end{cases}$$

Solving the first equation for y

$$y = \frac{2000}{x^2}$$

Substituting

$$3xy + 2x^2 = 800$$

$$3x\left(\frac{2000}{x^2}\right) + 2x^2 = 800$$

$$\frac{6000}{x} + 2x^2 = 800$$

$$x\left[\frac{6000}{x} + 2x^2\right] = x[800]$$

$$6000 + 2x^3 = 800x$$

$$2x^3 - 800x + 6000 = 0$$

Solving graphically by applying the x-intercept method with $x \geq 0$ yields:

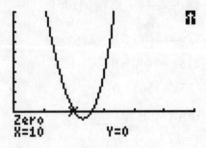

[0, 30] by [−250, 1000]

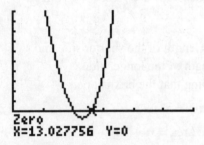

[0, 30] by [−250, 1000]

Substituting to find y

$$x = 10 \Rightarrow y = \frac{2000}{(10)^2} = 20$$

$$x \approx 13.03 \Rightarrow y \approx \frac{2000}{(13.03)^2} = 11.78$$

In the second case, x is not the smaller side. Therefore, the solution is a box with dimension 10 cm by 10 cm by 20 cm.

37. Let $t =$ time in years and $y =$ amount in dollars.

$$\begin{cases} y = 50{,}000(1.10)^t \\ y = 12{,}968.72t \end{cases}$$

$$50{,}000(1.10)^t = 12{,}968.72t$$

Solving graphically by applying the intersections of graphs method

Intersection
X=9.9998734   Y=129685.56

[0, 20] by [−50,000, 250,000]

In approximately ten years the trust fund will equal the amount of money received from the second account.

Chapter 5 Skills Check

1. $\begin{cases} 2x - 3y + z = 2 \\ 3x + 2y - z = 6 \\ x - 4y + 2z = 2 \end{cases} \xrightarrow{\substack{-3\,Eq3 + Eq2 \to Eq2 \\ -2\,Eq3 + Eq1 \to Eq3}}$

$\begin{cases} 2x - 3y + z = 2 \\ 14y - 7z = 0 \\ 5y - 3z = -2 \end{cases} \xrightarrow{-5\,Eq2 + 14\,Eq3 \to Eq3}$

$\begin{cases} 2x - 3y + z = 2 \\ 14y - 7z = 0 \\ -7z = -28 \end{cases} \xrightarrow{-\frac{1}{7}Eq3 \to Eq3}$

$\begin{cases} 2x - 3y + z = 2 \\ 14y - 7z = 0 \\ z = 4 \end{cases}$

Since z is isolated, back substitution yields

$14y - 7(4) = 0$

$14y - 28 = 0$

$14y = 28$

$y = 2$

and

$2x - 3(2) + (4) = 2$

$2x - 6 + 4 = 2$

$2x = 4$

$x = 2$

The solutions are $x = 2$, $y = 2$, $z = 4$.

Applying technology yields

```
rref([A])▶Frac
     [[1 0 0 2]
      [0 1 0 2]
      [0 0 1 4]]
```

2. $\begin{cases} 3x - 2y - 4z = 9 \\ x + 3y + 2z = -1 \\ 2x + 4y + 4z = 2 \end{cases} \xrightarrow{\substack{-3\,Eq2 + Eq1 \to Eq2 \\ -2\,Eq2 + Eq3 \to Eq3}}$

$\begin{cases} 3x - 2y - 4z = 9 \\ -11y - 10z = 12 \\ -2y = 4 \end{cases} \xrightarrow{-\frac{1}{2}Eq3 \to Eq3}$

$\begin{cases} 3x - 2y - 4z = 9 \\ -11y - 10z = 12 \\ y = -2 \end{cases}$

Since y is isolated, back substitution yields

$-11(-2) - 10z = 12$

$22 - 10z = 12$

$-10z = -10$

$z = 1$

and

$3x - 2(-2) - 4(1) = 9$

$3x + 4 - 4 = 9$

$3x = 9$

$x = 3$

The solutions are $x = 3$, $y = -2$, $z = 1$.

Applying technology yields

```
rref([A])▶Frac
     [[1 0 0 3 ]
      [0 1 0 -2]
      [0 0 1 1 ]]
```

3. $\begin{cases} 3x + 2y - z = 6 \\ 2x - 4y - 2z = 0 \\ 5x + 3y + 6z = 2 \end{cases} \xrightarrow[\substack{-3Eq2+2Eq1\to Eq2 \\ -5Eq2+2Eq3\to Eq3}]{}$

$\begin{cases} 3x + 2y - z = 6 \\ 16y + 4z = 12 \\ 26y + 22z = 4 \end{cases} \xrightarrow[\substack{\frac{1}{4}Eq2\to Eq2 \\ \frac{1}{2}Eq3\to Eq3}]{}$

$\begin{cases} 3x + 2y - z = 6 \\ 4y + z = 3 \\ 13y + 11z = 2 \end{cases} \xrightarrow{-13Eq2+4Eq3\to Eq3}$

$\begin{cases} 3x + 2y - z = 6 \\ 4y + z = 3 \\ 31z = -31 \end{cases} \xrightarrow{\frac{1}{31}Eq3\to Eq3}$

$\begin{cases} 3x + 2y - z = 6 \\ 4y + z = 3 \\ z = -1 \end{cases}$

Since z is isolated, back substitution yields
$4y + (-1) = 3$
$4y = 4$
$y = 1$
and
$3x + 2y - z = 6$
$3x + 2(1) - (-1) = 6$
$3x + 3 = 6$
$3x = 3$
$x = 1$
The solutions are $x = 1$, $y = 1$, $z = -1$.

Applying technology yields

```
rref([A])▶Frac
     [[1 0 0 1 ]
      [0 1 0 1 ]
      [0 0 1 -1]]
```

4. $\begin{cases} 3x + 6y + 9z = 27 \\ 2x + 3y - z = -2 \\ 4x + 5y + z = 6 \end{cases} \xrightarrow[\substack{-3Eq2+2Eq1\to Eq2 \\ -2Eq2+Eq3\to Eq3}]{}$

$\begin{cases} 3x + 6y + 9z = 27 \\ 3y + 21z = 60 \\ -y + 3z = 10 \end{cases} \xrightarrow{Eq2+3Eq3\to Eq3}$

$\begin{cases} 3x + 6y + 9z = 27 \\ 3y + 21z = 60 \\ 30z = 90 \end{cases} \xrightarrow{\frac{1}{30}Eq3\to Eq3}$

$\begin{cases} 3x + 6y + 9z = 27 \\ 3y + 21z = 60 \\ z = 3 \end{cases}$

Since z is isolated, back substitution yields
$3y + 21(3) = 60$
$3y + 63 = 60$
$3y = -3$
$y = -1$
and
$3x + 6y + 9z = 27$
$3x + 6(-1) + 9(3) = 27$
$3x + 21 = 27$
$3x = 6$
$x = 2$
The solutions are $x = 2$, $y = -1$, $z = 3$.

Applying technology yields

```
rref([A])▶Frac
     [[1 0 0 2 ]
      [0 1 0 -1]
      [0 0 1 3 ]]
```

5. Writing the system as an augmented matrix and reducing yields

$$\begin{bmatrix} 1 & 2 & -2 & | & 1 \\ 2 & -1 & 5 & | & 15 \\ 3 & -4 & 1 & | & 7 \end{bmatrix} \xrightarrow[-3R_1+R_3 \to R_3]{-2R_1+R_2 \to R_2}$$

$$\begin{bmatrix} 1 & 2 & -2 & | & 1 \\ 0 & -5 & 9 & | & 13 \\ 0 & -10 & 7 & | & 4 \end{bmatrix} \xrightarrow{-2R_2+R_3 \to R_3}$$

$$\begin{bmatrix} 1 & 2 & -2 & | & 1 \\ 0 & -5 & 9 & | & 13 \\ 0 & 0 & -11 & | & -22 \end{bmatrix} \xrightarrow[\left(-\frac{1}{11}\right)R_3 \to R_3]{\left(-\frac{1}{5}\right)R_2 \to R_2}$$

$$\begin{bmatrix} 1 & 2 & -2 & | & 1 \\ 0 & 1 & -\frac{9}{5} & | & -\frac{13}{5} \\ 0 & 0 & 1 & | & 2 \end{bmatrix}$$

$z = 2$

and

$$y - \frac{9}{5}(2) = -\frac{13}{5}$$
$$y - \frac{18}{5} = -\frac{13}{5}$$
$$y = \frac{5}{5} = 1$$

and

$$x + 2(1) - 2(2) = 1$$
$$x + 2 - 4 = 1$$
$$x - 2 = 1$$
$$x = 3$$

The solutions are $x = 3$, $y = 1$, $z = 2$.

Applying technology yields

```
rref([A])▶Frac
      [[1 0 0 3]
       [0 1 0 1]
       [0 0 1 2]]
```

6. Writing the system as an augmented matrix and reducing yields

$$\begin{bmatrix} -6 & 4 & -2 & | & 4 \\ 3 & -2 & 5 & | & -6 \\ 1 & -4 & 1 & | & -8 \end{bmatrix} \xrightarrow{R_1 \leftrightarrow R_3}$$

$$\begin{bmatrix} 1 & -4 & 1 & | & -8 \\ 3 & -2 & 5 & | & -6 \\ -6 & 4 & -2 & | & 4 \end{bmatrix} \xrightarrow[6R_1+R_3 \to R_3]{-3R_1+R_2 \to R_2}$$

$$\begin{bmatrix} 1 & -4 & 1 & | & -8 \\ 0 & 10 & 2 & | & 18 \\ 0 & -20 & 4 & | & -44 \end{bmatrix} \xrightarrow{2R_2+R_3 \to R_3}$$

$$\begin{bmatrix} 1 & -4 & 1 & | & -8 \\ 0 & 10 & 2 & | & 18 \\ 0 & 0 & 8 & | & -8 \end{bmatrix} \xrightarrow{\left(\frac{1}{8}\right)R_3 \to R_3}$$

$$\begin{bmatrix} 1 & -4 & 1 & | & -8 \\ 0 & 10 & 2 & | & 18 \\ 0 & 0 & 1 & | & -1 \end{bmatrix}$$

$z = -1$

and

$$10y + 2(-1) = 18$$
$$10y - 2 = 18$$
$$10y = 20$$
$$y = 2$$

and

$$x - 4y + z = -8$$
$$x - 4(2) + (-1) = -8$$
$$x - 8 - 1 = -8$$
$$x - 9 = -8$$
$$x = 1$$

The solutions are $x = 1$, $y = 2$, $z = -1$.

Applying technology yields

```
rref([A])▶Frac
      [[1 0 0 1 ]
       [0 1 0 2 ]
       [0 0 1 -1]]
```

7. Writing the system as an augmented matrix and reducing yields

$$\begin{bmatrix} 2 & 5 & 8 & | & 30 \\ 18 & 42 & 18 & | & 60 \end{bmatrix} \xrightarrow{-9R_1 + R_2 \to R_2}$$

$$\begin{bmatrix} 2 & 5 & 8 & | & 30 \\ 0 & -3 & -54 & | & -210 \end{bmatrix} \xrightarrow{\left(-\frac{1}{3}\right)R_2 \to R_2}$$

$$\begin{bmatrix} 2 & 5 & 8 & | & 30 \\ 0 & 1 & 18 & | & 70 \end{bmatrix}$$

Since the system has more variables than equations, the system is dependent and has infinitely many solutions.

Let $z = z =$ any real number

$y + 18z = 70$

$y = 70 - 18z$

and

$2x + 5y + 8z = 30$

$2x + 5(70 - 18z) + 8z = 30$

$2x + 350 - 90z + 8z = 30$

$2x + 350 - 82z = 30$

$2x = 82z - 320$

$x = 41z - 160$

The solution is

$x = 41z - 160, \ y = 70 - 18z, \ z = z.$

Applying technology yields

```
rref([B])▶Frac
[[1 0 -41 -160]
 [0 1 18  70  ]]
```

See the back substitution process above.

8. Writing the system as an augmented matrix and using technology to produce the reduced row-echelon form yields

$$\begin{bmatrix} 9 & 21 & 15 & | & 60 \\ 2 & 5 & 8 & | & 30 \\ 1 & 2 & -3 & | & -10 \end{bmatrix}$$

```
rref([A])▶Frac
[[1 0 -31 -110]
 [0 1 14  50 ]
 [0 0 0   0  ]]
```

Note that the system has a last row of all zeros. Therefore, the system is dependent and has infinitely many solutions.

Let $z = z$

$y + 14z = 50$

$y = 50 - 14z$

and

$x - 31z = -110$

$x = 31z - 110$

The solution is

$x = 31z - 110, \ y = 50 - 14z, \ z = z$

9. Writing the system as an augmented matrix and using technology to produce the reduced row-echelon form yields

$$\begin{bmatrix} 1 & 3 & 2 & | & 5 \\ 9 & 12 & 15 & | & 6 \\ 2 & 1 & 3 & | & -10 \end{bmatrix}$$

```
rref([A])
[[1 0 1.4 0]
 [0 1 .2  0]
 [0 0 0   1]]
```

Since the system has a last row of all zeros with a 1 as the augment, the system is inconsistent and has no solution.

10. Writing the system as an augmented matrix and using technology to produce the reduced row-echelon form yields

$$\begin{bmatrix} 5 & 7 & 10 & | & 6 \\ 2 & 5 & 6 & | & 1 \\ 3 & 2 & 4 & | & 6 \end{bmatrix}$$

```
rref([A])▶Frac
  [[1 0 8/11   0]
   [0 1 10/11  0]
   [0 0 0      1]]
```

Since the system has a last row of all zeros with a 1 as the augment, the system is inconsistent and has no solution.

11. Writing the system as an augmented matrix and using technology to produce the reduced row-echelon form yields

$$\begin{bmatrix} 3 & 2 & -1 & 1 & | & 12 \\ 1 & -4 & 3 & -1 & | & -18 \\ 1 & 1 & 3 & 2 & | & 0 \\ 2 & -1 & 3 & -3 & | & -10 \end{bmatrix}$$

```
rref([A])▶Frac
  [[1 0 0 0 1 ]
   [0 1 0 0 3 ]
   [0 0 1 0 -2]
   [0 0 0 1 1 ]]
```

The solutions are
$x = 1$, $y = 3$, $z = -2$, and $w = 1..$

12. Writing the system as an augmented matrix and using technology to produce the reduced row-echelon form yields

$$\begin{bmatrix} 2 & 1 & -3 & 4 & | & 7 \\ 1 & -2 & 1 & -2 & | & 0 \\ 3 & 1 & 4 & 1 & | & -2 \\ 1 & 3 & 2 & 2 & | & -1 \end{bmatrix}$$

```
rref([A])▶Frac
  [[1 0 0 0 2 ]
   [0 1 0 0 1 ]
   [0 0 1 0 -2]
   [0 0 0 1 -1]]
```

The solutions are
$x = 2$, $y = 1$, $z = -2$, and $w = -1..$

13. $B + D$

$$= \begin{bmatrix} 1 & 2 & 1 \\ 2 & -1 & 3 \end{bmatrix} + \begin{bmatrix} -2 & 3 & 1 \\ -3 & 2 & 2 \end{bmatrix}$$

$$= \begin{bmatrix} 1-2 & 2+3 & 1+1 \\ 2-3 & -1+2 & 3+2 \end{bmatrix}$$

$$= \begin{bmatrix} -1 & 5 & 2 \\ -1 & 1 & 5 \end{bmatrix}$$

14. $D - B$

$$= \begin{bmatrix} -2 & 3 & 1 \\ -3 & 2 & 2 \end{bmatrix} - \begin{bmatrix} 1 & 2 & 1 \\ 2 & -1 & 3 \end{bmatrix}$$

$$= \begin{bmatrix} -2-1 & 3-2 & 1-1 \\ -3-2 & 2-(-1) & 2-3 \end{bmatrix}$$

$$= \begin{bmatrix} -3 & 1 & 0 \\ -5 & 3 & -1 \end{bmatrix}$$

15. $5C$

$$= 5 \begin{bmatrix} 2 & 3 \\ -1 & 2 \\ 3 & -2 \end{bmatrix}$$

$$= \begin{bmatrix} 5(2) & 5(3) \\ 5(-1) & 5(2) \\ 5(3) & 5(-2) \end{bmatrix}$$

$$= \begin{bmatrix} 10 & 15 \\ -5 & 10 \\ 15 & -10 \end{bmatrix}$$

16. AB can not be calculated because the number of columns in matrix A is different from the number of rows in matrix B.

17. BA

$$= \begin{bmatrix} 1 & 2 & 1 \\ 2 & -1 & 3 \end{bmatrix} \begin{bmatrix} 1 & 3 & -3 \\ 2 & 4 & 1 \\ -1 & 3 & 2 \end{bmatrix}$$

$$= \begin{bmatrix} 1+4-1 & 3+8+3 & -3+2+2 \\ 2-2-3 & 6-4+9 & -6-1+6 \end{bmatrix}$$

$$= \begin{bmatrix} 4 & 14 & 1 \\ -3 & 11 & -1 \end{bmatrix}$$

19. DC

$$= \begin{bmatrix} -2 & 3 & 1 \\ -3 & 2 & 2 \end{bmatrix} \begin{bmatrix} 2 & 3 \\ -1 & 2 \\ 3 & -2 \end{bmatrix}$$

$$= \begin{bmatrix} -4-3+3 & -6+6-2 \\ -6-2+6 & -9+4-4 \end{bmatrix}$$

$$= \begin{bmatrix} -4 & -2 \\ -2 & -9 \end{bmatrix}$$

18. CD

$$= \begin{bmatrix} 2 & 3 \\ -1 & 2 \\ 3 & -2 \end{bmatrix} \begin{bmatrix} -2 & 3 & 1 \\ -3 & 2 & 2 \end{bmatrix}$$

$$= \begin{bmatrix} -4-9 & 6+6 & 2+6 \\ 2-6 & -3+4 & -1+4 \\ -6+6 & 9-4 & 3-4 \end{bmatrix}$$

$$= \begin{bmatrix} -13 & 12 & 8 \\ -4 & 1 & 3 \\ 0 & 5 & -1 \end{bmatrix}$$

20. A^2

$$= A \cdot A$$

$$= \begin{bmatrix} 1 & 3 & -3 \\ 2 & 4 & 1 \\ -1 & 3 & 2 \end{bmatrix} \begin{bmatrix} 1 & 3 & -3 \\ 2 & 4 & 1 \\ -1 & 3 & 2 \end{bmatrix}$$

$$= \begin{bmatrix} 1+6+3 & 3+12-9 & -3+3-6 \\ 2+8-1 & 6+16+3 & -6+4+2 \\ -1+6-2 & -3+12+6 & 3+3+4 \end{bmatrix}$$

$$= \begin{bmatrix} 10 & 6 & -6 \\ 9 & 25 & 0 \\ 3 & 15 & 10 \end{bmatrix}$$

21. A^{-1}

$$\begin{bmatrix} 1 & 3 & -3 & | & 1 & 0 & 0 \\ 2 & 4 & 1 & | & 0 & 1 & 0 \\ -1 & 3 & 2 & | & 0 & 0 & 1 \end{bmatrix} \xrightarrow[R_1+R_3 \to R_3]{-2R_1+R_2 \to R_2}$$

$$= \begin{bmatrix} 1 & 3 & -3 & | & 1 & 0 & 0 \\ 0 & -2 & 7 & | & -2 & 1 & 0 \\ 0 & 6 & -1 & | & 1 & 0 & 1 \end{bmatrix} \xrightarrow{3R_2+R_3 \to R_3}$$

$$= \begin{bmatrix} 1 & 3 & -3 & | & 1 & 0 & 0 \\ 0 & -2 & 7 & | & -2 & 1 & 0 \\ 0 & 0 & 20 & | & -5 & 3 & 1 \end{bmatrix} \xrightarrow{\left(\frac{1}{20}\right)R_3 \to R_3}$$

$$= \begin{bmatrix} 1 & 3 & -3 & | & 1 & 0 & 0 \\ 0 & -2 & 7 & | & -2 & 1 & 0 \\ 0 & 0 & 1 & | & -\dfrac{1}{4} & \dfrac{3}{20} & \dfrac{1}{20} \end{bmatrix} \xrightarrow[3R_3+R_1 \to R_1]{-7R_3+R_2 \to R_2}$$

$$= \begin{bmatrix} 1 & 3 & 0 & \frac{1}{4} & \frac{9}{20} & \frac{3}{20} \\ 0 & -2 & 0 & -\frac{1}{4} & -\frac{1}{20} & -\frac{7}{20} \\ 0 & 0 & 1 & -\frac{1}{4} & \frac{3}{20} & \frac{1}{20} \end{bmatrix} \xrightarrow{\left(-\frac{1}{2}\right)R_2 \rightarrow R_2}$$

$$= \begin{bmatrix} 1 & 3 & 0 & \frac{1}{4} & \frac{9}{20} & \frac{3}{20} \\ 0 & 1 & 0 & \frac{1}{8} & \frac{1}{40} & \frac{7}{40} \\ 0 & 0 & 1 & -\frac{1}{4} & \frac{3}{20} & \frac{1}{20} \end{bmatrix} \xrightarrow{(-3)R_2 + R_1 \rightarrow R_1}$$

$$= \begin{bmatrix} 1 & 0 & 0 & -\frac{1}{8} & \frac{3}{8} & -\frac{3}{8} \\ 0 & 1 & 0 & \frac{1}{8} & \frac{1}{40} & \frac{7}{40} \\ 0 & 0 & 1 & -\frac{1}{4} & \frac{3}{20} & \frac{1}{20} \end{bmatrix}$$

$$A^{-1} = \begin{bmatrix} -\frac{1}{8} & \frac{3}{8} & -\frac{3}{8} \\ \frac{1}{8} & \frac{1}{40} & \frac{7}{40} \\ -\frac{1}{4} & \frac{3}{20} & \frac{1}{20} \end{bmatrix}$$

22. Applying technology to calculate A^{-1} yields

```
[A]⁻¹▶Frac
[[4/3   1/3   -1]
 [-1    0     1 ]
 [-2/3  -2/3  1 ]]
```

23. Applying technology to calculate A^{-1} yields

```
[A]⁻¹▶Frac
 [[1    0   -1]
  [-1   1   1 ]
  [2    -2  -1]]
```

24. Applying technology to calculate A^{-1} yields

```
[C]⁻¹
[[-.4  .2  .2   0 …
 [-.8  .4  -.6  1 …
 [.6   .2  .2   -1…
 [0    0   0    1 …
```

$$A^{-1} = \begin{bmatrix} -0.4 & 0.2 & 0.2 & 0 \\ -0.8 & 0.4 & -0.6 & 1 \\ 0.6 & 0.2 & 0.2 & -1 \\ 0 & 0 & 0 & 1 \end{bmatrix}$$

25. $\begin{bmatrix} 1 & 1 & -3 \\ 2 & 4 & 1 \\ -1 & 3 & 2 \end{bmatrix} \begin{bmatrix} x \\ y \\ z \end{bmatrix} = \begin{bmatrix} 8 \\ 15 \\ 5 \end{bmatrix}$

Let $A = \begin{bmatrix} 1 & 1 & -3 \\ 2 & 4 & 1 \\ -1 & 3 & 2 \end{bmatrix}$. Applying technology to calculate A^{-1} yields $A^{-1} = \begin{bmatrix} -\dfrac{1}{6} & \dfrac{11}{30} & -\dfrac{13}{30} \\ \dfrac{1}{6} & \dfrac{1}{30} & \dfrac{7}{30} \\ -\dfrac{1}{3} & \dfrac{2}{15} & -\dfrac{1}{15} \end{bmatrix}$.

Multiplying both sides of the matrix equation by A^{-1} yields

$$\begin{bmatrix} -\dfrac{1}{6} & \dfrac{11}{30} & -\dfrac{13}{30} \\ \dfrac{1}{6} & \dfrac{1}{30} & \dfrac{7}{30} \\ -\dfrac{1}{3} & \dfrac{2}{15} & -\dfrac{1}{15} \end{bmatrix} \begin{bmatrix} 1 & 1 & -3 \\ 2 & 4 & 1 \\ -1 & 3 & 2 \end{bmatrix} \begin{bmatrix} x \\ y \\ z \end{bmatrix} = \begin{bmatrix} -\dfrac{1}{6} & \dfrac{11}{30} & -\dfrac{13}{30} \\ \dfrac{1}{6} & \dfrac{1}{30} & \dfrac{7}{30} \\ -\dfrac{1}{3} & \dfrac{2}{15} & -\dfrac{1}{15} \end{bmatrix} \begin{bmatrix} 8 \\ 15 \\ 5 \end{bmatrix}.$$

Using technology to carry out the multiplication yields:

$$\begin{bmatrix} 1 & 0 & 0 \\ 0 & 1 & 0 \\ 0 & 0 & 1 \end{bmatrix} \begin{bmatrix} x \\ y \\ z \end{bmatrix} = \begin{bmatrix} 2 \\ 3 \\ -1 \end{bmatrix}$$

$$\begin{bmatrix} x \\ y \\ z \end{bmatrix} = \begin{bmatrix} 2 \\ 3 \\ -1 \end{bmatrix}$$

$x = 2, \ y = 3, \ z = -1$

26. $\begin{bmatrix} 2 & 3 & 1 \\ 3 & 4 & -1 \\ 2 & 5 & 1 \end{bmatrix} \begin{bmatrix} x \\ y \\ z \end{bmatrix} = \begin{bmatrix} 20 \\ 40 \\ 60 \end{bmatrix}$

Let $A = \begin{bmatrix} 2 & 3 & 1 \\ 3 & 4 & -1 \\ 2 & 5 & 1 \end{bmatrix}$. Applying technology to calculate A^{-1} yields $A^{-1} = \begin{bmatrix} 0.9 & 0.2 & -0.7 \\ -0.5 & 0 & 0.5 \\ 0.7 & -0.4 & -0.1 \end{bmatrix}$.

Multiplying both sides of the matrix equation by A^{-1} yields

$$\begin{bmatrix} 0.9 & 0.2 & -0.7 \\ -0.5 & 0 & 0.5 \\ 0.7 & -0.4 & -0.1 \end{bmatrix} \begin{bmatrix} 2 & 3 & 1 \\ 3 & 4 & -1 \\ 2 & 5 & 1 \end{bmatrix} \begin{bmatrix} x \\ y \\ z \end{bmatrix} = \begin{bmatrix} 0.9 & 0.2 & -0.7 \\ -0.5 & 0 & 0.5 \\ 0.7 & -0.4 & -0.1 \end{bmatrix} \begin{bmatrix} 20 \\ 40 \\ 60 \end{bmatrix}.$$

Using technology to carry out the multiplication yields

$$\begin{bmatrix} 1 & 0 & 0 \\ 0 & 1 & 0 \\ 0 & 0 & 1 \end{bmatrix}\begin{bmatrix} x \\ y \\ z \end{bmatrix} = \begin{bmatrix} -16 \\ 20 \\ -8 \end{bmatrix}$$

$$\begin{bmatrix} x \\ y \\ z \end{bmatrix} = \begin{bmatrix} -16 \\ 20 \\ -8 \end{bmatrix}$$

$$x = -16, \ y = 20, \ z = -8$$

27.
$$\begin{bmatrix} -1 & 0 & 1 & 1 \\ 0 & 1 & 1 & 1 \\ 2 & 5 & 1 & 0 \\ 0 & 3 & 0 & 1 \end{bmatrix}\begin{bmatrix} x_1 \\ x_2 \\ x_3 \\ x_4 \end{bmatrix} = \begin{bmatrix} 6 \\ 12 \\ 20 \\ 24 \end{bmatrix}$$

Let $A = \begin{bmatrix} -1 & 0 & 1 & 1 \\ 0 & 1 & 1 & 1 \\ 2 & 5 & 1 & 0 \\ 0 & 3 & 0 & 1 \end{bmatrix}$. Applying technology to calculate A^{-1} yields

$$A^{-1} = \begin{bmatrix} -1.4 & 1.6 & -0.2 & -0.2 \\ 0.4 & -0.6 & 0.2 & 0.2 \\ 0.8 & -0.2 & 0.4 & -0.6 \\ -1.2 & 1.8 & -0.6 & 0.4 \end{bmatrix}.$$

Multiplying both sides of the matrix equation by A^{-1} yields

$$\begin{bmatrix} -1.4 & 1.6 & -0.2 & -0.2 \\ 0.4 & -0.6 & 0.2 & 0.2 \\ 0.8 & -0.2 & 0.4 & -0.6 \\ -1.2 & 1.8 & -0.6 & 0.4 \end{bmatrix}\begin{bmatrix} -1 & 0 & 1 & 1 \\ 0 & 1 & 1 & 1 \\ 2 & 5 & 1 & 0 \\ 0 & 3 & 0 & 1 \end{bmatrix}\begin{bmatrix} x_1 \\ x_2 \\ x_3 \\ x_4 \end{bmatrix} = \begin{bmatrix} -1.4 & 1.6 & -0.2 & -0.2 \\ 0.4 & -0.6 & 0.2 & 0.2 \\ 0.8 & -0.2 & 0.4 & -0.6 \\ -1.2 & 1.8 & -0.6 & 0.4 \end{bmatrix}\begin{bmatrix} 6 \\ 12 \\ 20 \\ 24 \end{bmatrix}.$$

Using technology to carry out the multiplication yields

$$\begin{bmatrix} 1 & 0 & 0 & 0 \\ 0 & 1 & 0 & 0 \\ 0 & 0 & 1 & 0 \\ 0 & 0 & 0 & 1 \end{bmatrix}\begin{bmatrix} x_1 \\ x_2 \\ x_3 \\ x_4 \end{bmatrix} = \begin{bmatrix} 2 \\ 4 \\ -4 \\ 12 \end{bmatrix}$$

$$\begin{bmatrix} x_1 \\ x_2 \\ x_3 \\ x_4 \end{bmatrix} = \begin{bmatrix} 2 \\ 4 \\ -4 \\ 12 \end{bmatrix}$$

$$x_1 = 2, \ x_2 = 4, \ x_3 = -4, \ x_4 = 12$$

28.
$$\begin{bmatrix} 2 & 2 & 0 & 1 \\ 2 & 1 & 0 & 2 \\ 1 & 1 & 1 & 1 \\ 0 & 1 & 0 & 1 \end{bmatrix}\begin{bmatrix} x_1 \\ x_2 \\ x_3 \\ x_4 \end{bmatrix} = \begin{bmatrix} 4 \\ 12 \\ 4 \\ 8 \end{bmatrix}$$

Let $A = \begin{bmatrix} 2 & 2 & 0 & 1 \\ 2 & 1 & 0 & 2 \\ 1 & 1 & 1 & 1 \\ 0 & 1 & 0 & 1 \end{bmatrix}$. Applying technology to calculate A^{-1} yields

$$A^{-1} = \begin{bmatrix} 0.25 & 0.25 & 0 & -0.75 \\ 0.5 & -0.5 & 0 & 0.5 \\ -0.25 & -0.25 & 1 & -0.25 \\ -0.5 & 0.5 & 0 & 0.5 \end{bmatrix}.$$

Multiplying both sides of the matrix equation by A^{-1} yields

$$\begin{bmatrix} 0.25 & 0.25 & 0 & -0.75 \\ 0.5 & -0.5 & 0 & 0.5 \\ -0.25 & -0.25 & 1 & -0.25 \\ -0.5 & 0.5 & 0 & 0.5 \end{bmatrix}\begin{bmatrix} 2 & 2 & 0 & 1 \\ 2 & 1 & 0 & 2 \\ 1 & 1 & 1 & 1 \\ 0 & 1 & 0 & 1 \end{bmatrix}\begin{bmatrix} x_1 \\ x_2 \\ x_3 \\ x_4 \end{bmatrix} = \begin{bmatrix} 0.25 & 0.25 & 0 & -0.75 \\ 0.5 & -0.5 & 0 & 0.5 \\ -0.25 & -0.25 & 1 & -0.25 \\ -0.5 & 0.5 & 0 & 0.5 \end{bmatrix}\begin{bmatrix} 4 \\ 12 \\ 4 \\ 8 \end{bmatrix}.$$

Using technology to carry out the multiplication yields

$$\begin{bmatrix} 1 & 0 & 0 & 0 \\ 0 & 1 & 0 & 0 \\ 0 & 0 & 1 & 0 \\ 0 & 0 & 0 & 1 \end{bmatrix}\begin{bmatrix} x_1 \\ x_2 \\ x_3 \\ x_4 \end{bmatrix} = \begin{bmatrix} -2 \\ 0 \\ -2 \\ 8 \end{bmatrix}$$

$$\begin{bmatrix} x_1 \\ x_2 \\ x_3 \\ x_4 \end{bmatrix} = \begin{bmatrix} -2 \\ 0 \\ -2 \\ 8 \end{bmatrix}$$

$x_1 = -2, x_2 = 0, x_3 = -2, x_4 = 8$

29. Isolating y in the second equation
$4x - y = 4$
$y = 4x - 4$

Substituting into the other equation
$x^2 - y = x$
$x^2 - (4x - 4) = x$
$x^2 - 4x + 4 = x$
$x^2 - 5x + 4 = 0$
$(x-4)(x-1) = 0$
$x = 4, x = 1$

Back substituting to calculate y

$x = 4 \Rightarrow y = 4(4) - 4 = 12$

$x = 1 \Rightarrow y = 4(1) - 4 = 0$

The solutions to the system are $(4, 12)$ and $(1, 0)$.

30. Isolating y in the second equation

$x^2 + 2y = 140$

$2y = 140 - x^2$

$y = \dfrac{140 - x^2}{2}$

Substituting into the other equation

$x^2 y = 2000$

$x^2 \left(\dfrac{140 - x^2}{2} \right) = 2000$

$2(x^2) \left(\dfrac{140 - x^2}{2} \right) = 2(2000)$

$140x^2 - x^4 = 4000$

$x^4 - 140x^2 + 4000 = 0$

Solving graphically by applying the x-intercept method

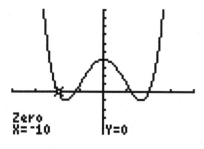

[–20, 20] by [–5000, 10,000]

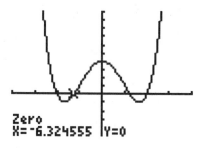

[–20, 20] by [–5000, 10,000]

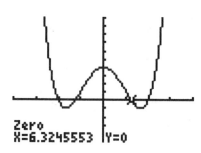

[–20, 20] by [–5000, 10,000]

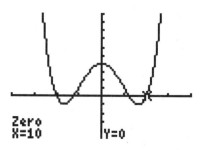

[–20, 20] by [–5000, 10,000]

Back substituting to calculate y

$x = -10 \Rightarrow y = \dfrac{2000}{x^2} = \dfrac{2000}{(-10)^2} = 20$

$x = 10 \Rightarrow y = \dfrac{2000}{x^2} = \dfrac{2000}{(10)^2} = 20$

$x \approx -6.32 \Rightarrow y = \dfrac{2000}{x^2} = \dfrac{2000}{(-6.32)^2} \approx 50.07$

$x \approx 6.32 \; y = \dfrac{2000}{x^2} = \dfrac{2000}{(6.32)^2} \approx 50.07$

The solutions to the system are $(-10, 20), (-6.32, 50.07), (6.32, 50.07),$ and $(10, 20)$.

Chapter 5 Review

31. Let x represent the number of \$40 tickets, y represent the number of \$60 tickets, and z represent the number of \$100 tickets.

$$\begin{cases} x+y+z=4000 \\ 40x+60y+100z=200,000 \\ y=\dfrac{1}{4}(x+z) \end{cases} \xrightarrow{4Eq3\to Eq3}$$

$$\begin{cases} x+y+z=4000 \\ 40x+60y+100z=200,000 \\ x-4y+z=0 \end{cases} \xrightarrow[\substack{-40Eq1+Eq2\to Eq2 \\ -1Eq1+Eq3\to Eq3}]{}$$

$$\begin{cases} x+y+z=4000 \\ 20y+60z=40,000 \\ -5y=-4000 \end{cases}$$

$-5y=-4000$

$y=800$

Substituting to find z

$20(800)+60z=40,000$

$16,000+60z=40,000$

$60z=24,000$

$z=400$

Substituting to find x

$x+(800)+(400)=4000$

$x=2800$

To generate \$200,000, the concert promoter needs to sell 2800 \$40 tickets, 800 \$60 tickets, and 400 \$100 tickets.

32. Let x represent the daily dosage of medication A, y represent the daily dosage of medication B, and z represent the daily dosage of medication C.

$$\begin{cases} 6x+2y+z=28.7 \\ z=\dfrac{1}{2}(x+y) \\ \dfrac{x}{y}=\dfrac{2}{3} \end{cases}$$

or

$$\begin{cases} 6x+2y+z=28.7 \\ x+y-2z=0 \\ 3x-2y=0 \end{cases}$$

$$\begin{cases} 6x + 2y + z = 28.7 \\ x + y - 2z = 0 \\ 3x - 2y = 0 \end{cases} \xrightarrow[\substack{Eq1-6Eq2\to Eq2 \\ -3Eq2+Eq3\to Eq3}]{}$$

$$\begin{cases} 6x + 2y + z = 28.7 \\ -4y + 13z = 28.7 \\ -5y + 6z = 0 \end{cases} \xrightarrow[\substack{-5Eq2+4Eq3\to Eq3}]{}$$

$$\begin{cases} 6x + 2y + z = 28.7 \\ -4y + 13z = 28.7 \\ -41z = -143.5 \end{cases} \xrightarrow[\substack{-\frac{1}{41}Eq3\to Eq3}]{}$$

$$\begin{cases} 6x + 2y + z = 28.7 \\ -4y + 13z = 28.7 \\ z = 3.5 \end{cases}$$

Substituting to find y

$-4y + 13(3.5) = 28.7$

$-4y + 45.5 = 28.7$

$-4y = -16.8$

$y = 4.2$

Substituting to find x

$6x + 2(4.2) + (3.5) = 28.7$

$6x + 8.4 + 3.5 = 28.7$

$6x = 16.8$

$x = 2.8$

Each dosage of medication A contains 2.8 mg, each dosage of medication B contains 4.2 mg, and each dosage of medication C contains 3.5 mg.

33. Let x represent the amount invested in property I (12%), y represent the amount invested in property II (15%), and z represent the amount invested in property III (10%).

$$\begin{cases} x + y + z = 750,000 \\ 0.12x + 0.15y + 0.10z = 89,500 \quad \text{or} \\ z = \dfrac{1}{2}(x + y) \end{cases} \qquad \begin{cases} x + y + z = 750,000 \\ 0.12x + 0.15y + 0.10z = 89,500 \\ x + y - 2z = 0 \end{cases}$$

or

$$\begin{bmatrix} 1 & 1 & 1 & | & 750,000 \\ 0.12 & 0.15 & 0.10 & | & 89,500 \\ 1 & 1 & -2 & | & 0 \end{bmatrix}$$

Using technology to solve the augmented matrix yields

```
rref([A])▶Frac
[[1 0 0 350000]
 [0 1 0 150000]
 [0 0 1 250000]]
```

To generate an annual return of $89,500, $350,000 needs to be invested in property I, $150,000 needs to be invested in property II, and $250,000 needs to be invested property III.

34. Let $x =$ the amount in 12% fund, $y =$ the amount in the 16% fund, and $z =$ the amount in the 8% fund.

$$\begin{cases} x+y+z=360,000 \\ 0.12x+0.16y+0.08z=35,200 \\ z=2(x+y) \end{cases} \quad \text{or} \quad \begin{cases} x+y+z=360,000 \\ 0.12x+0.16y+0.08z=35,200 \\ 2x+2y-z=0 \end{cases}$$

or

$$\begin{bmatrix} 1 & 1 & 1 & | & 360,000 \\ 0.12 & 0.16 & 0.08 & | & 35,200 \\ 2 & 2 & -1 & | & 0 \end{bmatrix}$$

Using technology to solve the augmented matrix yields

```
rref([A])▶Frac
[[1 0 0 80000 ]
 [0 1 0 40000 ]
 [0 0 1 240000]]
```

$80,000 is invested in the 12% fund, $40,000 is invested in the 16% fund, and $240,000 is invested in the 8% fund.

35. Let x represent the cost of property I, y represent the cost of property II, and z represent the cost of property III.

$$\begin{cases} x+y+z=1,180,000 \\ x=y+75,000 \\ z=3(x+y) \end{cases} \quad \text{or} \quad \begin{cases} x+y+z=1,180,000 \\ x-y=75,000 \\ 3x+3y-z=0 \end{cases}$$

or

$$\begin{bmatrix} 1 & 1 & 1 & | & 1,180,000 \\ 1 & -1 & 0 & | & 75,000 \\ 3 & 3 & -1 & | & 0 \end{bmatrix}$$

Using technology to solve the augmented matrix yields

```
rref([A])▶Frac
[[1 0 0 185000]
 [0 1 0 110000]
 [0 0 1 885000]]
```

Property I costs $185,000, property II costs $110,000, and property III costs $885,000.

36. Let $x =$ the number of Portfolio I units, $y =$ the number of Portfolio II units, and $z =$ the number of Portfolio III units.

$$\begin{cases} 10x + 12y + 10z = 290 \\ 2x + 8y + 4z = 138 \\ 3x + 5y + 8z = 161 \end{cases} \quad \text{or} \quad \begin{bmatrix} 10 & 12 & 10 & 290 \\ 2 & 8 & 4 & 138 \\ 3 & 5 & 8 & 161 \end{bmatrix}$$

Using the calculator to generate the reduced row-echelon form of the matrix yields

```
rref([A])
[[1 0 0 5 ]
 [0 1 0 10]
 [0 0 1 12]]
```

The solution to the system is $x = 5$, $y = 10$, $z = 12$. The client needs to purchase 5 units of Portfolio I, 10 units of Portfolio II, and 12 units of Portfolio III to achieve the investment objectives.

37. $\begin{cases} x + y + z = 375,000 \\ x = y + 50,000 \\ z = \dfrac{1}{2}(x + y) \end{cases}$

Note that $x = y + 50,000$ implies $x - y = 50,000$. Likewise $z = \dfrac{1}{2}(x + y)$ implies that

$$2z = 2\left(\frac{1}{2}(x + y) \right)$$

$$2z = x + y$$

$$x + y - 2z = 0.$$

Therefore, the system can be written as follows:

$$\begin{cases} x + y + z = 375,000 \\ x - y = 50,000 \\ x + y - 2z = 0 \end{cases}$$

Writing the system as an augmented matrix yields

$$\begin{bmatrix} 1 & 1 & 1 & | & 375{,}000 \\ 1 & -1 & 0 & | & 50{,}000 \\ 1 & 1 & -2 & | & 0 \end{bmatrix}$$

Applying technology to produce the reduced row-echelon form of the matrix yields

```
rref([A])
[[1 0 0 150000]
 [0 1 0 100000]
 [0 0 1 125000]]
```

Therefore, $x = 150{,}000$, $y = 100{,}000$, and $z = 125{,}000$. The first property costs \$150,000, the second property costs \$100,000, and the third property costs \$125,000.

38. Let x = grams of Food I, y = grams of Food II, and z = grams of Food III.

$$\begin{cases} 10\%x + 11\%y + 18\%z = 12.5 \\ 12\%x + 9\%y + 10\%z = 9.1 \\ 14\%x + 12\%y + 8\%z = 9.6 \end{cases} \quad \text{or} \quad \begin{cases} 0.10x + 0.11y + 0.18z = 12.5 \\ 0.12x + 0.09y + 0.10z = 9.1 \\ 0.14x + 0.12y + 0.08z = 9.6 \end{cases}$$

Converting the system to an augmented matrix yields

$$\begin{bmatrix} 0.10 & 0.11 & 0.18 & | & 12.5 \\ 0.12 & 0.09 & 0.10 & | & 9.1 \\ 0.14 & 0.12 & 0.08 & | & 9.6 \end{bmatrix}$$

Using the calculator to generate the reduced row-echelon form of the matrix yields

```
rref([A])
    [[1 0 0 20]
     [0 1 0 30]
     [0 0 1 40]]
```

The solution to the system is $x = 20$, $y = 30$, $z = 40$. The nutritionist recommends 20 grams of Food I, 30 grams of Food II, and 40 grams of Food III.

39. Let x = the number of passenger aircraft, y = the number of transport aircraft, and z = the number of jumbo aircraft.

$$\begin{cases} 200x + 200y + 200z = 2200 \\ 300x + 40y + 700z = 3860 \\ 40x + 130y + 70z = 920 \end{cases}$$

Converting the system to an augmented matrix yields

$$\begin{bmatrix} 200 & 200 & 200 & | & 2200 \\ 300 & 40 & 700 & | & 3860 \\ 40 & 130 & 70 & | & 920 \end{bmatrix}$$

Using the calculator to generate the reduced row-echelon form of the matrix yields

```
rref([A])
        [[1 0 0 3]
         [0 1 0 4]
         [0 0 1 4]]
```

The solution to the system is $x = 3$, $y = 4$, $z = 4$. The delivery service needs to schedule 3 passenger planes, 4 transport planes, and 4 jumbo planes.

40. Let $x =$ amount invested in the tech fund, $y =$ amount invested in the balanced fund, and $z =$ amount invested in the utility fund.

$$\begin{cases} 180x + 210y + 120z = 210,000 \\ 18x + 42y + 18z = 12\%(210,000) \end{cases} \quad \text{or} \quad \begin{cases} 180x + 210y + 120z = 210,000 \\ 18x + 42y + 18z = 0.12(210,000) \end{cases}$$

Converting the system to an augmented matrix yields

$$\begin{bmatrix} 180 & 210 & 120 & | & 210,000 \\ 18 & 42 & 18 & | & 25,200 \end{bmatrix}$$

Note that the system has more variables than equations. The system is dependent. Using the calculator to generate the reduced row-echelon form of the matrix yields

```
rref([A])▶Frac
...1 0 1/3 2800/3...
...0 1 2/7 200   ...
```

Let $z = z$.

$$y + \frac{2}{7}z = 200$$

$$y = 200 - \frac{2}{7}z$$

and

$$x + \frac{1}{3}z = \frac{2800}{3}$$

$$x = \frac{2800}{3} - \frac{1}{3}z$$

$$x = \frac{2800 - z}{3}$$

The solution is $x = \dfrac{2800 - z}{3}$, $y = 200 - \dfrac{2}{7}z$, $z = z$.

Since y must be greater than or equal to zero, z must not exceed 700. Therefore, $0 \leq z \leq 700$.

41. Let $x =$ the number of units of Product A, $y =$ the number of units of Product B, and $z =$ the number of units of Product C.

$$\begin{cases} 25x + 30y + 40z = 9260 \\ 30x + 36y + 60z = 12{,}000 \\ 150x + 180y + 200z = 52{,}600 \end{cases}$$

Converting the system to an augmented matrix yields

$$\begin{bmatrix} 25 & 30 & 40 & | & 9260 \\ 30 & 36 & 60 & | & 12{,}000 \\ 150 & 180 & 200 & | & 52{,}600 \end{bmatrix}$$

Using the calculator to generate the reduced row-echelon form of the matrix yields

```
rref([A])
  [[1 1.2 0 252]
   [0 0   1 74 ]
   [0 0   0 0  ]]
```

Since the last row consists entirely of zeros with zero as the augment, the system is dependent and has infinitely many solutions.

$z = 74$

Let $y = y$.

$x + 1.2y = 252$

$x = 252 - 1.2y$

The solution is $x = 252 - 1.2y, \ y = y, \ z = 74$.

Note that $x \geq 0$. Furthermore, $252 - 1.2y \geq 0$.

$$-1.2y \geq -252$$

$$y \leq \frac{-252}{-1.2}$$

$$y \leq 210$$

Therefore, $0 \leq y \leq 210$.

42. Let $x =$ the number of type A slugs, $y =$ the number of type B slugs, and $z =$ the number of type C slugs.

$$\begin{cases} 2x + 2y + 4z = 4000 & (\text{nutrient I}) \\ 6x + 8y + 20z = 16,000 & (\text{nutrient II}) \\ 2x + 4y + 12z = 8000 & (\text{nutrient III}) \end{cases}$$

Converting the system to an augmented matrix yields

$$\begin{bmatrix} 2 & 2 & 4 & | & 4000 \\ 6 & 8 & 20 & | & 16,000 \\ 2 & 4 & 12 & | & 8000 \end{bmatrix}$$

Using the calculator to generate the reduced row-echelon form of the matrix yields

```
rref([A])
  [[1 0 -2 0    ]
   [0 1 4  2000]
   [0 0 0  0   ]]
```

Since the last row consists entirely of zeros with zero as the augment, the system is dependent and has infinitely many solutions.

Let $z = z$

$y + 4z = 2000$

$y = 2000 - 4z$

and

$x - 2z = 0$

$x = 2z$

The solution is $x = 2z, \ y = 2000 - 4z, \ z = z$.

Note that $y \geq 0$. Furthermore, $2000 - 4z \geq 0$.

$$-4z \geq -2000$$

$$z \leq \frac{-2000}{-4}$$

$$z \leq 500$$

Therefore, $0 \leq z \leq 500$.

43. a. $A = \begin{bmatrix} 40,592 & 90,594 \\ 50,844 & 114,439 \\ 56,792 & 134,210 \\ 78,773 & 156,603 \\ 86,909 & 166,600 \end{bmatrix}$

b. $B = \begin{bmatrix} 36,211 & 98,630 \\ 49,494 & 128,406 \\ 74,297 & 155,893 \\ 94,629 & 173,256 \\ 109,721 & 198,711 \end{bmatrix}$

c. Trade balance = Exports − Imports

$$A - B = \begin{bmatrix} 40,592 & 90,594 \\ 50,844 & 114,439 \\ 56,792 & 134,210 \\ 78,773 & 156,603 \\ 86,909 & 166,600 \end{bmatrix} - \begin{bmatrix} 36,211 & 98,630 \\ 49,494 & 128,406 \\ 74,297 & 155,893 \\ 94,629 & 173,256 \\ 109,721 & 198,711 \end{bmatrix}$$

$$= \begin{bmatrix} 40,592 - 36,211 & 90,594 - 98,630 \\ 50,844 - 49,494 & 114,439 - 128,406 \\ 56,792 - 74,297 & 134,210 - 155,893 \\ 78,773 - 94,629 & 156,603 - 173,256 \\ 86,909 - 109,721 & 166,600 - 198,711 \end{bmatrix}$$

$$= \begin{bmatrix} 4381 & -8036 \\ 1350 & -13,967 \\ -17,505 & -21,683 \\ -15,856 & -16,653 \\ -22,812 & -32,111 \end{bmatrix}$$

d. No. The trend of the trade balances is getting worse for the U.S. over time.

e. In 1999, the final year represented in the matrix, the trade balance is worse with Canada.

44. a.   Intersection A

$x_4 + 2250 = x_1 + 2900$

Intersection B

$x_1 + 3050 = x_2 + 2500$

Intersection C

$x_2 + 4100 = x_3 + 2000$

Intersection D

$x_3 + 1800 = x_4 + 3800$

b.   Intersection A

$x_4 + 2250 = x_1 + 2900$

$x_1 - x_4 = -650$

Intersection B

$x_1 + 3050 = x_2 + 2500$

$x_1 - x_2 = -550$

Intersection C

$x_2 + 4100 = x_3 + 2000$

$x_2 - x_3 = -2100$

Intersection D

$x_3 + 1800 = x_4 + 3800$

$x_3 - x_4 = 2000$

The system is

$$\begin{cases} x_1 - x_4 = -650 \\ x_1 - x_2 = -550 \\ x_2 - x_3 = -2100 \\ x_3 - x_4 = 2000 \end{cases}$$

Writing the augmented matrix from the system yields

$$\begin{bmatrix} 1 & 0 & 0 & -1 & | & -650 \\ 1 & -1 & 0 & 0 & | & -550 \\ 0 & 1 & -1 & 0 & | & -2100 \\ 0 & 0 & 1 & -1 & | & 2000 \end{bmatrix}$$

Using technology to produce the reduced row-echelon form of the matrix yields

```
rref([A])
…1 0 0 -1  -650]
…0 1 0 -1  -100]
…0 0 1 -1  2000]
…0 0 0 0   0   ]]
```

Since the last row consists entirely of zeros with zero as the augment, the system is dependent and has infinitely many solutions.

Let $x_4 = x_4$.

$x_3 - x_4 = 2000$

$x_3 = 2000 + x_4$

and

$x_2 - x_4 = -100$

$x_2 = x_4 - 100$

and

$x_1 - x_4 = -650$

$x_1 = x_4 - 650$

The solution is

$x_1 = x_4 - 650, \ x_2 = x_4 - 100,$

$x_3 = 2000 + x_4, \ x_4 = x_4.$

To ensure that all the variables are positive, $x_1 \geq 0$, $x_4 - 650 \geq 0$, and

$x_4 \geq 650$.

45. a.

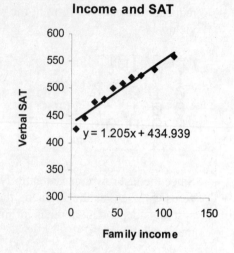

b.

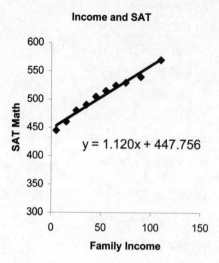

c. Solve the system
$$\begin{cases} y = 1.205x + 434.939 \\ y = 1.120x + 447.756 \end{cases}$$

Using the substitution method

$$1.205x + 434.939 = 1.120x + 447.756$$
$$1.205x - 1.120x + 434.939 = 1.120x - 1.120x + 447.756$$
$$0.085x + 434.939 = 447.756$$
$$0.085x + 434.939 - 434.939 = 447.756 - 434.939$$
$$0.085x = 12.817$$
$$x = \frac{12.817}{0.085}$$
$$x = 150.788$$

An income of approximately \$151,000 corresponds to equal SAT verbal and math scores.

Using the unrounded models and solving numerically yields the same solution.

X	Y1	Y2
148	613.28	613.53
149	614.49	614.65
150	615.69	615.77
151	616.9	616.89
152	618.1	618.01
153	619.31	619.13
154	620.51	620.25

X=151

46. a. $x + y = 10$

b. $20\%x + 5\%y = 15.5\%(10)$

or

$0.20x + 0.05y = 1.55$

c.   Applying the substitution method

$x + y = 10$

$x = 10 - y$

Substituting into the other equation

$0.20(10 - y) + 0.05y = 1.55$

$2 - 0.20y + 0.05y = 1.55$

$-0.15y + 2 = 1.55$

$-0.15y = -0.45$

$y = \dfrac{-0.45}{-0.15} = 3$

Substituting to find x

$x + y = 10$

$x + 3 = 10$

$x = 7$

To administer the desired concentration requires 7 cc of the 20% medication and 3 cc of the 5% medication.

47. $\begin{cases} p = q + 578 \\ p = 396 + q^2 \end{cases}$

$q + 578 = 396 + q^2$

$q^2 - q - 182 = 0$

$(q - 14)(q + 13) = 0$

$q = 14, \ q = -13$

Since $q \geq 0$ in the context of the question, then $q = 14$.

Substituting to calculate p

$p = 14 + 578$

$p = 592$

Equilibrium occurs when 14 units are produced and sold at a price of $592 per unit.

48. $C(x) = R(x)$

$2500x + x^2 + 27,550 = 3899x - 0.1x^2$

$1.1x^2 - 1399x + 27,550 = 0$

Solving graphically by applying the x-intercept method

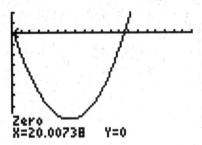

[0, 2000] by [−500,000, 100,000]

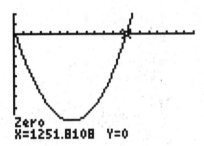

[0, 2000] by [−500,000, 100,000]

Break-even occurs when approximately 20 units are produced and sold or when approximately 1252 units are produced and sold.

Group Activity/Extended Application I

1. $A = \begin{bmatrix} 80,379 & 99,979 & 84,796 \\ 64,414 & 70,643 & 66,151 \\ 58,432 & 68,145 & 52,945 \\ 57,067 & 45,099 & 36,422 \end{bmatrix}$

   $B = \begin{bmatrix} 72,885 & 90,611 & 77,972 \\ 61,711 & 65,593 & 60,588 \\ 57,045 & 64,089 & 49,678 \\ 52,461 & 40,252 & 35,496 \end{bmatrix}$

2. $B - A$

   $= \begin{bmatrix} -7494 & -9368 & -6824 \\ -2703 & -5050 & -5563 \\ -1387 & -4056 & -3267 \\ -4606 & -4847 & -926 \end{bmatrix}$

3. Based on the matrix in part 2, at all institutions female professors are paid less than male professors. Gender bias seems to exist at the institutions.

4.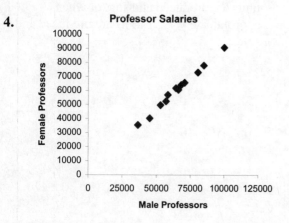

5.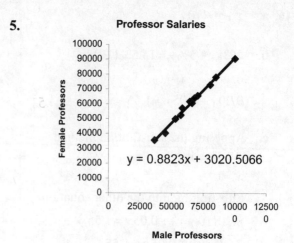

6. See part 5 above.

7.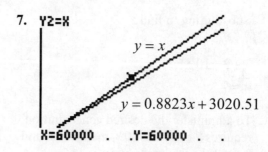

   [20,000, 100,000] by [20,000, 100,000]

8. Based on the graph in part 7, as salaries increase the gap between male and female salaries also increases.

9. Applying the intersection of graphs method:

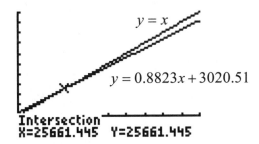

[0, 100,000] by [0, 100,000]

The male salaries exceed the female salaries
beginning at approximately $25,661.

Extended Application II

1. $D = \begin{bmatrix} 8 \\ 2 \\ 1 \\ 2 \\ 4 \\ 4 \\ 8 \end{bmatrix}$

2. $I - P = \begin{bmatrix} 1 & 0 & 0 & 0 & 0 & 0 & 0 \\ 0 & 1 & 0 & 0 & 0 & 0 & 0 \\ 0 & 0 & 1 & 0 & 0 & 0 & 0 \\ 0 & 0 & 0 & 1 & 0 & 0 & 0 \\ 0 & 0 & 0 & 0 & 1 & 0 & 0 \\ 0 & 0 & 0 & 0 & 0 & 1 & 0 \\ 0 & 0 & 0 & 0 & 0 & 0 & 1 \end{bmatrix} - \begin{bmatrix} 0 & 0 & 0 & 0 & 0 & 0 & 0 \\ 4 & 0 & 0 & 0 & 0 & 0 & 0 \\ 1 & 0 & 0 & 0 & 0 & 0 & 0 \\ 0 & 1 & 1 & 0 & 0 & 0 & 0 \\ 0 & 0 & 2 & 0 & 0 & 0 & 0 \\ 0 & 1 & 0 & 0 & 0 & 0 & 0 \\ 0 & 2 & 6 & 0 & 0 & 0 & 0 \end{bmatrix}$

$= \begin{bmatrix} 1 & 0 & 0 & 0 & 0 & 0 & 0 \\ -4 & 1 & 0 & 0 & 0 & 0 & 0 \\ -1 & 0 & 1 & 0 & 0 & 0 & 0 \\ 0 & -1 & -1 & 1 & 0 & 0 & 0 \\ 0 & 0 & -2 & 0 & 1 & 0 & 0 \\ 0 & -1 & 0 & 0 & 0 & 1 & 0 \\ 0 & -2 & -6 & 0 & 0 & 0 & 1 \end{bmatrix}$

3. Applying technology to find the inverse of $I - P$:

```
[[1   0  0  0  0  0 ...
 [4   1  0  0  0  0 ...
 [1   0  1  0  0  0 ...
 [5   1  1  1  0  0 ...
 [2   0  2  0  1  0 ...
 [4   1  0  0  0  1 ...
 [14  2  6  0  0  0 ...

...  0  0  0  0  0  0]
...  1  0  0  0  0  0]
...  0  1  0  0  0  0]
...  1  1  1  0  0  0]
...  0  2  0  1  0  0]
...  1  0  0  0  1  0]
...4 2  6  0  0  0  1]]
```

Note that the inverse matrix is shown in two graphics with several repeated columns. The inverse matrix is

$$(I-P)^{-1} = \begin{bmatrix} 1 & 0 & 0 & 0 & 0 & 0 & 0 \\ 4 & 1 & 0 & 0 & 0 & 0 & 0 \\ 1 & 0 & 1 & 0 & 0 & 0 & 0 \\ 5 & 1 & 1 & 1 & 0 & 0 & 0 \\ 2 & 0 & 2 & 0 & 1 & 0 & 0 \\ 4 & 1 & 0 & 0 & 0 & 1 & 0 \\ 14 & 2 & 6 & 0 & 0 & 0 & 1 \end{bmatrix}$$

4. $(I-P)X = D$

$(I-P)^{-1}(I-P)X = (I-P)^{-1}D$

$$X = \begin{bmatrix} 1 & 0 & 0 & 0 & 0 & 0 & 0 \\ 4 & 1 & 0 & 0 & 0 & 0 & 0 \\ 1 & 0 & 1 & 0 & 0 & 0 & 0 \\ 5 & 1 & 1 & 1 & 0 & 0 & 0 \\ 2 & 0 & 2 & 0 & 1 & 0 & 0 \\ 4 & 1 & 0 & 0 & 0 & 1 & 0 \\ 14 & 2 & 6 & 0 & 0 & 0 & 1 \end{bmatrix} \begin{bmatrix} 8 \\ 2 \\ 1 \\ 2 \\ 4 \\ 4 \\ 8 \end{bmatrix}$$

Applying technology to carry out the multiplication yields

$$X = \begin{bmatrix} 8 \\ 34 \\ 9 \\ 45 \\ 22 \\ 38 \\ 130 \end{bmatrix}$$

5. Based on the calculations in part 4, the required number of pipes is 45, the required number of clamps is 22, the required number of braces is 38, and the required number of bolts is 130.

Chapter 6
Special Topics

Section 6.1 Skills Check

1. $y \le 5x - 4$

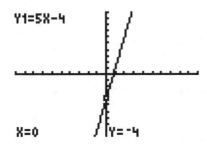

[−10, 10] by [−10, 10]

Note that the line is solid because of the "equal to" in the given inequality.

Test (0, 0).
$$0 \le 5(0) - 4$$
$$0 \le -4$$

Since the statement is false, the region not containing (0, 0) is the solution to the inequality.

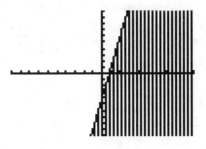

[−10, 10] by [−10, 10]

3. $6x - 3y \ge 12$
$$-3y \ge -6x + 12$$
$$\frac{-3y}{-3} \le \frac{-6x + 12}{-3}$$
$$y \le 2x - 4$$

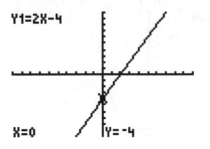

[−10, 10] by [−10, 10]

Note that the line is solid because of the "equal to" in the given inequality.

Test (0, 0).
$$0 \le 2(0) - 4$$
$$0 \le -4$$

Since the statement is false, the region not containing (0, 0) is the solution to the inequality.

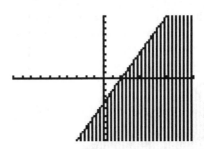

[−10, 10] by [−10, 10]

5. $4x + 5y \le 20$
$$5y \le 20 - 4x$$
$$y \le \frac{20 - 4x}{5}$$
$$y \le 4 - \frac{4}{5}x$$

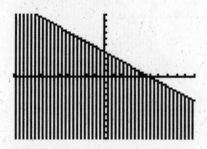

Y1=4-(4/5)X

X=0   Y=4

[–10, 10] by [–10, 40]

Note that the line is solid because of the "equal to" in the given inequality.

Test (0, 0).

$$y \le 4 - \frac{4}{5}x$$

$$0 \le 4 - \frac{4}{5}(0)$$

$$0 \le 4$$

Since the statement is true, the region containing (0, 0) is the solution to the inequality.

[–10, 10] by [–10, 40]

7. To determine the solution region and the corners of the solution region, pick a point in a potential solution region and test it. For example, pick $(2,2)$.

$$x + y \le 5$$
$$2 + 2 \le 5$$
$$4 \le 5$$
$$\text{True statement}$$

$$2x + y \le 8$$
$$2(2) + 2 \le 8$$
$$4 + 2 \le 8$$
$$6 \le 8$$
$$\text{True statement}$$

$$x \ge 0$$
$$2 \ge 0$$
$$\text{True statement}$$

$$y \ge 0$$
$$2 \ge 0$$
$$\text{True statement}$$

Since all the inequalities are true at the point $(2,2)$, the region that contains $(2,2)$ is the solution region. The corners of the region are $(0,0), (4,0), (3,2),$ and $(0,5)$.

9. To determine the solution region and the corners of the solution region, pick a point in a potential solution region and test it. Pick $(2,4)$.

$$4x + 2y > 8$$
$$4(2) + 2(8) > 8$$
$$24 > 8$$
$$\text{True statement}$$
$$3x + y > 5$$
$$3(2) + 4 > 5$$
$$10 > 5$$
$$\text{True statement}$$

$x \geq 0$

$2 \geq 0$

True statement

$y \geq 0$

$4 \geq 0$

True statement

Since all the inequalities are true at the point $(2,4)$, the region that contains $(2,4)$ is the solution region. The corners of the region are $(2,0),(0,5)$, and $(1,2)$.

11. To determine the solution region and the corners of the solution region, pick a point in a potential solution region and test it. Pick $(3,3)$.

$2x + 6y \geq 12$

$2(3) + 6(3) \geq 12$

$6 + 18 \geq 12$

$24 \geq 12$

True statement

$3x + y \geq 5$

$3(3) + 3 \geq 5$

$12 \geq 5$

True statement

$x + 2y \geq 5$

$3 + 2(3) \geq 5$

$9 \geq 5$

True statement

$x \geq 0$

$3 \geq 0$

True statement

$y \geq 0$

$3 \geq 0$

True statement

Since all the inequalities are true at the point $(3,3)$, the region that contains $(3,3)$ is the solution region. The corners of the region are $(0,5),(1,2),(3,1)$, and $(6,0)$.

13. The graph of the system is

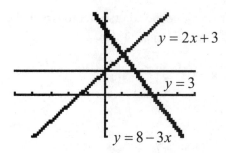

[–5, 5] by [–5, 10]

Note that $y = 3$ is a dashed line. The other two lines are solid. The intersection point between $y = 8 - 3x$ and $y = 2x + 3$ is $(1,5)$.

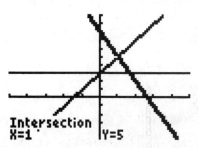

The intersection point between $y = 3$ and $y = 2x + 3$ is $(0,3)$.

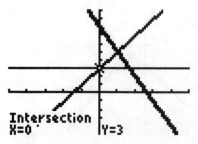

The intersection point between $y = 3$ and $y = 8 - 3x$ is $\left(\dfrac{5}{3}, 3 \right)$.

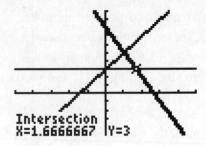

Intersection
X=1.6666667 Y=3

To determine the solution region, pick a
point to test. Pick $(1,4)$.

$y \le 8 - 3x$

$4 \le 8 - 3(1)$

$4 \le 5$

True statement

$y \le 2x + 3$

$4 \le 2(1) + 3$

$4 \le 5$

True statement

$y > 3$

$4 > 3$

True statement

Since all the inequalities that form the
system are true at the point $(1,4)$, the region
that contains $(1,4)$ is the solution region.
The corners of the region are

$(0,3), (1,5),$ and $\left(\dfrac{5}{3}, 3\right)$.

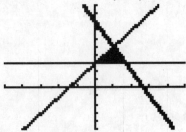

[–10, 10] by [–10, 10]

Recall that $y = 3$ is dashed. The other two
lines are solid.

15. Rewriting the inequalities:

$2x + y < 5$

$\qquad y < -2x + 5$

$\qquad$ and

$2x - y > -1$

$\qquad -y > -1 - 2x$

$\qquad \dfrac{-y}{-1} < \dfrac{-1 - 2x}{-1}$

$\qquad y < 2x + 1$

The new system is

$\begin{cases} y < -2x + 5 \\ y < 2x + 1 \\ x \ge 0, y \ge 0 \end{cases}$

The graph of the system is

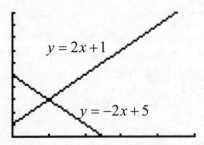

$y = 2x + 1$

$y = -2x + 5$

[0, 5] by [0, 10]

Note that both lines are dashed. The
intersection point between the two lines is
$(1,3)$.

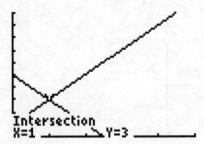

Intersection
X=1   Y=3

To determine the solution region, pick a
point to test. Pick $(1,2)$.

$y < -2x + 5$

$2 < -2(1) + 5$

$2 < 3$

True statement

$y < 2x + 1$

$2 < 2(1) + 1$

$2 < 3$

True statement

$x \geq 0$

$1 \geq 0$

True statement

$y \geq 0$

$2 \geq 0$

True statement

Since all the inequalities that form the system are true at the point $(1, 2)$, the region that contains $(1, 2)$ is the solution region. Considering the graph, the corners of the region are $(0, 0), (0, 1), (2.5, 0)$ and $(1, 3)$.

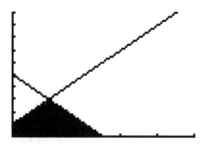

[0, 5] by [0, 10]

Recall that both lines are dashed.

17. Rewriting the system:

$$x + 2y \geq 4$$

$$2y \geq 4 - x$$

$$y \geq \frac{4 - x}{2}$$

and

$$x + y \leq 5$$

$$y \leq 5 - x$$

and

$$2x + y \leq 8$$

$$y \leq 8 - 2x$$

The new system is

$$\begin{cases} y \geq \dfrac{4 - x}{2} \\ y \leq 5 - x \\ y \leq 8 - 2x \\ x \geq 0, y \geq 0 \end{cases}$$

The graph of the system is

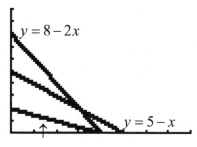

$$y = \frac{4 - x}{2}$$

[0, 8] by [0, 10]

Note that all the lines are solid.

Determining the intersection point between $y = 8 - 2x$ and $y = 5 - x$:

$8-2x=5-x$

$-x=-3$

$x=3$

Substituting to find y

$y=5-x$

$y=5-3$

$y=2$

The intersection point is $(3,2)$.

Determining the intersection point between $y=8-2x$ and $y=\dfrac{4-x}{2}$:

$8-2x=\dfrac{4-x}{2}$

$2(8-2x)=2\left(\dfrac{4-x}{2}\right)$

$16-4x=4-x$

$-3x=-12$

$x=4$

Substituting to find y

$y=8-2x$

$y=8-2(4)$

$y=0$

The intersection point is $(4,0)$.

To determine the solution region, pick a point to test. Pick $(1,3)$.

$y\ge\dfrac{4-x}{2}$

$3\ge\dfrac{4-1}{2}$

$3\ge\dfrac{3}{2}$

True statement

$y\le5-x$

$3\le5-1$

$3\le4$

True statement

$y\le8-2x$

$3\le8-2(1)$

$3\le6$

True statement

$x\ge0$

$1\ge0$

True statement

$y\ge0$

$3\ge0$

True statement

Since all the inequalities that form the system are true at the point $(1,3)$, the region that contains $(1,3)$ is the solution region. Considering the graph of the system, the corners of the region are $(0,2),(0,5),(3,2),$ and $(4,0)$.

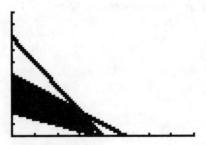

[0, 8] by [0, 10]

Recall that all the lines are solid.

19. Rewriting the system:

$$x^2 - 3y < 4$$

$$-3y < 4 - x^2$$

$$y > \frac{4 - x^2}{-3}$$

$$y > \frac{x^2 - 4}{3}$$

and

$$2x + 3y < 4$$

$$3y < 4 - 2x$$

$$y < \frac{4 - 2x}{3}$$

The new system is

$$\begin{cases} y > \dfrac{x^2 - 4}{3} \\ y < \dfrac{4 - 2x}{3} \end{cases}$$

The graph of the system is

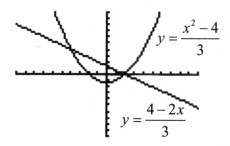

[−10, 10] by [−10, 10]

Note that all the lines are dashed.

Determining the intersection points between the two functions graphically yields

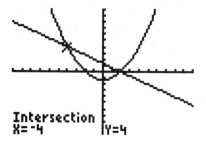

Intersection
X=-4   Y=4

[−10, 10] by [−10, 10]

and

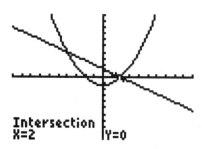

Intersection
X=2   Y=0

[−10, 10] by [−10, 10]

The intersection points are $(-4, 4)$ and $(2, 0)$.

To determine the solution region, pick a point to test. Pick $(0, 0)$.

$$y > \frac{x^2 - 4}{3}$$

$$0 > \frac{(0)^2 - 4}{3}$$

$$0 > -\frac{4}{3}$$

True statement

$$y < \frac{4 - 2x}{3}$$

$$0 < \frac{4 - 2(0)}{3}$$

$$0 < \frac{4}{3}$$

True statement

Since all the inequalities that form the system are true at the point $(0,0)$, the region that contains $(0,0)$ is the solution region. Considering the graph of the system, the corners of the region are $(-4,4)$ and $(2,0)$.

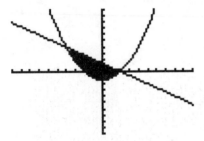

[–10, 10] by [–10, 10]

Recall that all the lines are dashed.

21. Rewriting the system:

$$x^2 - y - 8x \le -6$$
$$-y \le -x^2 + 8x - 6$$
$$y \ge x^2 - 8x + 6$$
$$\text{and}$$
$$y + 9x \le 18$$
$$y \le 18 - 9x$$

The new system is
$$\begin{cases} y \ge x^2 - 8x + 6 \\ y \le 18 - 9x \end{cases}$$

The graph of the system is

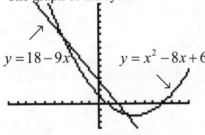

$y = 18 - 9x$ $y = x^2 - 8x + 6$

[–10, 10] by [–20, 80]

Note that all the lines are solid.

Determining the intersection points between the two functions graphically yields

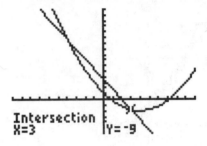

[–10, 10] by [–20, 80]

and

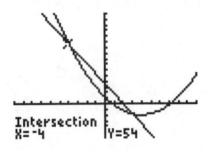

[–10, 10] by [–20, 80]

The intersection points are $(3,-9)$ and $(-4,54)$.

To determine the solution region, pick a point to test. Pick $(0,10)$.

$$y \ge x^2 - 8x + 6$$
$$10 \ge (0)^2 - 8(0) + 6$$
$$10 \ge 6$$
True statement
$$y \le 18 - 9x$$
$$10 \le 18 - 9(0)$$
$$10 \le 18$$
True statement

Since all the inequalities that form the system are true at the point $(0,10)$, the region that contains $(0,10)$ is the solution region. Considering the graph of the system, the corners of the region are $(3,-9)$ and $(-4,54)$.

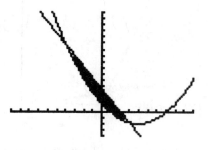

[–10, 10] by [–20, 80]

Recall that all the lines are solid.

Section 6.1 Exercises

23. a. Let x = the number of Turbo blowers, and y = the number of Tornado blowers.

$x + y \geq 780$

b. $x + y \geq 780$

$y \geq 780 - x$

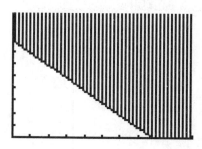

[0, 1000] by [0, 1000]

Note that the line is solid since an "equal to" is part of the inequality.

25. a. Let x = minutes of cable television time, and y = minutes of radio time.

$240x + 150y \leq 36{,}000$

b. $240x + 150y \leq 36{,}000$

$150y \leq 36{,}000 - 240x$

$y \leq \dfrac{36{,}000 - 240x}{150}$

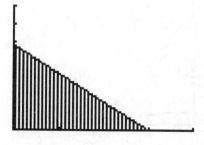

[0, 200] by [0, 350]

Note that the line is solid since an "equal to" is part of the inequality.

27. a. Let x = minutes of television time, and y = minutes of radio time. Then, $0.12x$ represents the number, in millions, of registered voters reached by television advertising, and $0.009y$ represents the number, in millions, of registered votersreached by radio advertising. The corresponding inequalities are

$$\begin{cases} x + y \geq 100 \\ 0.12x + 0.009y \geq 7.56 \\ x \geq 0, y \geq 0 \end{cases}$$

b. Rewriting the system:

$x + y \geq 100$

$y \geq 100 - x$

and

$0.12x + 0.009y \geq 7.56$

$0.009y \geq 7.56 - 0.12x$

$y \geq \dfrac{7.56 - 0.12x}{0.009}$

The new system is

$$\begin{cases} y \geq 100 - x \\ y \geq \dfrac{7.56 - 0.12x}{0.009} \\ x \geq 0, y \geq 0 \end{cases}$$

The graph of the system is

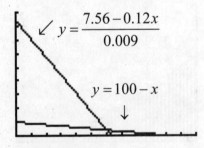

$y = \dfrac{7.56 - 0.12x}{0.009}$

$y = 100 - x$

[0, 120] by [0, 900]

The intersection point between the two lines is $(60, 40)$. Both lines are solid

since there is an "equal to" in both inequalities.

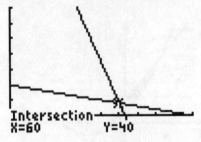

Intersection
X=60   Y=40

To determine the solution region, pick a point to test. Pick $(60, 100)$.

$x + y \geq 100$

$60 + 100 \geq 100$

$160 \geq 100$

True statement

$0.12x + 0.009y \geq 7.56$

$0.12(60) + 0.009(100) \geq 7.56$

$7.2 + 0.9 \geq 7.56$

$8.1 \geq 7.56$

True statement

$x \geq 0$

$60 \geq 0$

True statement

$y \geq 0$

$100 \geq 0$

True statement

Since all the inequalities that form the system are true at the point $(60, 100)$, the region that contains $(60, 100)$ is the solution region. The graph of the solution is

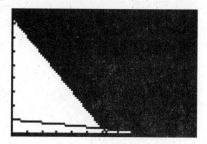

[0, 120] by [0, 900]

Recall that both lines are solid.

One of the corner points is the intersection point of the two lines, $(60,40)$. Another corner point occurs where $y = \dfrac{7.56 - 0.12x}{0.009}$ crosses the y-axis. A third corner point occurs where $y = 100 - x$ crosses the x-axis.

To find the y-intercept, let $x = 0$.
$$y = \frac{7.56 - 0.12(0)}{0.009}$$
$$y = \frac{7.56}{0.009}$$
$$y = 840$$
$$(0,840)$$

To find the x-intercept, let $y = 0$.
$$0 = 100 - x$$
$$x = 100$$
$$(100,0)$$

Therefore, the corner points of the solution region are $(60,40), (100,0)$, and $(0,840)$.

29. The system of inequalities is
$$\begin{cases} x + y \geq 780 \\ 78x + 117y \leq 76{,}050 \\ x \geq 0, y \geq 0 \end{cases}$$

Rewriting the system:
$$x + y \geq 780$$
$$y \geq 780 - x$$
and
$$78x + 117y \leq 76{,}050$$
$$117y \leq 76{,}050 - 78x$$
$$y \leq \frac{76{,}050 - 78x}{117}$$

The new system is
$$\begin{cases} y \geq 780 - x \\ y \leq \dfrac{76{,}050 - 78x}{117} \\ x \geq 0, y \geq 0 \end{cases}$$

The graph of the system is

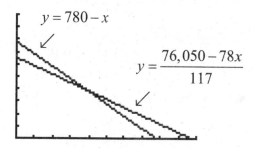

[0, 1000] by [0, 1000]

The intersection point between the two lines is $(390,390)$. Both lines are solid since there is an "equal to" in both inequalities.

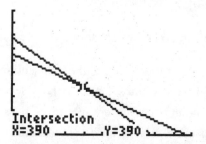

To determine the solution region, pick a point to test. Pick $(700,100)$.

$x + y \geq 780$

$700 + 100 \geq 780$

$800 \geq 780$

True statement

$78x + 117y \leq 76,050$

$78(700) + 117(100) \leq 76,050$

$54,600 + 11,700 \leq 76,050$

$66,300 \leq 76,050$

True statement

$x \geq 0$

$700 \geq 0$

True statement

$y \geq 0$

$100 \geq 0$

True statement

Since all the inequalities that form the system are true at the point $(700,100)$, the region that contains $(700,100)$ is the solution region. The graph of the solution is

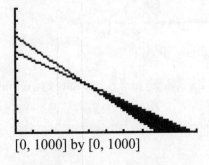

[0, 1000] by [0, 1000]

Recall that both lines are solid.

One of the corner points is the intersection point of the two lines, $(390,390)$. A second corner point occurs where $y = 780 - x$ crosses the x-axis. A third corner point occurs where $y = \dfrac{76,050 - 78x}{117}$ crosses the x-axis.

To find the x-intercept, let $y = 0$.

$y = 780 - x$

$0 = 780 - x$

$x = 780$

$(780,0)$

To find the x-intercept, let $y = 0$.

$0 = \dfrac{76,050 - 78x}{117}$

$0 = 76,050 - 78x$

$78x = 76,050$

$x = \dfrac{76,050}{78} = 975$

$(975,0)$

Therefore, the corner points of the solution region are $(390,390), (780,0)$ and $(975,0)$.

31. a. Let x = number of manufacturing days on assembly line 1, and y = number of manufacturing days on assembly line 2.

Then, the system of inequalities is

$\begin{cases} 80x + 40y \geq 3200 \\ 20x + 20y \geq 1000 \\ 100x + 40y \geq 3400 \\ x \geq 0, y \geq 0 \end{cases}$

Rewriting the system:
$$80x + 40y \geq 3200$$
$$40y \geq 3200 - 80x$$
$$y \geq \frac{3200 - 80x}{40}$$
$$y \geq 80 - 2x$$

and

$$20x + 20y \geq 1000$$
$$20y \geq 1000 - 20x$$
$$y \geq \frac{1000 - 20x}{20}$$
$$y \geq 50 - x$$

and

$$100x + 40y \geq 3400$$
$$40y \geq 3400 - 100x$$
$$y \geq \frac{3400 - 100x}{40}$$
$$y \geq 85 - 2.5x$$

The new system is
$$\begin{cases} y \geq 80 - 2x \\ y \geq 50 - x \\ y \geq 85 - 2.5x \\ x \geq 0, y \geq 0 \end{cases}$$

b. To solve the system, first consider the graph of the system

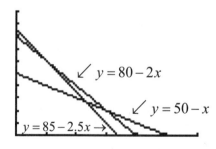

$y = 80 - 2x$

$y = 50 - x$

$y = 85 - 2.5x \rightarrow$

[0, 60] by [0, 100]

The intersection point between
$y = 80 - 2x$ and $y = 50 - x$ is $(30, 20)$.

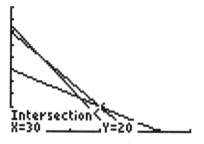

The intersection point between
$y = 80 - 2x$ and $y = 85 - 2.5x$ is
$(10, 60)$.

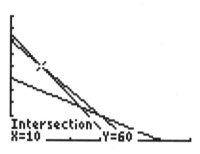

The intersection point between
$y = 50 - x$ and $y = 85 - 2.5x$ is
$(23.\overline{3}, 26.\overline{6})$.

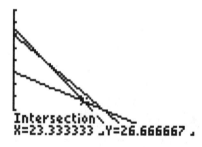

Note that all the lines are solid because there is an "equal to" as part of all the inequalities.

To determine the solution region, pick a point to test. Pick $(25, 40)$.

$80x + 40y \geq 3200$

$80(25) + 40(40) \geq 3200$

$3600 \geq 3200$

True statement

$20x + 20y \geq 1000$

$20(25) + 20(40) \geq 1000$

$500 + 800 \geq 1000$

$1300 \geq 1000$

True statement

$100x + 40y \geq 3400$

$100(25) + 40(40) \geq 3400$

$4100 \geq 3400$

True statement

$x \geq 0$

$25 \geq 0$

True statement

$y \geq 0$

$40 \geq 0$

True statement

Since all the inequalities that form the system are true at the point $(25, 40)$, the region that contains $(25, 40)$ is the solution region. The graph of the solution is

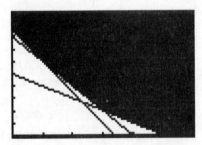

[0, 60] by [0, 100]

Recall that all lines are solid.

One of the corner points is the intersection point between $y = 80 - 2x$ and $y = 85 - 2.5x$, which is $(10, 60)$. A second corner point is the intersection

point between $y = 80 - 2x$ and $y = 50 - x$, which is $(30, 20)$. A third corner point occurs where $y = 50 - x$ crosses the x-axis. A fourth corner point occurs where $y = 85 - 2.5x$ crosses the y-axis. Therefore, to find the x-intercept, let $y = 0$.

$0 = 50 - x$

$x = 50$

$(50, 0)$

To find the y-intercept, let $x = 0$.

$y = 85 - 2.5(0)$

$y = 85$

$(0, 85)$

Therefore, the corner points of the solution region are

$(10, 60), (30, 20), (50, 0)$ and $(0, 85)$.

33. Let x = the number of bass, and y = the number of trout.

Then, the system of inequalities is
$$\begin{cases} 4x + 10y \leq 1600 \\ 6x + 7y \leq 1600 \\ x \geq 0, y \geq 0 \end{cases}$$

Rewriting the system:

$4x + 10y \leq 1600$

$\qquad 10y \leq 1600 - 4x$

$\qquad\qquad y \leq \dfrac{1600 - 4x}{10}$

$\qquad$ and

$6x + 7y \leq 1600$

$\qquad 7y \leq 1600 - 6x$

$\qquad\qquad y \leq \dfrac{1600 - 6x}{7}$

The new system is

$$\begin{cases} y \le \dfrac{1600-4x}{10} \\ y \le \dfrac{1600-6x}{7} \\ x \ge 0, y \ge 0 \end{cases}$$

To solve the system, first consider the graph of the system

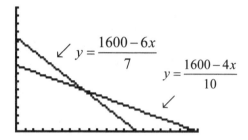

[0, 400] by [0, 300]

The intersection point between the two lines is $(150,100)$.

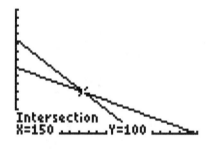

Note that the lines are solid because there is an "equal to" as part of all the inequalities.

To determine the solution region, pick a point to test. Pick $(100,75)$.

$$y \le \dfrac{1600-4x}{10}$$

$$75 \le \dfrac{1600-4(100)}{10}$$

$$75 \le \dfrac{1200}{10}$$

$$75 \le 120$$

True statement

$$y \le \dfrac{1600-6x}{7}$$

$$75 \le \dfrac{1600-6(100)}{7}$$

$$75 \le \dfrac{1000}{7}$$

$$75 \le\sim 142$$

True statement

$$x \ge 0$$

$$100 \ge 0$$

True statement

$$y \ge 0$$

$$75 \ge 0$$

True statement

Since all the inequalities that form the system are true at the point $(100,75)$, the region that contains $(100,75)$ is the solution region. The graph of the solution is

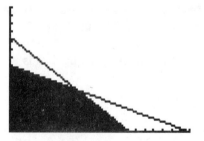

[0, 400] by [0, 300]

Recall that all lines are solid.

One of the corner points is the intersection point between the two lines, which is $(150,100)$. A second corner point occurs where $y = \dfrac{1600 - 4x}{10}$ crosses the y-axis. A third corner point occurs where $y = \dfrac{1600 - 6x}{7}$ crosses the x-axis.

To find the y-intercept, let $x = 0$.

$$y = \frac{1600 - 4(0)}{10}$$

$$y = \frac{1600}{10}$$

$$y = 160$$

$$(0, 160)$$

To find the x-intercept, let $y = 0$.

$$0 = \frac{1600 - 6x}{7}$$

$$0 = 1600 - 6x$$

$$6x = 1600$$

$$x = \frac{1600}{6} = 266.\overline{6}$$

$$\left(266.\overline{6}, 0\right)$$

A fourth corner point occurs at the origin, $(0, 0)$. Therefore, the corner points of the solution region are

$$(0, 0), (150, 100), (0, 160) \text{ and } \left(266.\overline{6}, 0\right).$$

35. a. Let x = number of Standard chairs, and let y = number of Deluxe chairs.

Then, the system of inequalities is

$$\begin{cases} 4x + 6y \le 480 \\ 2x + 6y \le 300 \\ x \ge 0, y \ge 0 \end{cases}$$

Rewriting the system:

$$4x + 6y \le 480$$

$$6y \le 480 - 4x$$

$$y \le \frac{480 - 4x}{6}$$

and

$$2x + 6y \le 300$$

$$6y \le 300 - 2x$$

$$y \le \frac{300 - 2x}{6}$$

The new system is

$$\begin{cases} y \le \dfrac{480 - 4x}{6} \\ y \le \dfrac{300 - 2x}{6} \\ x \ge 0, y \ge 0 \end{cases}$$

b. To solve the system, first consider the graph of the system

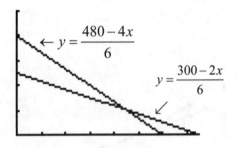

[0, 150] by [0, 100]

The intersection point between the two lines is $(90, 20)$.

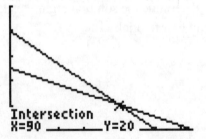

Note that the lines are solid because there is an "equal to" as part of all the inequalities.

To determine the solution region, pick a point to test. Pick $(5,10)$.

$4x + 6y \leq 480$

$4(5) + 6(10) \leq 480$

$80 \leq 480$

True statement

$2x + 6y \leq 300$

$2(5) + 6(10) \leq 300$

$70 \leq 300$

True statement

$x \geq 0$

$5 \geq 0$

True statement

$y \geq 0$

$10 \geq 0$

True statement

Since all the inequalities that form the system are true at the point $(5,10)$, the region that contains $(5,10)$ is the solution region. The graph of the solution is

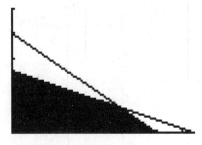

[0, 150] by [0, 100]

Recall that all lines are solid.

One of the corner points is the intersection point between the two lines, which is $(90,20)$. A second corner point occurs where $y = \dfrac{480 - 4x}{6}$ crosses the x-axis. A third corner point

occurs where $y = \dfrac{300 - 2x}{6}$ crosses the y-axis.

To find the x-intercept, let $y = 0$.

$0 = \dfrac{480 - 4x}{6}$

$0 = 480 - 4x$

$4x = 480$

$x = \dfrac{480}{4}$

$(120, 0)$

To find the y-intercept, let $x = 0$.

$y = \dfrac{300 - 2(0)}{6}$

$y = \dfrac{300}{6}$

$y = 50$

$(0, 50)$

A fourth corner point occurs at the origin, $(0,0)$. Therefore, the corner points of the solution region are $(0,0), (90,20), (120,0)$ and $(0,50)$.

37. a. Let x = number of commercial heating systems, and y = number of domestic heating systems.

Then, the system of inequalities is
$$\begin{cases} x + y \leq 1400 \\ x \geq 500 \\ y \geq 750 \end{cases}$$

Rewriting the system:
$x + y \leq 1400$
    $y \leq 1400 - x$

The new system is
$$\begin{cases} y \leq 1400 - x \\ x \geq 500 \\ y \geq 750 \end{cases}$$

b. To solve the system, first consider the graph of the system

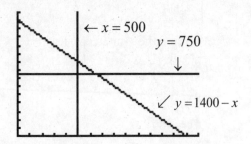

[0, 1500] by [0, 1500]

The intersection point between the $x = 500$ and $y = 1400 - x$ is $(500, 900)$. Note that solving by the substitution method yields $y = 1400 - 500 = 900$.

The intersection point between $y = 750$ and $y = 1400 - x$ is $(650, 750)$.

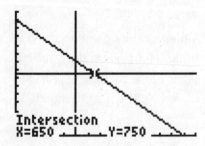

The intersection point between $x = 500$ and $y = 750$ is $(500, 750)$.

Note that the lines are solid because there is an "equal to" as part of all the inequalities.

To determine the solution region, pick a point to test. Pick $(600, 775)$.

$y \leq 1400 - x$

$775 \leq 1400 - 600$

$775 \leq 800$

True statement

$x \geq 500$

$600 \geq 500$

True statement

$y \geq 750$

$775 \geq 750$

True statement

Since all the inequalities that form the system are true at the point $(600, 775)$, the region that contains $(600, 775)$ is the solution region. The graph of the solution is

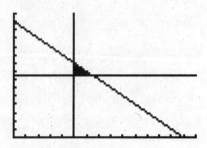

[0, 1500] by [0, 1500]

Recall that all lines are solid.

One of the corner points is the intersection point between $x = 500$ and $y = 1400 - x$, which is $(500, 900)$.

Another corner point is the intersection point between $y = 750$ and $y = 1400 - x$, which is $(650, 750)$.

A third corner point occurs at the intersection between $x = 500$ and $y = 750$, which is $(500, 750)$.

Therefore, the corner points of the solution region are $(500, 750), (500, 900),$ and $(650, 750)$.

Section 6.2 Skills Check

1. Test the corner points of the feasible region in the objective function.

   At $(0,0)$: $f = 4(0) + 9(0) = 0$

   At $(0,40)$: $f = 4(0) + 9(40) = 360$

   At $(67,0)$: $f = 4(67) + 9(0) = 268$

   At $(10,38)$: $f = 4(10) + 9(38) = 382$

   The maximum value is 382 occurring at $(10,38)$, and the minimum value is 0 occurring at $(0,0)$.

3. Test the corner points of the feasible region in the objective function.

   At $(0,0)$: $f = 4(0) + 2(0) = 0$

   At $(0,2)$: $f = 4(0) + 2(2) = 4$

   At $(2,4)$: $f = 4(2) + 2(4) = 16$

   At $(4,3)$: $f = 4(4) + 2(3) = 22$

   At $(5,0)$: $f = 4(5) + 2(0) = 20$

   The maximum value is 22 occurring at $(4,3)$, and the minimum value is 0 occurring at $(0,0)$.

5. a. The graph of the system is

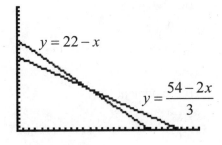

   [0, 30] by [0, 30]

   The intersection point between the two lines is $(12,10)$.

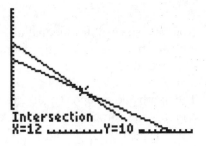

To determine the solution region, pick a point to test. Pick $(1,1)$.

$$y \le \frac{54 - 2x}{3}$$

$$1 \le \frac{54 - 2(1)}{3}$$

$$1 \le \frac{52}{3}$$

$$1 \le 17.\overline{3}$$

True statement

$$y \le 22 - x$$

$$1 \le 22 - 1$$

$$1 \le 21$$

True statement

$$x \ge 0$$

$$1 \ge 0$$

True statement

$$y \ge 0$$

$$1 \ge 0$$

True statement

Since all the inequalities that form the system are true at the point $(1,1)$, the region that contains $(1,1)$ is the solution region. The graph of the solution is

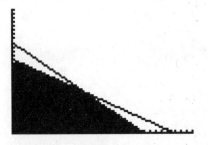

[0, 30] by [0, 30]

Note that since all the inequalities contain an "equal to", all the boundary lines are solid.

The graph above represents the feasible region. The corners of the region are $(0,0), (12,10), (22,0)$, and $(0,18)$.

b. Testing the corner points of the feasible region in the objective function yields

At $(0,0)$: $f = 3(0) + 5(0) = 0$

At $(0,18)$: $f = 3(0) + 5(18) = 90$

At $(22,0)$: $f = 3(22) + 5(0) = 66$

At $(12,10)$: $f = 3(12) + 5(10) = 86$

The maximum value is 90 occurring at $(0,18)$.

7. a. To solve the system and determine the feasible region, first graph the system. To graph the system, rewrite the system by solving for y in the inequalities.

$$x + 2y \geq 15$$
$$2y \geq 15 - x$$
$$y \geq \frac{15 - x}{2}$$
and
$$x + y \geq 10$$
$$y \geq 10 - x$$

The new system is
$$\begin{cases} y \geq \dfrac{15 - x}{2} \\ y \geq 10 - x \\ x \geq 0, y \geq 0 \end{cases}$$

The graph of the system is

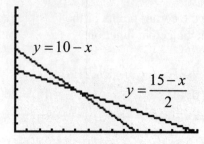

[0, 15] by [0, 15]

The intersection point between the two lines is $(5,5)$.

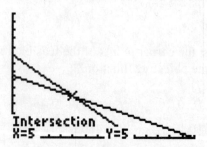

To determine the solution region, pick a point to test. Pick $(5,6)$.

$$y \geq \frac{15 - x}{2}$$
$$6 \geq \frac{15 - 5}{2}$$
$$6 \geq \frac{10}{2}$$
$$6 \geq 5$$
True statement
$$y \geq 10 - x$$
$$6 \geq 10 - 5$$
$$6 \geq 5$$
True statement
$$x \geq 0$$
$$5 \geq 0$$
True statement
$$y \geq 0$$
$$6 \geq 0$$
True statement

Since all the inequalities that form the system are true at the point $(5,6)$, the region that contains $(5,6)$ is the solution region. The graph of the solution is

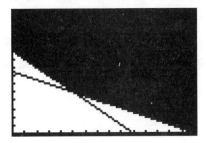

[0, 15] by [0, 15]

Note that since all the inequalities contain an "equal to", all the boundary lines are solid.

The graph above represents the feasible region. The corners of the region are $(5,5),(0,10)$, and $(15,0)$.

b. Testing the corner points of the feasible region in the objective function yields

At $(5,5)$: $g = 4(5)+2(5)=30$

At $(0,10)$: $g = 4(0)+2(10)=20$

At $(15,0)$: $g = 4(15)+2(0)=60$

The minimum value is 20 occurring at $(0,10)$.

9. Graphing the system yields

$$\begin{cases} y \le -\dfrac{1}{2}x+4 \\ y \le -x+6 \\ y \le -\dfrac{1}{3}x+4 \\ x \ge 0, y \ge 0 \end{cases}$$

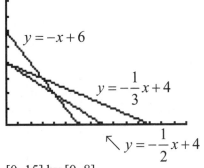

[0, 15] by [0, 8]

To determine the solution region, pick a point to test. Pick $(1,1)$. When substituted into the inequalities that form the system, the point $(1,1)$ creates true statements in all cases.

Since all the inequalities that form the system are true at the point $(1,1)$, the region that contains $(1,1)$ is the solution region. The solution represents the feasible region. The graph of the feasible region is

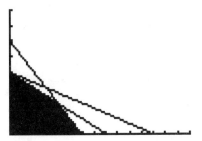

[0, 15] by [0, 8]

Note that since all the inequalities contain an "equal to", all the boundary lines are solid.

The corner points of the feasible region are $(0,0),(0,4),(6,0),$ and $(4,2)$.

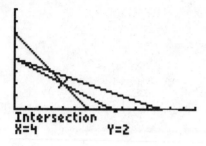

Testing the corner points of the feasible region to maximize the objective function, $f = 20x + 30y$, yields

At $(0,0)$: $f = 20(0) + 30(0) = 0$

At $(0,4)$: $f = 20(0) + 30(4) = 120$

At $(6,0)$: $f = 20(6) + 30(0) = 120$

At $(4,2)$: $f = 20(4) + 30(2) = 140$

The maximum value is 140 occurring at $(4,2)$.

11. Rewriting the system of constraints and graphing the system yields

$$\begin{cases} y \le \dfrac{32 - 3x}{4} \\ y \le \dfrac{15 - x}{2} \\ y \le 18 - 2x \\ x \ge 0, y \ge 0 \end{cases}$$

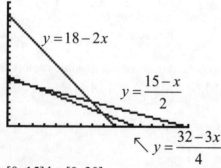

[0, 15] by [0, 20]

To determine the solution region, pick a point to test. Pick $(1,1)$. When substituted

into the inequalities that form the system, the point $(1,1)$ creates true statements in all cases.

Since all the inequalities that form the system are true at the point $(1,1)$, the region that contains $(1,1)$ is the solution region. The solution represents the feasible region. The graph of the feasible region is

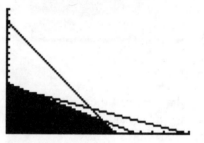

[0, 15] by [0, 20]

Note that since all the inequalities contain an "equal to", all the boundary lines are solid.

The corner points of the feasible region are $(0,0), (0,7.5), (9,0), (2,6.5)$, and $(8,2)$.

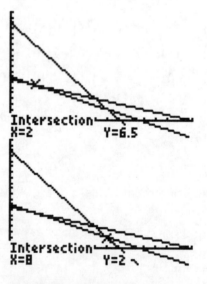

Testing the corner points of the feasible region to maximize the objective function, $f = 80x + 160y$, yields

At $(0,0)$: $f = 80(0)+160(0) = 0$

At $(0,7.5)$: $f = 80(0)+160(7.5) = 1200$

At $(9,0)$: $f = 80(9)+160(0) = 720$

At $(2,6.5)$: $f = 80(2)+160(6.5) = 1200$

At $(8,2)$: $f = 80(8)+160(2) = 960$

The maximum value is 1200 occurring at $(0,7.5)$ and $(2,6.5)$. Since the maximum occurs at two corner points, the maximum also occurs at all points along the line segment connecting $(0,7.5)$ and $(2,6.5)$.

13. Rewriting the system of constraints and graphing the system yields

$$\begin{cases} y \geq 6-2x \\ y \geq 8-4x \\ y \geq \dfrac{6-x}{2} \\ x \geq 0, y \geq 0 \end{cases}$$

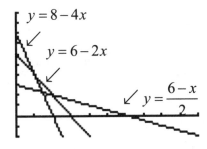

[0, 10] by [–2, 10]

To determine the solution region, pick a point to test. Pick $(2,4)$. When substituted into the inequalities that form the system, the point $(2,4)$ creates true statements in all cases.

Since all the inequalities that form the system are true at the point $(2,4)$, the region that contains $(2,4)$ is the solution region.

The solution represents the feasible region. The graph of the feasible region is

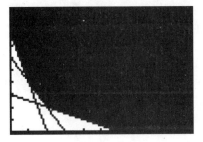

[0, 10] by [0, 10]

Note that since all the inequalities contain an "equal to", all the boundary lines are solid.

The corner points of the feasible region are $(0,8),(6,0),(1,4)$ and $(2,2)$.

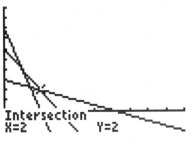

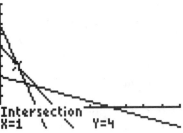

Testing the corner points of the feasible region to minimize the objective function, $g = 40x+30y$, yields

At $(0,8)$: $g = 40(0)+30(8) = 240$

At $(6,0)$: $g = 40(6)+30(0) = 240$

At $(1,4)$: $g = 40(1)+30(4) = 160$

At $(2,2)$: $g = 40(2)+30(2) = 140$

The minimum value is 140 occurring at $(2,2)$.

15. Rewriting the system of constraints and graphing the system yields

$$\begin{cases} y \ge 6 - 3x \\ y \ge 4 - x \\ y \le \dfrac{8-x}{5} \\ x \ge 0, y \ge 0 \end{cases}$$

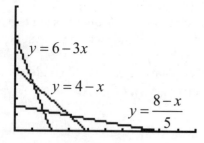

[0, 5] by [0, 8]

To determine the solution region, pick a point to test. Pick $\left(4, \dfrac{1}{2}\right)$. When substituted into the inequalities that form the system, the point $\left(4, \dfrac{1}{2}\right)$ creates true statements in all cases.

Since all the inequalities that form the system are true at the point $\left(4, \dfrac{1}{2}\right)$, the region that contains $\left(4, \dfrac{1}{2}\right)$ is the solution region. The solution represents the feasible region. The graph of the feasible region is

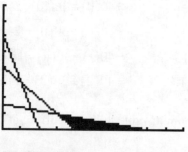

[0, 5] by [0, 8]

Note that since all the inequalities contain an "equal to", all the boundary lines are solid.

The corner points of the feasible region are $(4, 0), (8, 0),$ and $(3, 1)$.

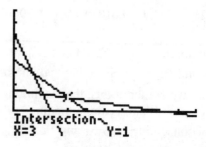

Testing the corner points of the feasible region to minimize the objective function, $g = 46x + 23y$, yields

At $(4, 0)$: $g = 46(4) + 23(0) = 184$
At $(8, 0)$: $g = 46(8) + 23(0) = 368$
At $(3, 1)$: $g = 46(3) + 23(1) = 161$

The minimum value is 161 occurring at $(3, 1)$.

17. Rewriting the system of constraints and graphing the system yields

$$\begin{cases} y \ge \dfrac{12 - 3x}{2} \\ y \ge 7 - 2x \\ x \ge 0, y \ge 0 \end{cases}$$

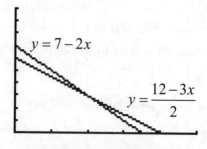

[0, 5] by [0, 10]

To determine the solution region, pick a point to test. Pick $(1,10)$. When substituted into the inequalities that form the system, the point $(1,10)$ creates true statements in all cases.

Since all the inequalities that form the system are true at the point $(1,10)$, the region that contains $(1,10)$ is the solution region. The solution represents the feasible region. The graph of the feasible region is

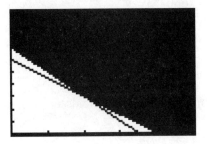

[0, 5] by [0, 10]

Note that since all the inequalities contain an "equal to", all the boundary lines are solid.

The corner points of the feasible region are $(0,7),(4,0),$ and $(2,3)$.

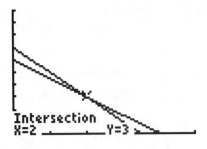

Testing the corner points of the feasible region to minimize the objective function, $g = 60x + 10y$, yields

At $(0,7)$: $\qquad g = 60(0)+10(7)=70$

At $(4,0)$: $\qquad g = 60(4)+10(0)=240$

At $(2,3)$: $\qquad g = 60(2)+10(3)=150$

The minimum value is 70 occurring at $(0,7)$.

Section 6.2 Exercises

19. See the solution to Section 6.1, question 29. The corner points of the feasible region are $(390,390),(780,0)$ and $(975,0)$.

Testing the corner points of the feasible region to maximize the objective function, $f = 32x + 45y$, yields

At $(390,390)$: $f = 32(390)+45(390)$
$$= \$30,030$$

At $(780,0)$: $f = 32(780)+45(0)$
$$= \$24,960$$

At $(975,0)$: $f = 32(975)+45(0)$
$$= \$31,200$$

The maximum is \$31,200 occurring at $(975,0)$. To maximize profit, the company needs to manufacture 975 Turbo models and 0 Tornado models.

21. See the solution to Section 6.1, question 36. The corner points of the feasible region are $(0,0),(9,6),(12,0)$ and $(0,12)$.

Testing the corner points of the feasible region to maximize the objective function, $f = 24x + 30y$, yields

At $(0,0)$: $f = 24(0)+30(0) = \$0$

At $(9,6)$: $f = 24(9)+30(6) = \$396$

At $(12,0)$: $f = 24(12)+30(0) = \$288$

At $(0,12)$: $f = 24(0)+30(12) = \$360$

The maximum is \$396 occurring at $(9,6)$.

To maximize profit, the company needs to sell 9 Safecut chainsaws and 6 Deluxe chainsaws.

23. a. Determine the feasible region by solving and graphing the system of inequalities that represent the constraints. Let x represent the number of cable television commercials, and let y represent the number of radio commercials.

$$\begin{cases} 240x + 150y \le 36,000 \\ \dfrac{1}{4}x + \dfrac{1}{10}y \ge 33 \\ x \ge 0, y \ge 0 \end{cases}$$

Rewriting yields
$$\begin{cases} y \le \dfrac{36,000 - 240x}{150} \\ y \ge \dfrac{660 - 5x}{2} \\ x \ge 0, y \ge 0 \end{cases}$$

Graphing the system yields

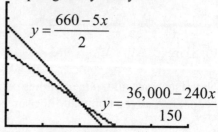

$[0, 250]$ by $[0, 400]$

To determine the solution region, pick a point to test. Pick $(125,30)$. When substituted into the inequalities that form the system, the point $(125,30)$ creates true statements in all cases.

Since all the inequalities that form the system are true at the point $(125,30)$, the region that contains $(125,30)$ is the solution region. The solution represents the feasible region. The graph of the feasible region is

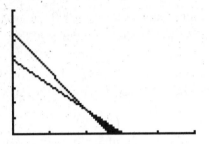

[0, 250] by [0, 400]

Note that since all the inequalities contain an "equal to", all the boundary lines are solid.

The corner points of the feasible region are $(150,0),(132,0),$ and $(100,80).$

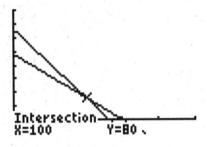

Intersection
X=100   Y=80

Testing the corner points of the feasible region to maximize the objective function, $f = 500x + 550y$, yields

At $(132,0)$: $f = 500(132) + 550(0)$
$= \$66,000$

At $(150,0)$: $f = 500(150) + 550(0)$
$= \$75,000$

At $(100,80)$: $f = 500(100) + 550(80)$
$= \$94,000$

The maximum value is \$94,000 occurring at $(100,80).$ To produce a maximum profit the company needs to buy 100 minutes on cable television and 80 minutes on radio.

b. See part a) above. The maximum value is \$94,000.

25. Determine the feasible region by solving and graphing the system of inequalities that represent the constraints. Let x represent the number of assembly line 1 days, and let y represent the number assembly line 2 days.

$$\begin{cases} 80x + 40y \geq 3200 \\ 20x + 20y \geq 1000 \\ 100x + 40y \geq 3400 \\ x \geq 0, y \geq 0 \end{cases}$$

Rewriting yields
$$\begin{cases} y \geq \dfrac{3200 - 80x}{40} \\ y \geq \dfrac{1000 - 20x}{20} \\ y \geq \dfrac{3400 - 100x}{40} \\ x \geq 0, y \geq 0 \end{cases}$$

Graphing the system yields

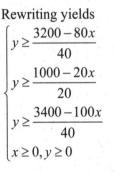

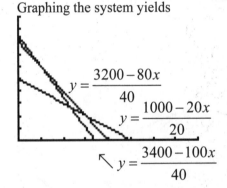

$y = \dfrac{3200 - 80x}{40}$

$y = \dfrac{1000 - 20x}{20}$

$y = \dfrac{3400 - 100x}{40}$

[0, 80] by [0, 100]

To determine the solution region, pick a point to test. Pick $(25,40).$ When substituted into the inequalities that form the system, the point $(25,40)$ creates true statements in all cases.

Since all the inequalities that form the system are true at the point $(25,40)$, the region that contains $(25,40)$ is the solution region. The solution represents the feasible region. The graph of the feasible region is

[0, 80] by [0, 100]

Note that since all the inequalities contain an "equal to", all the boundary lines are solid. The corner points of the feasible region are $(0,85),(50,0),(30,20)$ and $(10,60)$.

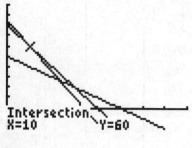

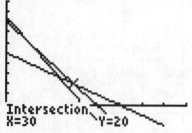

Testing the corner points of the feasible region to minimize the objective function, $g = 20,000x + 40,000y$, yields

At $(0,85)$: $g = 20,000(0) + 40,000(85)$
$= \$3,400,000$

At $(50,0)$: $g = 20,000(50) + 40,000(0)$
$= \$1,000,000$

At $(10,60)$: $g = 20,000(10) + 40,000(60)$
$= \$2,600,000$

At $(30,20)$: $g = 20,000(30) + 40,000(20)$
$= \$1,400,000$

The minimum value is \$1,000,000 occurring at $(50,0)$. To minimize the cost, the company needs to manufacture the television sets on assembly line 1 for 50 days and assembly line 2 for zero days.

27. Determine the feasible region by solving and graphing the system of inequalities that represent the constraints. Let x represent the number of Van Buren models, and let y represent the number of Jefferson models.

$$\begin{cases} 200x + 500y \le 5000 \\ 240,000x + 300,000y \le 3,600,000 \\ x \ge 0, y \ge 0 \end{cases}$$

Rewriting yields
$$\begin{cases} y \le \dfrac{50 - 2x}{5} \\ y \le \dfrac{360 - 24x}{30} \\ x \ge 0, y \ge 0 \end{cases}$$

Graphing the system yields

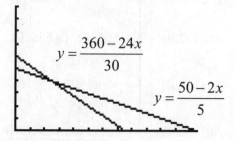

[0, 25] by [0, 20]

To determine the solution region, pick a point to test. Pick $(1,1)$. When substituted into the inequalities that form the system, the point $(1,1)$ creates true statements in all cases.

Since all the inequalities that form the system are true at the point $(1,1)$, the region that contains $(1,1)$ is the solution region.

The solution represents the feasible region. The graph of the feasible region is

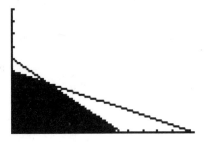

$[0, 25]$ by $[0, 20]$

Note that since all the inequalities contain an "equal to", all the boundary lines are solid.

The corner points of the feasible region are $(0,0),(15,0),(0,10),$ and $(5,8)$.

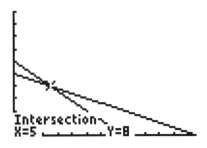

Testing the corner points of the feasible region to maximize the objective function, $f = 60,000x + 75,000y$, yields

At $(0,0)$: $f = 60,000(0) + 75,000(0)$
$= \$0$
At $(15,0)$: $f = 60,000(15) + 75,000(0)$
$= \$900,000$
At $(0,10)$: $f = 60,000(0) + 75,000(10)$
$= \$750,000$
At $(5,8)$: $f = 60,000(5) + 75,000(8)$
$= \$900,000$

The maximum profit is \$900,000 occurring at $(15,0)$ and $(5,8)$. Since there are two corner points that maximize the objective function, any point along the line segment

connecting $(15,0)$ and $(5,8)$ also maximizes the objective function. Therefore, the contractor can build 15 Van Buren models and 0 Jefferson models, 5 Van Buren models and 8 Jefferson models, or any other whole number combination that lies along the line segment connecting $(15,0)$ and $(5,8)$. In particular the possible solutions are $(10,4),(15,0)$ and $(5,8)$.

29. Determine the feasible region by solving and graphing the system of inequalities that represent the constraints. Let x represent the number of weeks operating facility 1, and let y represent the number of weeks operating facility 2.

$$\begin{cases} 400x + 400y \geq 4000 \\ 300x + 100y \geq 1800 \\ 200x + 400y \geq 2400 \\ x \geq 0, y \geq 0 \end{cases}$$

Rewriting yields
$$\begin{cases} y \geq 10 - x \\ y \geq 18 - 3x \\ y \geq \dfrac{12 - x}{2} \\ x \geq 0, y \geq 0 \end{cases}$$

Graphing the system yields

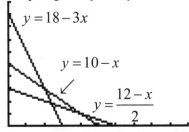

$[0, 20]$ by $[0, 20]$

To determine the solution region, pick a point to test. Pick $(10,5)$. When substituted into the inequalities that form the system, the point $(10,5)$ creates true statements in all cases.

Since all the inequalities that form the system are true at the point $(10,5)$, the region that contains $(10,5)$ is the solution region. The solution represents the feasible region. The graph of the feasible region is

[0, 20] by [0, 20]

Note that since all the inequalities contain an "equal to", all the boundary lines are solid.

The corner points of the feasible region are $(0,18),(12,0),(4,6)$ and $(8,2)$.

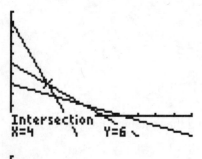

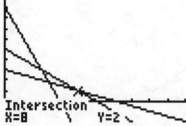

Testing the corner points of the feasible region to minimize the objective function, $g = 15,000x + 20,000y$, yields

At $(0,18)$: $g = 15,000(0) + 20,000(18)$
$= \$360,000$

At $(12,0)$: $g = 15,000(12) + 20,000(0)$
$= \$180,000$

At $(4,6)$: $g = 15,000(4) + 20,000(6)$
$= \$180,000$

At $(8,2)$: $g = 15,000(8) + 20,000(2)$
$= \$160,000$

The minimum value is \$160,000 occurring at $(8,2)$. Operating facility 1 for 8 weeks and facility 2 for 2 weeks yields a minimum cost of \$160,000 for filling the orders.

31. Determine the feasible region by solving and graphing the system of inequalities that represent the constraints. Let x represent the number of servings of Diet A, and let y represent the number of servings of Diet B.

$$\begin{cases} 2x + 1y \geq 18 \\ 4x + 1y \geq 26 \\ x \geq 0, y \geq 0 \end{cases}$$

Rewriting yields
$$\begin{cases} y \geq 18 - 2x \\ y \geq 26 - 4x \\ x \geq 0, y \geq 0 \end{cases}$$

Graphing the system yields

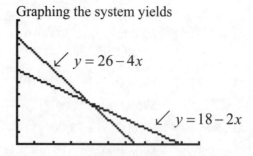

[0, 10] by [0, 30]

To determine the solution region, pick a point to test. Pick $(5,10)$. When substituted into the inequalities that form the system, the point $(5,10)$ creates true statements in all cases.

Since all the inequalities that form the system are true at the point $(5,10)$, the region that contains $(5,10)$ is the solution region. The solution represents the feasible region. The graph of the feasible region is

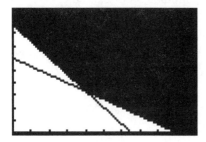

[0, 10] by [0, 30]

Note that since all the inequalities contain an "equal to", all the boundary lines are solid.

The corner points of the feasible region are $(0,26), (9,0),$ and $(4,10)$.

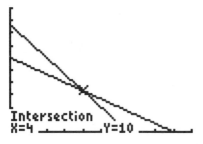

Testing the corner points of the feasible region to minimize the objective function, $g = 0.09x + 0.035y$, yields

At $(0,26): g = 0.09(0) + 0.035(26) = 0.91$
At $(9,0): \quad g = 0.09(9) + 0.035(0) = 0.81$
At $(4,10): g = 0.09(4) + 0.035(10) = 0.71$

The minimum value is 0.71 occurring at $(4,10)$. Four servings of Diet A and ten servings of Diet B yields the minimum amount of detrimental substance, 0.71 oz.

Section 6.3 Skills Check

1. $f(1) = 2(1) + 3 = 5$
    $f(2) = 2(2) + 3 = 7$
    $f(3) = 2(3) + 3 = 9$
    $f(4) = 2(4) + 3 = 11$
    $f(5) = 2(5) + 3 = 13$
    $f(6) = 2(6) + 3 = 15$
    The first six terms are 5, 7, 9, 11, 13, and 15.

3. $a_1 = \dfrac{10}{1} = 10$
    $a_2 = \dfrac{10}{2} = 5$
    $a_3 = \dfrac{10}{3}$
    $a_4 = \dfrac{10}{4} = \dfrac{5}{2}$
    $a_5 = \dfrac{10}{5} = 2$

    The first five terms are $10, 5, \dfrac{10}{3}, \dfrac{5}{2},$ and 2.

5. To move from one term in the sequence to the next term, add two. Therefore the next three terms are 9, 11, and 13.

7. $a_n = a_1 + (n-1)d$
    $a_n = -3 + (n-1)(4)$
    $a_n = -3 + 4n - 4$
    $a_n = 4n - 7$
    $a_8 = 4(8) - 7 = 25$

9. To move from one term in the sequence to the next term, multiply by 2. Therefore the next four terms are 24, 48, 96, and 192.

11. $a_n = a_1 r^{n-1}$
    $a_n = 10(3)^{n-1}$
    $a_6 = 10(3)^{6-1} = 10(3)^5 = 2430$

13. $a_n = a_{n-1} - 2$ implies that the next term in the sequence is the previous term in the sequence minus 2. For example, $a_2 = a_{2-1} - 2 = a_1 - 2$. The second term is the first term minus 2. Therefore, if the first term is 5, then the first four terms will be 5, 3, 1, and -1.

15. $a_n = 2a_{n-1} + 3$
    $a_1 = 2$
    $a_2 = 2a_{2-1} + 3 = 2a_1 + 3 = 2(2) + 3 = 7$
    $a_3 = 2a_{3-1} + 3 = 2a_2 + 3 = 2(7) + 3 = 17$
    $a_4 = 2a_{4-1} + 3 = 2a_3 + 3 = 2(17) + 3 = 37$
    The first four terms are 2, 7, 17, and 37.

Section 6.3 Exercises

17. Let a_1 represent the starting salary

$$a_n = a_1 + (n-1)d$$
$$a_n = a_1 + (n-1)1500$$
$$a_8 = a_1 + (8-1)1500$$
$$a_8 = a_1 + (7)1500 = a_1 + 10,500$$

The salary would increase by \$10,500.

19. a. $f(n) = 300 + 60n$

b. $f(1) = 300 + 60(1) = 360$
$f(2) = 300 + 60(2) = 420$
$f(3) = 300 + 60(3) = 480$
$f(4) = 300 + 60(4) = 540$
$f(5) = 300 + 60(5) = 600$
$f(6) = 300 + 60(6) = 660$

The first six terms are 360, 420, 480, 540, 600, and 660.

21. a. Job 1
$$a_n = a_1 + (n-1)d$$
$$a_n = 40,000 + (n-1)2000$$
$$a_n = 2000n + 38,000$$
Job 2
$$a_n = a_1 + (n-1)d$$
$$a_n = 36,000 + (n-1)2400$$
$$a_n = 2400n + 33,600$$
After 5 years, $n = 6$.
Job 1:
$$a_6 = 2000(6) + 38,000 = 50,000$$
Job 2:
$$a_6 = 2400(6) + 33,600 = 48,000$$
After 5 years, Job 1 pays \$2000 more than Job 2.

b. After 10 years, $n = 11$.
Job 1:
$$a_{11} = 2000(11) + 38,000 = 60,000$$
Job 2:
$$a_{11} = 2400(11) + 33,600 = 60,000$$
The salaries are the same.

c. To decide between the jobs on the basis of salary requires an analysis of the length of time you plan to stay in the job. If you plan to stay in the job less than 10 years, Job 1 is clearly better.

23. a. Year 1: $S = P + Prt$
$$S = 1000 + 1000(5\%)(1)$$
$$S = 1000 + 1000(0.05)(1)$$
$$S = 1000 + 50$$
$$S = 1050$$
Year 2: $S = P + Prt$
$$S = 1050 + 1050(5\%)(1)$$
$$S = 1050 + 1050(0.05)(1)$$
$$S = 1050 + 52.5$$
$$S = 1102.50$$
Year 3: $S = P + Prt$
$$S = 1102.50 + 1102.50(5\%)(1)$$
$$S = 1102.50 + 1102.50(0.05)(1)$$
$$S = 1102.50 + 55.13$$
$$S = 1157.63$$
Year 4: $S = P + Prt$
$$S = 1157.63 + 1157.63(5\%)(1)$$
$$S = 1157.63 + 1157.63(0.05)(1)$$
$$S = 1157.63 + 57.88$$
$$S = 1215.51$$
The sequence is 1050, 1102.50, 1157.63, and 1215.51.

25. a. $20\%(50{,}000)=(0.20)(50{,}000)$

$=10{,}000$

$50{,}000-3(10{,}000)=\$20{,}000$

The value after 3 years is $20,000.

b. $a_n = a_1 + (n-1)d$

$a_n = 40{,}000 + (n-1)(-10{,}000)$

$a_n = 40{,}000 - 10{,}000n + 10{,}000$

$a_n = 50{,}000 - 10{,}000n$

Note that n represents the number of years of depreciation.

c.   The value of the car depreciates as follows: 50,000, 40,000, 30,000, 20,000, 10,000, and 0.

27. $a_n = a_1 r^{n-1}$

$a_n = 5000(2)^{n-1}$

Since $n=1$ represents zero hours having passed, then after six hours $n=7$.

$a_7 = 5000(2)^{7-1}$

$a_7 = 5000(2)^6 = 320{,}000$

After 6 hours the number of bacteria in the culture is 320,000.

29. $s(n)=128\left(\dfrac{1}{4}\right)^n$, where $n=$ the number

of bounces, and $s=$ the height of the bounce.

$s(4)=128\left(\dfrac{1}{4}\right)^4=\dfrac{1}{2}$

The height after the fourth bounce is $\dfrac{1}{2}$ foot.

31. If a company loses 2% of it's profit, then 98% of it's profit remains. Therefore,

$f(n)=8{,}000{,}000(0.98)^n$, where

$n=$ the number of years and $f(n)=$ the company's profit.

$f(5)=8{,}000{,}000(0.98)^5$

$f(5)=7{,}231{,}366.37$

After five years the company's projected profit is $7,231,366.37.

33. $a_n = a_1 + (n-1)d$

$a_n = 54{,}000 + (n-1)(3600)$

$a_n = 54{,}000 + 3600n - 3600$

$a_n = 50{,}400 + 3600n$

X	Y1
11	90000
12	93600
13	97200
14	100800
15	104400
16	108000
17	111600

X=16

When $n=16$ or after 15 years, the salary is twice the original amount of $54,000. Therefore the salary doubles after 15 years.

35. $A = P\left(1+\dfrac{r}{n}\right)^{nt}$

$A = 10{,}000\left(1+\dfrac{0.08}{365}\right)^{365(10)}$

$A = 10{,}000(1.000219178)^{3650}$

$A = 22{,}253.46$

The future value of $10,000 compounded daily at 8% is $22,253.46.

37. The next number in the sequence is the sum of the previous two numbers in the sequence. Therefore, the next four numbers in the sequence are 13, 21, 34, and 55.

39. Making a payment of 1% interest plus 10% of the balance at the beginning of the month implies that she will make a payment of 11% of balance each month. However, the balance will decrease by 10%, since 1% of the payment is interest.

Month 1

Payment:
$$10{,}000(0.11) = 1100$$

Balance after

Payment:
$$10{,}000(1.01) - 1100 = 9000$$

Month 2

Payment:
$$9000(0.11) = 990$$

Balance after

Payment:
$$9000(1.01) - 990 = 8100$$

Month 3

Payment:
$$8100(0.11) = 891$$

Balance after

Payment:
$$8100(1.01) - 891 = 7290$$

Month 4

Payment:
$$7290(0.11) = 801.90$$

The sequence of payments is $1100, $990, $891, and $801.90.

Section 6.4 Skills Check

1. Note that $r = \dfrac{3}{9} = \dfrac{1}{3}$.

$$s_n = \frac{a_1\left(1-r^n\right)}{1-r}$$

$$s_6 = \frac{9\left(1-\left(\dfrac{1}{3}\right)^6\right)}{1-\left(\dfrac{1}{3}\right)} = \frac{364}{27}$$

3. Note that $d = 10-7 = 13-10 = 3$.

$$a_n = a_1 + (n-1)d$$
$$a_n = 7 + (n-1)(3)$$
$$a_n = 3n+4$$
$$a_{10} = 3(10)+4 = 34$$
$$s_n = \frac{n(a_1 + a_n)}{2}$$
$$s_{10} = \frac{10(a_1 + a_{10})}{2} = \frac{10(7+34)}{2} = 205$$

5. $a_n = a_1 + (n-1)d$
$$a_n = -4 + (n-1)(2)$$
$$a_n = 2n-6$$
$$a_{15} = 2(15)-6 = 24$$
$$s_n = \frac{n(a_1 + a_n)}{2}$$
$$s_{15} = \frac{15(a_1 + a_{15})}{2} = \frac{15(-4+24)}{2} = 150$$

7. $s_n = \dfrac{a_1\left(1-r^n\right)}{1-r}$

$$s_{15} = \frac{3\left(1-(2)^{15}\right)}{1-(2)} = 98{,}301$$

9. Note that $r = \dfrac{10}{5} = \dfrac{20}{10} = 2$.

$$s_n = \frac{a_1\left(1-r^n\right)}{1-r}$$

$$s_{10} = \frac{5\left(1-(2)^{10}\right)}{1-(2)} = 5115$$

11. Note that $r = \dfrac{9}{81} = \dfrac{1}{9} = \dfrac{1}{9}$.

$$S = \frac{a_1}{1-r}$$

$$S = \frac{81}{1-\left(\dfrac{1}{9}\right)} = \frac{729}{8}$$

13. $\displaystyle\sum_{i=1}^{6} 2^i = 2^1 + 2^2 + 2^3 + 2^4 + 2^5 + 2^6$
$$= 2+4+8+16+32+64 = 126$$

15. $\displaystyle\sum_{i=1}^{4}\left(\frac{1+i}{i}\right) = 2 + \frac{3}{2} + \frac{4}{3} + \frac{5}{4} = \frac{73}{12}$

17. $S = \dfrac{a_1}{1-r} = \dfrac{\dfrac{3}{4}}{1-\dfrac{3}{4}} = 3$

19. Since $r = \dfrac{4}{3} \geq 1$, the infinite sum does not exist. No solution.

Section 6.4 Exercises

21. Note that $d = 400$ and $a_1 = -2000$.

$$a_n = a_1 + (n-1)d$$
$$a_n = -2000 + (n-1)(400)$$
$$a_n = 400n - 2400$$
$$a_{12} = 400(12) - 2400 = 2400$$
$$s_n = \frac{n(a_1 + a_n)}{2}$$
$$s_{12} = \frac{12(a_1 + a_{12})}{2}$$
$$= \frac{12(-2000 + 2400)}{2} = 2400$$

The profit for the year is $2400.

23. $1 + 2 + 3 + 5 = 11$. A male bee has 11 ancestors through four generations.

Job 1 produces $21,600 more income over a 12-year period.

c. The length of time a person stays in the job determines which job is best from a salary point of view. For the first 5 or 12 years, Job 1 produces more income. However, since the raises for Job 1 are greater, eventually the income generated from Job 2 will exceed the income generated from Job 1.

25. a.
$$a_n = a_1 + (n-1)d$$
$$a_n = 1 + (n-1)(1)$$
$$a_n = n$$
$$a_{12} = 12$$
$$s_n = \frac{n(a_1 + a_n)}{2}$$
$$s_{12} = \frac{12(1+12)}{2} = 78$$

The clock chimes 78 times in a 12-hour period.

b. In a 24-hour period, the clock chimes twice as many times as in a 12-hour period. Therefore, the clock chimes 156 times every 24 hours.

27. Note that the sequence is geometric with a common ratio of 1.10. Therefore,

$$s_n = \frac{a_1(1 - r^n)}{1 - r}$$
$$s_{12} = \frac{2000\left(1 - (1.10)^{12}\right)}{1 - (1.10)} \approx 42{,}768.57$$

The total profit over the first 12 months is $42,768.57.

29. If the pump removes $\frac{1}{3}$ of the water with each stroke, then $\frac{2}{3}$ of the water remains after each stroke. Therefore, a geometric sequence can be created with a common ratio of $\frac{2}{3}$.

$$a_1 = 81\left(\frac{2}{3}\right) = 54$$
$$a_n = a_1 r^{n-1}$$
$$a_n = 54\left(\frac{2}{3}\right)^{n-1}$$
$$a_4 = 54\left(\frac{2}{3}\right)^{4-1} = 54\left(\frac{2}{3}\right)^3 = 16\,\text{cm}^3$$

The amount of water in the container after four strokes is 16 cm^3.

31. a. 5

b. $5 \cdot 5 = 25$

c. $5^3 = 125$
$$5^4 = 625$$

d. The sequence is geometric with a common ratio of 5.

33. Since 10% of the balance is paid each month, then 90% of the balance remains. Note that the situation is modeled by a geometric series with a common ratio of 0.90.

$$a_n = 10,000(0.90)^n$$
$$a_{12} = 10,000(0.90)^{12} = 2824.30$$

35. Note that the situation is modeled by a geometric function having a common ratio of $\frac{1}{4}$.

$$a_1 = 128\left(\frac{1}{4}\right) = 32$$

$$a_n = 32\left(\frac{1}{4}\right)^{n-1}$$, where n represents the number of bounces and a_n represents the height after that bounce. Note that when the ball hits the ground for the fifth time, it has bounced four times.

$$s_n = \frac{a_1(1-r^n)}{1-r}$$

$$s_4 = \frac{32\left(1-\left(\frac{1}{4}\right)^4\right)}{1-\left(\frac{1}{4}\right)}$$

$$s_4 = 42.5$$

Note that s_4 is the sum of the rebound heights. To calculate the distance the ball actually travels, add in the initial distance before the first rebound of 128 feet and double s_4 to take into consideration that the ball rises and falls with each bounce. Therefore the total distance traveled is $128 + 2(42.5) = 213$ feet.

37. a. Note that if the car loses 16% of its value each year, then it retains 84% of its value. The value of the car is modeled by a geometric sequence with a common ratio of 0.84.

The value of the car after n years is given by $a_n = 35,000(0.84)^n$. The amount of depreciation after n years is given by the original value of the car less the depreciated value. Therefore,

$$s_n = 35,000 - 35,000(0.84)^n$$
$$s_n = 35,000\left(1-(0.84)^n\right)$$

b. $a_n = 35,000(0.84)^n$

39. $s_n = \frac{a_1(1-r^n)}{1-r}$

$$s_{96} = \frac{100\left(1-\left(1+\frac{0.12}{12}\right)^{96}\right)}{1-\left(1+\frac{0.12}{12}\right)}$$

$$s_{96} = \frac{100\left(1-(1.01)^{96}\right)}{1-(1.01)}$$

$$s_{96} \approx 15,992.73$$

The future value of the annuity is $15,992.73.

41. $s_n = \dfrac{a_1\left(1-r^n\right)}{1-r}$

$s_n = \dfrac{\left(R(1+i)^{-n}\right)\left(1-(1+i)^n\right)}{1-(1+i)}$

$s_n = \dfrac{R(1+i)^{-n}\left(1-(1+i)^n\right)}{1-1-i}$

$s_n = \dfrac{R\left[(1+i)^{-n}(1)-(1+i)^{-n}(1+i)^n\right]}{-i}$

$s_n = \dfrac{R\left[(1+i)^{-n}-(1+i)^{-n+n}\right]}{-i}$

$s_n = \dfrac{R\left[(1+i)^{-n}-(1+i)^{0}\right]}{-i}$

$s_n = \dfrac{R\left[(1+i)^{-n}-1\right]}{-i}$

$s_n = R\left[\dfrac{1-(1+i)^{-n}}{i}\right]$

Section 6.5 Skills Check

1. $f'(x) = (4x^4 + 6x^2 + 2x) +$
$\qquad (6x^4 + 3x^2 + 24x^2 + 12)$
$\quad f'(x) = 10x^4 + 33x^2 + 2x + 12$

3. $f'(x) = \dfrac{3x^4 - (2x^4 - 6x)}{x^4}$

$\quad f'(x) = \dfrac{x^4 + 6x}{x^4} = \dfrac{x(x^3 + 6)}{x^4} = \dfrac{x^3 + 6}{x^3}$

5. $f'(2) = 4(2)^3 - 3(2)^2 + 4(2) - 2 = 26$

7. $f'(3) = \dfrac{\left((3)^2 - 1\right)(2) + (2(3))(2(3))}{\left((3)^2 - 1\right)^2}$

$\qquad = \dfrac{(8)(2) + (6)(6)}{(8)^2}$

$\qquad = \dfrac{16 + 36}{64}$

$\qquad = \dfrac{52}{64} = \dfrac{13}{16}$

9. $y - y_1 = m(x - x_1)$
$\quad y - 8 = 12(x - 2)$
$\quad y - 8 = 12x - 24$
$\quad y = 12x - 16$

11. a. $f(x + h) = 4(x + h) + 5$

b. $f(x + h) - f(x)$
$\quad = \left[4(x + h) + 5\right] - \left[4x + 5\right]$
$\quad = 4x + 4h + 5 - 4x - 5$
$\quad = 4h$

c. $\dfrac{f(x + h) - f(x)}{h}$

$\quad = \dfrac{4h}{h}$

$\quad = 4$

13. $8x^2 + 4y = 12$
$\qquad 4y = 12 - 8x^2$
$\qquad y = \dfrac{12 - 8x^2}{4}$
$\qquad y = \dfrac{12}{4} - \dfrac{8x^2}{4}$
$\qquad y = 3 - 2x^2$

15. $9x^3 + 5y = 18$
$\qquad 5y = 18 - 9x^3$
$\qquad y = \dfrac{18 - 9x^3}{5}$

17. a. $0 = 2x - 2$
$\qquad 2x = 2$
$\qquad x = 1$

b. Recall that the x-coordinate of the vertex is given by $\dfrac{-b}{2a}$.

$\dfrac{-b}{2a} = \dfrac{-(-2)}{2(1)} = \dfrac{2}{2} = 1$

The solution to part a) and the x-coordinate of the vertex are equal.

c. $y = (1)^2 - 2(1) + 5 = 4$
$\quad (1, 4)$

19. a. $0 = 12x - 24$
$\qquad 12x = 24$
$\qquad x = 2$

b. Recall that the x-coordinate of the vertex is given by $\dfrac{-b}{2a}$.

$\dfrac{-b}{2a} = \dfrac{-(-24)}{2(6)} = \dfrac{24}{12} = 2$

The solution to part a) and the x-coordinate of the vertex are equal.

c. $y = 6(2)^2 - 24(2) + 15$

$y = 24 - 48 + 15 = -9$

$(2, -9)$

21. a. $3x^2 - 18x - 48 = 0$

$3(x^2 - 6x - 16) = 0$

$3(x - 8)(x + 2) = 0$

$x = 8,\ x = -2$

b. $x = 8$

$y = (8)^3 - 9(8)^2 - 48(8) + 15$

$y = -433$

$(8, -433)$

$x = -2$

$y = (-2)^3 - 9(-2)^2 - 48(-2) + 15$

$y = 67$

$(-2, 67)$

c. $x = -2$ produces the maximum.

23. $y = 3\sqrt{x} = 3\sqrt[2]{x^1} = 3x^{\frac{1}{2}}$

25. $y = 2\sqrt[3]{x^2} = 2x^{\frac{2}{3}}$

27. $y = \sqrt[3]{x^2 + 1} = (x^2 + 1)^{\frac{1}{3}}$

29. $y = \sqrt[3]{(x^3 - 2)^2} = (x^3 - 2)^{\frac{2}{3}}$

31. $y = \sqrt{x} + \sqrt[3]{2x}$

$y = x^{\frac{1}{2}} + (2x)^{\frac{1}{3}}$

33. $y^2 + 4x - 3 = 0$

$y^2 = 3 - 4x$

$\sqrt{y^2} = \pm\sqrt{3 - 4x}$

$y = \pm\sqrt{3 - 4x}$

35. $y^2 + y - 6x = 0$

$a = 1,\ b = 1,\ c = -6x$

$y = \dfrac{-b \pm \sqrt{b^2 - 4ac}}{2a}$

$y = \dfrac{-1 \pm \sqrt{(1)^2 - 4(1)(-6x)}}{2(1)}$

$y = \dfrac{-1 \pm \sqrt{1 + 24x}}{2}$

37. $y = u^2$

$u = 4x^3 + 5$

39. $u = x^3 + x$

$y = \sqrt{x^3 + x}$

$y = \sqrt{u} = u^{\frac{1}{2}}$

41. $\dfrac{3}{2}u^{\frac{1}{2}} \cdot 2x$

$= \dfrac{3}{2}(x^2 - 1)^{\frac{1}{2}} \cdot 2x$

$= 3x\sqrt{x^2 - 1}$

43. $y = 3x^{-1} - 4x^{-2} - 6$

45. $y = \left(x^2 - 3\right)^{-3}$

47. $f'(x) = -6\left(\dfrac{1}{x^4}\right) + 4\left(\dfrac{1}{x^2}\right) + \dfrac{1}{x}$

$f'(x) = \dfrac{-6}{x^4} + \dfrac{4}{x^2} + \dfrac{1}{x}$

49. $f'(x) = \left(4x^2 - 3\right)^{-\frac{1}{2}}(8x)$

$f'(x) = \dfrac{8x}{\left(4x^2 - 3\right)^{\frac{1}{2}}}$

$f'(x) = \dfrac{8x}{\sqrt{4x^2 - 3}}$

51. $\log\left[x^3 \left(3x - 4\right)^5 \right]$

$\log\left(x^3\right) + \log\left[\left(3x - 4\right)^5\right]$

$3\log x + 5\log\left(3x - 4\right)$

53. $f'(x) = 0$

$3x^2 - 3x = 0$

$3x(x - 1) = 0$

$3x = 0, \ x - 1 = 0$

$x = 0, \ x = 1$

55. $f'(x) = 0$

$\left(x^2 - 4\right)3\left(x - 3\right)^2 + \left(x - 3\right)^3(2x) = 0$

$\left(x - 3\right)^2 \left[3\left(x^2 - 4\right) + \left(x - 3\right)(2x) \right] = 0$

$\left(x - 3\right)^2 \left[3x^2 - 12 + 2x^2 - 6x \right] = 0$

$\left(x - 3\right)^2 \left[5x^2 - 6x - 12 \right] = 0$

$\left(x - 3\right)^2 = 0$ implies $x = 3$

$5x^2 - 6x - 12 = 0$

$a = 5, \ b = -6, \ c = -12$

$x = \dfrac{-b \pm \sqrt{b^2 - 4ac}}{2a}$

$x = \dfrac{-(-6) \pm \sqrt{(-6)^2 - 4(5)(-12)}}{2(5)}$

$x = \dfrac{6 \pm \sqrt{36 + 240}}{10}$

$x = \dfrac{6 \pm \sqrt{276}}{10}$

$x = \dfrac{6 \pm \sqrt{4 \cdot 69}}{10}$

$x = \dfrac{6 \pm 2\sqrt{69}}{10}$

$x = \dfrac{3 \pm \sqrt{69}}{5}$

$x = 3, \ x = \dfrac{3 + \sqrt{69}}{5}, \ x = \dfrac{3 - \sqrt{69}}{5}$

Chapter 6 Skills Check

1. $5x + 2y \leq 10$

$$2y \leq 10 - 5x$$

$$y \leq \frac{10 - 5x}{2}$$

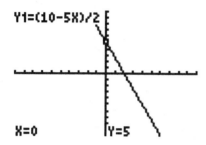

[–10, 10] by [–10, 10]

Note that the line is solid because of the "equal to" in the given inequality.

Test (0, 0).

$$y \leq \frac{10 - 5x}{2}$$

$$0 \leq \frac{10 - 5(0)}{2}$$

$$0 \leq 5$$

Since the statement is true, the region containing (0, 0) is the solution to the inequality.

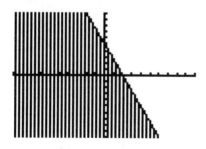

[–10, 10] by [–10, 10]

2. $5x - 4y > 12$

$$-4y > 12 - 5x$$

$$y < \frac{12 - 5x}{-4}$$

$$y < \frac{5x - 12}{4}$$

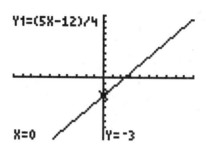

[–10, 10] by [–10, 10]

Note that the line is dashed because there is no "equal to" in the given inequality.

Test (0, 0).

$$y < \frac{5x - 12}{4}$$

$$0 < \frac{5(0) - 12}{4}$$

$$0 < \frac{-12}{4}$$

$$0 < -3$$

Since the statement is false, the region not containing (0, 0) is the solution to the inequality.

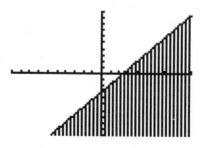

[–10, 10] by [–10, 10]

3. Rewriting the system yields
$$2x + y \leq 3$$
$$y \leq 3 - 2x$$
and
$$x + y \leq 2$$
$$y \leq 2 - x$$

The new system is
$$\begin{cases} y \leq 3 - 2x \\ y \leq 2 - x \\ x \geq 0, y \geq 0 \end{cases}$$

The graph of the system is

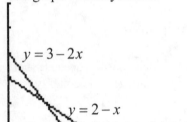

$y = 3 - 2x$

$y = 2 - x$

[0, 5] by [0, 5]

Note that all the lines are solid.
Determining the intersection point between $y = 3 - 2x$ and $y = 2 - x$:

$$3 - 2x = 2 - x$$
$$-x = -1$$
$$x = 1$$
Substituting to find y
$$y = 2 - x$$
$$y = 2 - 1$$
$$y = 1$$

The intersection point is $(1,1)$.

To determine the solution region, pick a point to test. Pick $\left(\frac{1}{2}, 1\right)$.

$$y \leq 3 - 2x$$
$$1 \leq 3 - 2\left(\frac{1}{2}\right)$$
$$1 \leq 2$$
True statement
$$y \leq 2 - x$$
$$1 \leq 2 - \frac{1}{2}$$
$$1 \leq \frac{3}{2}$$
True statement
$$x \geq 0$$
$$\frac{1}{2} \geq 0$$
True statement
$$y \geq 0$$
$$1 \geq 0$$
True statement

Since all the inequalities that form the system are true at the point $\left(\frac{1}{2}, 1\right)$, the region that contains $\left(\frac{1}{2}, 1\right)$ is the solution region. Considering the graph of the system, the corners of the region are
$$(0,0), (0,2), (1,1), \text{and } \left(\frac{3}{2}, 0\right).$$

[0, 5] by [0, 5]

Recall that all the lines are solid.

4. Rewriting the system:
$$3x + 2y \le 6$$
$$2y \le 6 - 3x$$
$$y \le \frac{6-3x}{2}$$
and
$$3x + 6y \le 12$$
$$6y \le 12 - 3x$$
$$y \le \frac{12-3x}{6}$$

The new system is
$$\begin{cases} y \le \dfrac{6-3x}{2} \\ y \le \dfrac{12-3x}{6} \\ x \ge 0, y \ge 0 \end{cases}$$

The graph of the system is

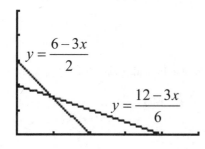

[0, 5] by [0, 5]

Note that all the lines are solid. The intersection point between the two lines is $(1,1.5)$.

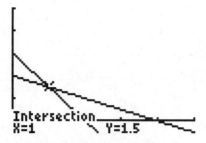

To determine the solution region, pick a point to test. Pick $(1,1)$.

$$y \le \frac{6-3x}{2}$$
$$1 \le \frac{6-3(1)}{2}$$
$$1 \le \frac{3}{2}$$
True statement
$$y \le \frac{12-3x}{6}$$
$$1 \le \frac{12-3(1)}{6}$$
$$1 \le \frac{3}{2}$$
True statement
$$x \ge 0$$
$$1 \ge 0$$
True statement
$$y \ge 0$$
$$1 \ge 0$$
True statement

Since all the inequalities that form the system are true at the point $(1,1)$, the region that contains $(1,1)$ is the solution region. Considering the graph of the system, the corners of the region are $(0,0),(0,2),(1,1.5),$ and $(2,0)$.

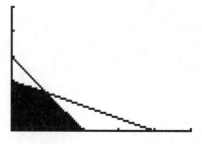

[0, 5] by [0, 5]

Recall that all the lines are solid.

5. Rewriting the system yields

$$2x + y \leq 30$$
$$y \leq 30 - 2x$$

and

$$x + y \leq 19$$
$$y \leq 19 - x$$

and

$$x + 2y \leq 30$$
$$2y \leq 30 - x$$
$$y \leq \frac{30 - x}{2}$$

The new system is

$$\begin{cases} y \leq 30 - 2x \\ y \leq 19 - x \\ y \leq \dfrac{30 - x}{2} \\ x \geq 0, y \geq 0 \end{cases}$$

The graph of the system is

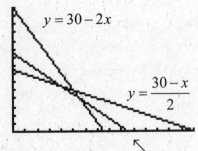

$$y = 30 - 2x$$
$$y = \frac{30 - x}{2}$$

[0, 30] by [0, 30]

$$y = 19 - x$$

The intersection point between $y = 30 - 2x$ and $y = 19 - x$ is $(11, 8)$.

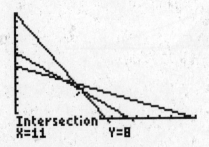

The intersection point between $y = 30 - 2x$ and $y = \dfrac{30 - x}{2}$ is $(10, 10)$.

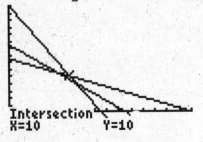

The intersection point between $y = 19 - x$ and $y = \dfrac{30 - x}{2}$ is $(8, 11)$.

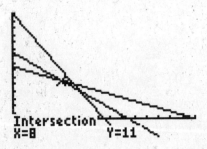

Note that all the lines are solid because there is an "equal to" as part of all the inequalities.

To determine the solution region, pick a point to test. Pick $(1, 1)$.

$$2x + y \leq 30$$
$$2(1) + (1) \leq 30$$
$$3 \leq 30$$

True statement

$$x + y \leq 19$$
$$(1) + (1) \leq 19$$
$$2 \leq 19$$

True statement

$x + 2y \leq 30$

$(1) + 2(1) \leq 30$

$3 \leq 30$

True statement

$x \geq 0$

$1 \geq 0$

True statement

$y \geq 0$

$1 \geq 0$

True statement

Since all the inequalities that form the system are true at the point $(1,1)$, the region that contains $(1,1)$ is the solution region. The graph of the solution is

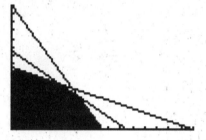

[0, 30] by [0, 30]

Recall that all lines are solid.

One of the corner points is the intersection point between $y = 30 - 2x$ and $y = 19 - x$, which is $(11,8)$. A second corner point is the intersection point between $y = 19 - x$ and $y = \dfrac{30 - x}{2}$, which is $(8,11)$. A third corner point occurs where $y = 30 - 2x$ crosses the x-axis. A fourth corner point occurs where $y = \dfrac{30 - x}{2}$ crosses the y-axis.

To find the x-intercept, let $y = 0$.

$0 = 30 - 2x$

$2x = 30$

$x = 15$

$(15,0)$

To find the y-intercept, let $x = 0$.

$y = \dfrac{30 - x}{2}$

$y = \dfrac{30 - (0)}{2}$

$y = 15$

$(0,15)$

Therefore, the corner points of the solution region are

$(8,11), (11,8), (15,0), (0,15),$ and $(0,0)$.

6. Rewriting the system:

$2x + y \leq 10$

$\qquad y \leq 10 - 2x$

$\qquad$ and

$x + 2y \leq 11$

$\qquad 2y \leq 11 - x$

$\qquad y \leq \dfrac{11 - x}{2}$

The new system is

$$\begin{cases} y \leq 10 - 2x \\ y \leq \dfrac{11 - x}{2} \\ x \geq 0, y \geq 0 \end{cases}$$

The graph of the system is

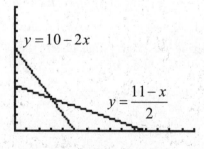

[0, 15] by [0, 15]

Note that all the lines are solid.

The intersection point between the two lines is $(3,4)$.

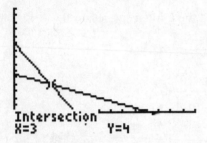

To determine the solution region, pick a point to test. Pick $(1,1)$.

$y \leq 10 - 2x$

$1 \leq 10 - 2(1)$

$1 \leq 8$

True statement

$y \leq \dfrac{11 - x}{2}$

$y \leq \dfrac{11 - (1)}{2}$

$1 \leq 5$

True statement

$x \geq 0$

$1 \geq 0$

True statement

$y \geq 0$

$1 \geq 0$

True statement

Since all the inequalities that form the system are true at the point $(1,1)$, the region that contains $(1,1)$ is the solution region. Considering the graph of the system, the corners of the region are

$(0,0), \left(0, \dfrac{11}{2}\right), (3,4),$ and $(5,0)$.

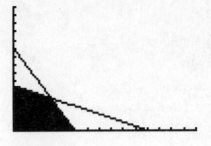

[0, 15] by [0, 15]

Recall that all the lines are solid.

7. Rewriting the system:

$15x - x^2 - y \geq 0$

$-y \geq x^2 - 15x$

$y \leq 15x - x^2$

and

$y - \dfrac{44x + 60}{x} \geq 0$

$y \leq \dfrac{44x + 60}{x}$

The new system is

$$\begin{cases} y \leq 15x - x^2 \\ y \geq \dfrac{44x + 60}{x} \\ x \geq 0, y \geq 0 \end{cases}$$

The graph of the system is

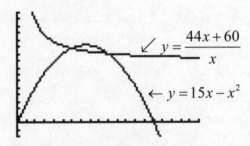

[0, 20] by [−10, 80]

Note that all the lines are solid.
Determining the intersection points between the two functions graphically yields

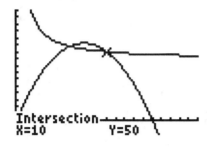

[0, 20] by [−10, 80]

and

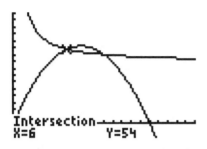

[0, 20] by [−10, 80]

The intersection points are $(10, 50)$ and $(6, 54)$.

To determine the solution region, pick a point to test. Pick $(8, 54)$.

$y \leq 15x - x^2$

$54 \leq 15(8) - (8)^2$

$54 \leq 56$

True statement

$y \geq \dfrac{44x + 60}{x}$

$54 \geq \dfrac{44(8) + 60}{(8)}$

$54 \geq 51.5$

True statement

$x \geq 0$

$8 \geq 0$

True statement

$y \geq 0$

$54 \geq 0$

True statement

Since all the inequalities that form the system are true at the point $(8, 54)$, the region that contains $(8, 54)$ is the solution region. Considering the graph of the system, the corners of the region are $(10, 50)$ and $(6, 54)$.

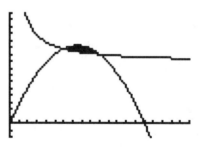

[0, 20] by [−10, 80]

Recall that all the lines are solid.

8. Rewriting the system yields

$$x^3 - 26x + 100 - y \geq 0$$

$$-y \geq -x^3 + 26x - 100$$

$$y \leq x^3 - 26x + 100$$

and

$$19x^2 - 20 - y \leq 0$$

$$-y \leq 20 - 19x^2$$

$$y \geq 19x^2 - 20$$

The new system is

$$\begin{cases} y \leq x^3 - 26x + 100 \\ y \geq 19x^2 - 20 \\ x \geq 0, y \geq 0 \end{cases}$$

The graph of the system is

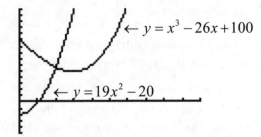

$\leftarrow y = x^3 - 26x + 100$

$\leftarrow y = 19x^2 - 20$

[0, 10] by [−50, 150]

Note that all the lines are solid.
Determining the intersection point between the two functions graphically yields

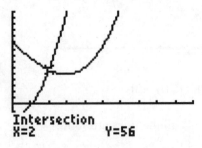

Intersection
X=2   Y=56

[0, 10] by [−50, 150]

The intersection point is $(2, 56)$.

To determine the solution region, pick a point to test. Pick $(1, 1)$.

$$y \leq x^3 - 26x + 100$$

$$1 \leq (1)^3 - 26(1) + 100$$

$$1 \leq 75$$

True statement

$$y \geq 19x^2 - 20$$

$$1 \geq 19(1)^2 - 20$$

$$1 \geq -1$$

True statement

$$x \geq 0$$

$$1 \geq 0$$

True statement

$$y \geq 0$$

$$1 \geq 0$$

True statement

Since all the inequalities that form the system are true at the point $(1, 1)$, the region that contains $(1, 1)$ is the solution region.
Considering the graph of the system, the corners of the region are $(2, 56), (0, 100)$, and $(0, -20)$.

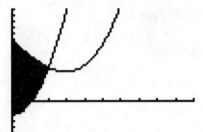

[0, 10] by [−50, 150]

Recall that all the lines are solid.

9. Rewriting the system of constraints and graphing the system yields

$$\begin{cases} y \geq 8 - 3x \\ y \geq \dfrac{14 - 2x}{5} \\ x \geq 0, y \geq 0 \end{cases}$$

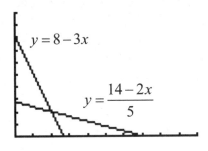

$y = 8 - 3x$

$y = \dfrac{14 - 2x}{5}$

[0, 10] by [0, 10]

To determine the solution region, pick a point to test. Pick $(2,3)$. When substituted into the inequalities that form the system, the point $(2,3)$ creates true statements in all cases.

Since all the inequalities that form the system are true at the point $(2,3)$, the region that contains $(2,3)$ is the solution region. The solution represents the feasible region. The graph of the feasible region is

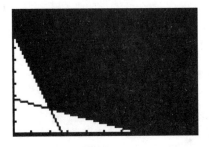

[0, 10] by [0, 10]

Note that since all the inequalities contain an "equal to", all the boundary lines are solid.

The corner points of the feasible region are $(0,8),(7,0),$ and $(2,2)$.

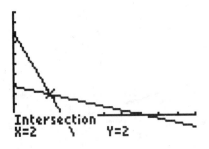

Intersection
X=2   Y=2

Testing the corner points of the feasible region to minimize the objective function, $g = 3x + 4y$, yields

At $(0,8)$: $g = 3(0) + 4(8) = 32$

At $(7,0)$: $g = 3(7) + 4(0) = 21$

At $(2,2)$: $g = 3(2) + 4(2) = 14$

The minimum value is 14 occurring at $(2,2)$.

10. Rewriting the system of constraints and graphing the system yields

$$\begin{cases} y \le \dfrac{105 - 7x}{3} \\ y \le \dfrac{59 - 2x}{5} \\ y \le \dfrac{70 - x}{7} \\ x \ge 0, y \ge 0 \end{cases}$$

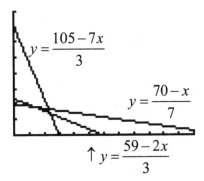

$y = \dfrac{105 - 7x}{3}$

$y = \dfrac{70 - x}{7}$

$\uparrow y = \dfrac{59 - 2x}{3}$

[0, 60] by [0, 40]

To determine the solution region, pick a point to test. Pick $(2,3)$. When substituted into the inequalities that form the system, the point $(2,3)$ creates true statements in all cases.

Since all the inequalities that form the system are true at the point $(2,3)$, the region that contains $(2,3)$ is the solution region.

The solution represents the feasible region. The graph of the feasible region is

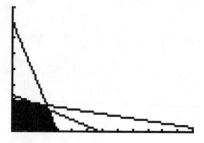

[0, 60] by [0, 40]

Note that since all the inequalities contain an "equal to", all the boundary lines are solid.

The corner points of the feasible region are $(0,0),(0,10),(15,0),(12,7),$ and $(7,9)$.

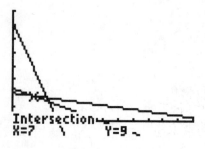

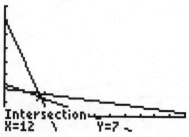

Testing the corner points of the feasible region to maximize the objective function, $f = 7x + 12y$, yields

At $(0,0)$: $f = 7(0) + 12(0) = 0$

At $(0,10)$: $f = 7(0) + 12(10) = 120$

At $(15,0)$: $f = 7(15) + 12(0) = 105$

At $(12,7)$: $f = 7(12) + 12(7) = 168$

At $(7,9)$: $f = 7(7) + 12(9) = 157$

The maximum value is 168 occurring at $(12,7)$.

11. Rewriting the system:
$$2x + y \le 30$$
$$y \le 30 - 2x$$
$$\text{and}$$
$$x + y \le 19$$
$$y \le 19 - x$$
$$\text{and}$$
$$x + 2y \le 30$$
$$2y \le 30 - x$$
$$y \le \frac{30 - x}{2}$$

The new system is
$$\begin{cases} y \le 30 - 2x \\ y \le 19 - x \\ y \le \dfrac{30 - x}{2} \\ x \ge 0, y \ge 0 \end{cases}$$

The graph of the system is

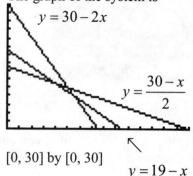

[0, 30] by [0, 30]

$$y = 19 - x$$

The intersection point between $y = 30 - 2x$ and $y = 19 - x$ is $(11,8)$.

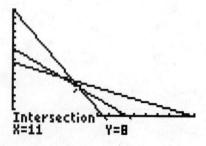

The intersection point between $y = 30 - 2x$

and $y = \dfrac{30 - x}{2}$ is $(10,10)$.

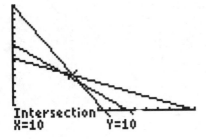

The intersection point between $y = 19 - x$

and $y = \dfrac{30 - x}{2}$ is $(8,11)$.

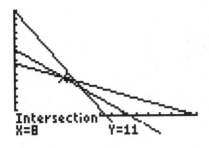

Note that all the lines are solid because there is an "equal to" as part of all the inequalities.

To determine the solution region, pick a point to test. Pick $(1,1)$.

$2x + y \le 30$

$2(1) + (1) \le 30$

$3 \le 30$

True statement

$x + y \le 19$

$(1) + (1) \le 19$

$2 \le 19$

True statement

$x + 2y \le 30$

$(1) + 2(1) \le 30$

$3 \le 30$

True statement

$x \ge 0$

$1 \ge 0$

True statement

$y \ge 0$

$1 \ge 0$

True statement

Since all the inequalities that form the system are true at the point $(1,1)$, the region that contains $(1,1)$ is the solution region. The graph of the solution is

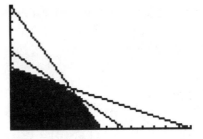

[0, 30] by [0, 30]

Recall that all lines are solid.

One of the corner points is the intersection point between $y = 30 - 2x$ and $y = 19 - x$, which is $(11,8)$. A second corner point is the intersection point between $y = 19 - x$

and $y = \dfrac{30 - x}{2}$, which is $(8,11)$. A third corner point occurs where $y = 30 - 2x$ crosses the x-axis. A fourth corner point

occurs where $y = \dfrac{30 - x}{2}$ crosses the y-axis.

To find the x-intercept, let $y = 0$.

$0 = 30 - 2x$

$2x = 30$

$x = 15$

$(15, 0)$

To find the y-intercept, let $x = 0$.

$y = \dfrac{30 - x}{2}$

$y = \dfrac{30 - (0)}{2}$

$y = 15$

$(0, 15)$

Therefore, the corner points of the solution region are $(8, 11), (11, 8), (15, 0), (0, 15),$ and $(0, 0)$.

Testing the corner points of the feasible region to maximize the objective function, $f = 3x + 5y$, yields

At $(0, 0)$: $\quad f = 3(0) + 5(0) = 0$

At $(0, 15)$: $\quad f = 3(0) + 5(15) = 75$

At $(15, 0)$: $\quad f = 3(15) + 5(0) = 45$

At $(8, 11)$: $\quad f = 3(8) + 5(11) = 79$

At $(11, 8)$: $\quad f = 3(11) + 5(8) = 73$

The maximum value is 79 occurring at $(8, 11)$.

12. Geometric, with a common ratio of 6.

$\dfrac{\frac{2}{3}}{\frac{1}{9}} = 6, \dfrac{4}{\frac{2}{3}} = 6, \dfrac{24}{4} = 6, \text{etc.}$

13. Arithmetic, with a common difference of 12.
$16 - 4 = 12, 28 - 16 = 12, \text{etc.}$

14. Geometric, with a common ratio of $-\dfrac{3}{4}$.

$\dfrac{-12}{16} = -\dfrac{3}{4}, \dfrac{9}{-12} = -\dfrac{3}{4}, \text{etc.}$

15. $a_n = a_1 r^{n-1}$

$a_n = 64 r^{n-1}$

$a_8 = 64 r^{8-1} = \dfrac{1}{2}$

$2(64 r^7) = 2\left(\dfrac{1}{2}\right)$

$128 r^7 = 1$

$r^7 = \dfrac{1}{128}$

$r = \sqrt[7]{\dfrac{1}{128}} = \dfrac{1}{2}$

$a_n = 64\left(\dfrac{1}{2}\right)^{n-1}$

$a_5 = 64\left(\dfrac{1}{2}\right)^{5-1} = 64\left(\dfrac{1}{2}\right)^4 = 4$

16. Note that the common ratio is 3 and that the first term is $\dfrac{1}{9}$.

$a_n = a_1 r^{n-1}$

$a_n = \dfrac{1}{9}(3)^{n-1}$

$a_6 = \dfrac{1}{9}(3)^{6-1} = \dfrac{1}{9}(3)^5 = 27$

17. $s_n = \dfrac{a_1\left(1 - r^n\right)}{1 - r}$

$s_{10} = \dfrac{5\left(1 - (-2)^{10}\right)}{1 - (-2)}$

$s_{10} = \dfrac{5(-1023)}{3} = -1705$

18. Note that the sequence is arithmetic with a common difference of 3.

$a_n = a_1 + (n-1)d$

$a_n = 3 + (n-1)(3)$

$a_n = 3n$

$a_{12} = 3(12) = 36$

$s_n = \dfrac{n(a_1 + a_n)}{2}$

$s_{12} = \dfrac{12(a_1 + a_{12})}{2} = \dfrac{12(3+36)}{2} = 234$

24. $f'(x) = 0$

$2x^3 + x^2 - x = 0$

$x(2x^2 + x - 1) = 0$

$x(2x - 1)(x + 1) = 0$

$x = 0,\ 2x - 1 = 0,\ x + 1 = 0$

$x = 0,\ x = \dfrac{1}{2},\ x = -1$

19. $a_1 = \left(\dfrac{4}{5}\right)^1 = \dfrac{4}{5}$

$r = \dfrac{4}{5}$

$S = \dfrac{a_1}{1-r} = \dfrac{\frac{4}{5}}{1 - \frac{4}{5}} = \dfrac{\frac{4}{5}}{\frac{1}{5}} = 4$

20. $\left(x^3 + 2x + 3\right)(2x) + \left(x^2 - 5\right)\left(3x^2 + 2\right)$

$= 2x^4 + 4x^2 + 6x + \left(3x^4 + 2x^2 - 15x^2 - 10\right)$

$= 5x^4 - 9x^2 + 6x - 10$

21. $x^3 - 8x^2 - 9x = 0$

$x\left(x^2 - 8x - 9\right) = 0$

$x(x - 9)(x + 1) = 0$

$x = 0,\ x - 9 = 0,\ x + 1 = 0$

$x = 0,\ x = 9,\ x = -1$

22. $y = x + x^{-2} - \dfrac{2}{x^{\frac{3}{2}}} + \dfrac{3}{x^{\frac{1}{3}}}$

$y = x + x^{-2} - 2x^{-\frac{3}{2}} + 3x^{-\frac{1}{3}}$

23. $f'(x) = \dfrac{8}{x^3} + \dfrac{5}{x} + 4x$

Chapter 6 Review

25. Determine the feasible region by solving and graphing the system of inequalities that represent the constraints. Let x represent the number of Deluxe model DVD players, and let y represent the number of Superior model DVD players.

$$\begin{cases} 3x + 2y \leq 1800 \\ 40x + 60y \leq 36,000 \\ x \geq 0, y \geq 0 \end{cases}$$

Rewriting yields

$$\begin{cases} y \leq \dfrac{1800 - 3x}{2} \\ y \leq \dfrac{36,000 - 40x}{60} \\ x \geq 0, y \geq 0 \end{cases}$$

Graphing the system yields

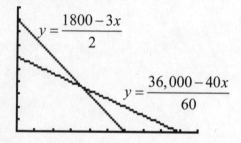

$$y = \frac{1800 - 3x}{2}$$

$$y = \frac{36,000 - 40x}{60}$$

[0, 1000] by [0, 1000]

To determine the solution region, pick a point to test. Pick $(1,1)$. When substituted into the inequalities that form the system, the point $(1,1)$ creates true statements in all cases.

Since all the inequalities that form the system are true at the point $(1,1)$, the region that contains $(1,1)$ is the solution region.

The solution represents the feasible region. The graph of the feasible region is

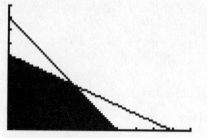

[0, 1000] by [0, 1000]

Note that since all the inequalities contain an "equal to", all the boundary lines are solid.

The corner points of the feasible region are $(0,0), (0,600), (600,0),$ and $(360,360)$.

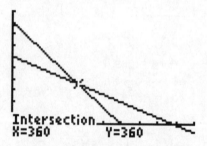

Intersection
X=360   Y=360

Testing the corner points of the feasible region to maximize the objective function, $f = 30x + 40y$, yields

At $(0,0)$:
$f = 30(0) + 40(0) = \$0$
At $(600,0)$:
$f = 30(600) + 40(0) = \$18,000$
At $(0,600)$:
$f = 30(0) + 40(600) = \$24,000$
At $(360,360)$:
$f = 30(360) + 40(360) = \$25,200$

The maximum value is \$25,200 occurring at $(360,360)$. To produce a maximum profit the company needs to produce 360 Deluxe models and 360 Superior models.

26. a. Determine the feasible region by solving and graphing the system of inequalities that represent the constraints. Let x represent the number of days of production at the Pottstown plant, and let y represent the number of days of production at the Ethica plant.

$$\begin{cases} 20x + 40y \geq 1600 \\ 60x + 40y \geq 2400 \\ x \geq 0, y \geq 0 \end{cases}$$

Rewriting yields

$$\begin{cases} y \geq \dfrac{1600 - 20x}{40} \\ y \geq \dfrac{2400 - 60x}{40} \\ x \geq 0, y \geq 0 \end{cases}$$

Graphing the system yields

$$y = \frac{2400 - 60x}{40}$$

$$y = \frac{1600 - 20x}{40}$$

[0, 100] by [0, 100]

To determine the solution region, pick a point to test. Pick $(20, 40)$. When substituted into the inequalities that form the system, the point $(20, 40)$ creates true statements in all cases.

Since all the inequalities that form the system are true at the point $(20, 40)$, the region that contains $(20, 40)$ is the solution region. The solution represents the feasible region. The graph of the feasible region is

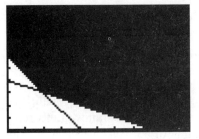

[0, 100] by [0, 100]

Note that since all the inequalities contain an "equal to", all the boundary lines are solid.

The corner points of the feasible region are $(80, 0), (0, 60),$ and $(20, 30)$.

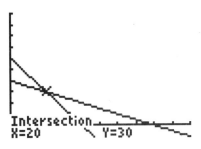

Intersection
X=20   Y=30

Testing the corner points of the feasible region to minimize the objective function, $g = 20,000x + 24,000y$, yields

At $(80, 0)$:
$g = 20,000(80) + 24,000(0)$
$\quad = \$1,600,000$

At $(0, 60)$:
$g = 20,000(0) + 24,000(60)$
$\quad = \$1,440,000$

At $(20, 30)$:
$g = 20,000(20) + 24,000(30)$
$\quad = \$1,120,000$

The minimum cost is \$1,120,000 occurring at $(20, 30)$. Therefore, operating the Pottstown plant for 20 days and the Ethica plant for 30 days produces the minimum manufacturing cost of \$1,120,000.

b. Refer to part a). The minimum cost is $1,200,000.

27. Determine the feasible region by solving and graphing the system of inequalities that represent the constraints. Let x represent the number of pounds of Feed A, and let y represent the number of pounds of Feed B.

$$\begin{cases} 2x+8y \geq 80 \\ 4x+2y \geq 132 \\ x \geq 0, y \geq 0 \end{cases}$$

Rewriting yields

$$\begin{cases} y \geq \dfrac{80-2x}{8} \\ y \geq 66-2x \\ x \geq 0, y \geq 0 \end{cases}$$

Graphing the system yields

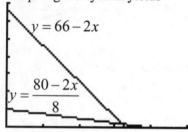

$y = 66 - 2x$

$y = \dfrac{80 - 2x}{8}$

[0, 50] by [0, 70]

To determine the solution region, pick a point to test. Pick $(25,50)$. When substituted into the inequalities that form the system, the point $(25,50)$ creates true statements in all cases.

Since all the inequalities that form the system are true at the point $(25,50)$, the region that contains $(25,50)$ is the solution region. The solution represents the feasible region. The graph of the feasible region is

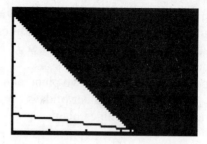

[0, 50] by [0, 70]

Note that since all the inequalities contain an "equal to", all the boundary lines are solid.

The corner points of the feasible region are $(0,66),(32,2)$ and $(40,0)$.

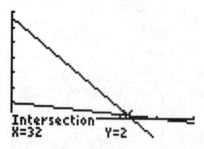

Intersection
X=32   Y=2

Testing the corner points of the feasible region to minimize the objective function, $g = 1.40x + 1.60y$, yields

At $(0,66)$:
$g = 1.40(0) + 1.60(66) = \$105.60$
At $(32,2)$:
$g = 1.40(32) + 1.60(2) = \$48$
At $(40,0)$:
$g = 1.40(40) + 1.60(0) = \$56$

The minimum value is $48 occurring at $(32,2)$. To minimize the cost of the feed, the laboratory needs to use 32 pounds of Feed A and 2 pounds of Feed B.

28. Determine the feasible region by solving and graphing the system of inequalities that represent the constraints. Let x represent the number of leaf blowers, and let y represent the number of weed wackers.

$$\begin{cases} x + y \le 460 \\ x \le 260 \\ y \le 240 \\ x \ge 0, y \ge 0 \end{cases}$$

Rewriting yields

$$\begin{cases} y \le 460 - x \\ x \le 260 \\ y \le 240 \\ x \ge 0, y \ge 0 \end{cases}$$

Graphing the system yields

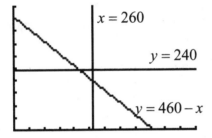

[0, 600] by [0, 500]

To determine the solution region, pick a point to test. Pick $(1,1)$. When substituted into the inequalities that form the system, the point $(1,1)$ creates true statements in all cases.

Since all the inequalities that form the system are true at the point $(1,1)$, the region that contains $(1,1)$ is the solution region.

The solution represents the feasible region. The graph of the feasible region is

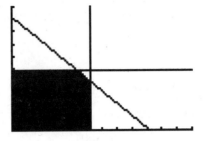

[0, 600] by [0, 500]

Note that since all the inequalities contain an "equal to", all the boundary lines are solid.

The corner points of the feasible region are $(0,0), (0,240), (260,0), (220,240)$, and $(260,200)$.

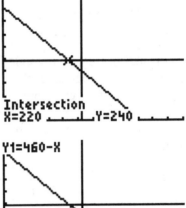

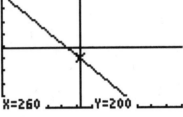

Testing the corner points of the feasible region to maximize the objective function, $f = 5x + 10y$, yields

At $(0,0)$:

$f = 5(0) + 10(0) = \$0$

At $(0,240)$:

$f = 5(0) + 10(240) = \$2400$

At $(260,0)$:

$f = 5(260) + 10(0) = \$1300$

At $(220,240)$:

$f = 5(220) + 10(240) = \$3500$

At $(260,200)$:

$f = 5(260) + 10(200) = \$3300$

The maximum value is \$3500 occurring at $(220,240)$. To produce a maximum profit the company needs to produce 220 leaf blowers and 240 weed wackers.

29. Determine the feasible region by solving and graphing the system of inequalities that represent the constraints. Let x represent the two-bedroom apartments rented, and let y represent the number of three bedroom apartments rented.

$$\begin{cases} 2x + 3y \leq 180 \\ x \leq 60 \\ y \leq 40 \\ x \geq 0, y \geq 0 \end{cases}$$

Rewriting yields

$$\begin{cases} y \leq \dfrac{180 - 2x}{3} \\ x \leq 60 \\ y \leq 40 \\ x \geq 0, y \geq 0 \end{cases}$$

Graphing the system yields

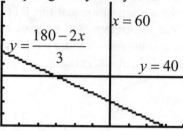

[0, 100] by [0, 100]

To determine the solution region, pick a point to test. Pick $(1,1)$. When substituted into the inequalities that form the system, the point $(1,1)$ creates true statements in all cases.

Since all the inequalities that form the system are true at the point $(1,1)$, the region that contains $(1,1)$ is the solution region. The solution represents the feasible region. The graph of the feasible region is

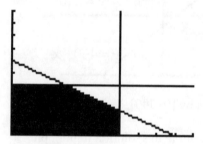

[0, 100] by [0, 100]

Note that since all the inequalities contain an "equal to", all the boundary lines are solid.

The corner points of the feasible region are $(0,0),(0,40),(60,0),(30,40)$, and $(60,20)$.

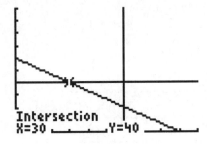

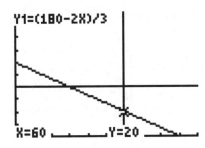

Testing the corner points of the feasible region to maximize the objective function, $f = 800x + 1150y$, yields

At $(0,0)$:

$f = 800(0) + 1150(0) = \$0$

At $(0,40)$:

$f = 800(0) + 1150(40) = \$46{,}000$

At $(60,0)$:

$f = 800(60) + 1150(0) = \$48{,}000$

At $(30,40)$:

$f = 800(30) + 1150(40) = \$70{,}000$

At $(60,20)$:

$f = 800(60) + 1150(20) = \$71{,}000$

The maximum value is \$71,000 occurring at $(60,20)$. To produce a maximum profit the woman needs to rent sixty two-bedroom apartments and twenty 3-bedroom apartments.

30. Determine the feasible region by solving and graphing the system of inequalities that represent the constraints. Let x represent the amount of auto loans in millions of dollars, and let y represent the amount of home equity loans in millions of dollars.

$\begin{cases} x + y \le 30 \\ x \ge 2y \\ x \ge 0, y \ge 0 \end{cases}$

Rewriting yields

$\begin{cases} y \le 30 - x \\ y \le \dfrac{x}{2} \\ x \ge 0, y \ge 0 \end{cases}$

Graphing the system yields

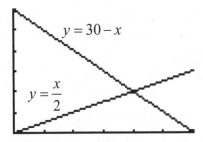

[0, 30] by [0, 30]

To determine the solution region, pick a point to test. Pick $(15,1)$. When substituted into the inequalities that form the system, the point $(15,1)$ creates true statements in all cases.

Since all the inequalities that form the system are true at the point $(15,1)$, the region that contains $(15,1)$ is the solution region. The solution represents the feasible region. The graph of the feasible region is

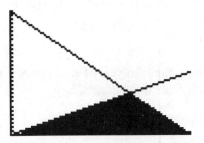

[0, 30] by [0, 30]

Note that since all the inequalities contain an "equal to", all the boundary lines are solid.

The corner points of the feasible region are $(0,0),(30,0),$ and $(20,10)$.

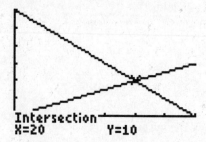

Intersection
X=20   Y=10

Testing the corner points of the feasible region to maximize the objective function, $f = 0.08x + 0.07y$, yields

At $(0,0)$: $f = 0.08(0) + 0.07(0) = \$0$

At $(30,0)$: $f = 0.08(30) + 0.07(0) = \$2.4$

At $(20,10)$: $f = 0.08(20) + 0.07(10) = \$2.3$

The maximum value is \$2.4 million occurring at $(30,0)$. To produce a maximum profit the finance company should make \$30 million in auto loans and no home equity loans.

31. a. Job 1: $a_n = a_1 + (n-1)d$

$a_n = 20,000 + (n-1)(1000)$

$a_n = 1000n + 19,000$

$a_5 = 1000(5) + 19,000 = 24,000$

Job 2: $a_n = a_1 + (n-1)d$

$a_n = 18,000 + (n-1)(1600)$

$a_n = 1600n + 16,400$

$a_5 = 1600(5) + 16,400 = 24,400$

During the fifth year of employment, job two produces a higher salary.

b. Job 1: $s_n = \dfrac{n(a_1 + a_n)}{2}$

$s_5 = \dfrac{5(a_1 + a_5)}{2}$

$s_5 = \dfrac{5(20,000 + 24,000)}{2}$

$s_5 = 110,000$

Job 2: $s_n = \dfrac{n(a_1 + a_n)}{2}$

$s_5 = \dfrac{5(a_1 + a_5)}{2}$

$s_5 = \dfrac{5(18,000 + 24,400)}{2}$

$s_5 = 106,000$

Over the first five years of employment, the total salary earned from job one is higher than the total salary earned for job two.

32. Note that the given series is geometric with a common ratio of 0.6.

$s_n = \dfrac{a_1(1-r^n)}{1-r}$

$s_{21} = \dfrac{400\left(1-(0.6)^{21}\right)}{1-(0.6)}$

$s_{21} = 999.978063$

The level of the drug in the bloodstream after 21 doses over 21 days is approximately 999.98 mg.

33. a. Note that the question is modeled by a geometric series with a common ratio of 2 and an initial value of 1.

$a_1 = 1, r = 2$

$a_n = a_1 r^{n-1}$

$a_n = 1(2)^{n-1}$

$a_{64} = 1(2)^{64-1} = (2)^{63} \approx 9.22337 \times 10^{18}$

b. $a_1 = 1, r = 2$

$$s_n = \frac{a_1(1-r^n)}{1-r}$$

$$s_{64} = \frac{1(1-(2)^{64})}{1-(2)}$$

$$s_{64} \approx 1.84467 \times 10^{19}$$

34. $a_1 = 20,000, r = 1.06$

$a_n = a_1 r^{n-1}$

Note that after 5 years, $n = 6$.

$a_6 = 20,000(1.06)^{6-1}$

$a_6 = 20,000(1.06)^5$

$a_6 = 26,764.51$

After earning interest for five years, the value of the investment is \$26,764.51.

35. $s_n = \frac{a_1(1-r^n)}{1-r}$

$$s_{60} = \frac{300\left(1-\left(1+\frac{0.12}{12}\right)^{60}\right)}{1-\left(1+\frac{0.12}{12}\right)}$$

$$s_{60} = \frac{300\left(1-(1.01)^{60}\right)}{1-(1.01)}$$

$$s_{60} \approx 24,500.90$$

The future value of the annuity is \$24,500.90.

36. Recall that to earn a profit, revenue must exceed cost.

$R(x) > C(x)$

$177.50x > 3x^2 + 1228$

Graphing the functions yields

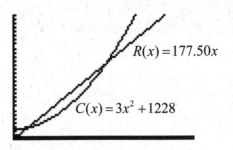

$[0, 100]$ by $[0, 15,000]$

The two intersection points are approximately $(51.18, 9082.08)$ and $(8, 1420)$

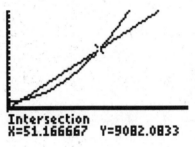

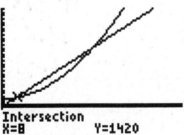

If $8 < x < 52$, $R(x) > C(x)$ and profit is achieved.

Chapter 6 Extended Application

Year	Period (months)	Salary, Plan I	Salary, Plan II
1	0–6	20,000	20,000
	6–12	20,000	20,300
2	12–18	20,500	20,600
	18–24	20,500	20,900
3	24–30	21,000	21,200
	30–36	21,000	21,500
Total for 3 years		$123,000	$124,500

1. The raises in Plan I total $3000.

2. The raises in Plan II total $4,500.

3. Plan II is clearly better for the employee. It yields an extra $1500 over the 3-year period.

4. See complete table above.

Year	Period (months)	Salary, Plan I	Salary, Plan II
Total for 3 years		$123,000	$124,500
Year 3	36–42	21,500	21,800
	42–48	21,500	22,100
Total for 4 years		$166,000	$168,400

Plan II continues to be better. It yields $8,400 in raises, whereas Plan I yields $6000 in raises.

5. For at least the first four years, the school board will not save money by awarding $300 raises every six months instead of $1000 annual raises. The school board did not take into consideration that the semi-annual raises would be awarded more frequently and therefore would compound faster than annual raises.

6. The difference in the raises over four years is $8400 - 6000 = \$2400$.

    200 teachers × $2400 per teacher
    = $480,000

    The school district will spend an additional $480,000 over the first four years.